CONTEMPORARY MATHEMATICS

Titles in this Series

CONTEMPORARY MATHEMATICS

Volume 8

Ordered Fields and Real Algebraic Geometry

AMERICAN MATHEMATICAL SOCIETY
Providence • Rhode Island

PROCEEDINGS OF THE SPECIAL SESSION ON ORDERED FIELDS AND REAL ALGEBRAIC GEOMETRY 87TH ANNUAL MEETING OF THE AMERICAN MATHEMATICAL SOCIETY

HELD IN SAN FRANCISCO, CALIFORNIA

JANUARY 7–11, 1981

EDITED BY

D. W. DUBOIS AND T. RECIO

Library of Congress Cataloging in Publication Data

Special Session on Ordered Fields and Real Algebraic Geometry (1981: San Francisco, Calif.)
Ordered fields and real algebraic geometry.
(Contemporary mathematics, ISSN 0271–4132; v. 8)
"Proceedings of the Special Session on Ordered Fields and Real Algebraic Geometry, 87th Annual Meeting of the American Mathematical Society, held at San Francisco, California, January 7–11, 1982"—T.p. verso.
Bibliography: p.
1. Ordered fields—Congresses. 2. Geometry, Algebraic—Congresses. I. Dubois, D. W. (Donald Ward), 1923– . II. Recio, T. (Tómas), 1949– . III. American Mathematical Society. Meeting (87th: 1981: San Francisco, Calif.) IV. Title. V. Series: Contemporary mathematics (American Mathematical Society); v. 8.
QA247.S67 1981 512'.3 82-3951
ISBN 0-8218-5007-5 AACR2

1980 Mathematics Subject Classification.
Primary 12D15, 12J15, 32C05, 58A07, 10C04, 13J25, 06F25, 03B52.

CONTENTS

INTRODUCTION

The contents of this book comprise, with only minor editorial changes, papers submitted by those mathematicians who accepted our invitation to speak at the American Mathematical Society's 1981 Special Session on Ordered Fields and Real Algebraic Geometry in San Francisco. We wish to thank all those who made possible the Special Session and this publication. Special thanks are offered to Professors Kenneth Ross and James Milgram, and to our incredibly patient typist, Rose Henry.

D. W. Dubois
T. Recio

Contemporary Mathematics
Volume 8, 1982

NORMAL DECOMPOSITIONS OF SEMI-ALGEBRAIC SETS

C. Andradas[1]

INTRODUCTION

Extensive studies have been done lately on semi-algebraic sets. However, many times they are considered as a particular case of the semi-analytic sets and so, the definitions used in this study are of analytic nature. This is the case, for example, of the definitions of regular points: a point p is regular of dimension r if locally the semi-algebraic set is defined by analytic functions with linearly independent differentials at p. It seems interesting to replace here the analytic functions by polynomials in such a way that the new defintion extends the classical regularity in a real algebraic variety given by Whitney [Whitney 1957]. This is our Definition 1.1. Then, we prove, Theorem 2.12, the analogue to Lojasiewicz's theorem on semi-analyticity of the sets of regular points, for the semi-algebraic case. We ask, also, for the semi-algebraic version of Galbiati's coherence theorem of semi-analytic sets (Prop. 4.7). This, as well as Th. 2.12, is done by means of appropriate decompositions of the semi-algebraic sets: the normal decompositions and normal stratifications. These decompositions are studied in §2 and §3. In §3 the problem of finding a canonical representation (in the sense that it is univoquely determined by T) is also treated. We define the notion of index of a point in a semi-algebraic set and look for a unique standard normal stratification with good properties with respect to the index function defined by it on T. It is characterized as the maximum of a suitable order defined in the set of normal stratifications.

I wish to thank Professor T. Recio who proposed the problem to me. The definitions 1.1 and the index notion were given by him. Also, some of the results had been proved by him in earlier papers but in a different way and sometimes using the analytic results. Here the proof is completely algebraic.

1980 Mathematics Subject Classification 14G30, 32C05, 58A07.

[1]During the preparation of this paper the author was supported by a grant from the I.N.A.P.E.

0271-4132/81/0000-0671/$05.50

1. SOME RESULTS ON REAL ALGEBRAIC SETS.

We recall in this section some basic results for the algebraic case. Proofs can be obtained in [Recio 1977a] and the notes [S.G.A.R.M. 1980].

Given $E \subset \mathbb{R}^n$, let $I(E)$ be the ideal of polynomials vanishing on E. Let V be an algebraic set in $\mathbb{R}^n$, $\Gamma(V)$ its coordinate ring and $d = \dim \Gamma(V) = \dim V$. Let $I(V) = (f_1,\dots,f_s)$ and $J(f_1,\dots,f_s)$ its jacobian matrix. The topology considered in the paper is the standard one on $\mathbb{R}^n$.

DEFINITION 1.1. A point p of V is regular of dimension r or r-regular if there is an open neighborhood U of p and polynomials $g_{r+1},\dots,g_n$ with linearly independent differentials at p such that:

$$U \cap V = \{x \in U \mid g_{r+1}(x) = \dots = g_n(x) = 0\}.$$

DEFINITION 1.2. [Brumfiel 1979] We define the ideal of degeneracy of V at a point p to be the ideal of the polynomials vanishing on $U \cap V$ for some open neighborhood U of p. We denote it by $J_p(V)$.

Note that there is an open neighborhood W of p such that $J_p(V) = I(W \cap V)$.

THEOREM 1.3. (a) A point p is regular of dimension r if and only if $\mathbb{R}[\underline{x}]_p/J_p(V)\,\mathbb{R}[\underline{x}]_p$ is a regular local ring of dimension r.

(b) A point p is regular of dimension d if and only if the rank of the Jacobian matrix at p is $n-d$.

COROLLARY 1.4. (a) If p is r-regular, $0 \le r \le d$, $J_p(V)$ is prime.

(b) There are no regular points of dimension higher than d.

(c) If V is irreducible and p is regular of dimension d, $I(V) = J_p(V)$.

From now on, we denote by $R_k(V)$ the set of regular points of dimension k, and $R(V) = \bigcup_{k=0}^{d} R_k(V)$.

PROPOSITION 1.5. If V is irreducible then $V' = V - R_d(V)$ is an algebraic set and $\dim V' < d$.

Note that Proposition 1.5 does not hold if V is reducible.

PROPOSITION 1.6. Let V be an algebraic set. The set $R(V)$ of regular points is dense in V.

Given a set M, M_p means the germ of M at the point p.

DEFINITION 1.7. Let V be an algebraic set, p a point of V. The dimension of V_p is k if there is an open neighborhood of p in which there are no regular points of V of dimension higher than k and in every neighborhood of p there are regular points of V of dimension k.

PROPOSITION 1.8. Let V, W be algebraic sets, p a point of $V \cup W$. Then $\dim (V \cup W)_p = \max \{\dim V_p, \dim W_p\}$.

2. NORMAL DECOMPOSITIONS.

Definitions 1.1, 1.2 and 1.7 hold word by word in the case of a semi-algebraic set T in $\mathbb{R}^n$. A decomposition of a semialgebraic set T of $\mathbb{R}^n$ is a finite family of sets Γ of the form $\Gamma = \Gamma_o \cap \Gamma_c$ where Γ_o is an open semialgebraic set and Γ_c is an algebraic set. The dimension of T is the maximum of the dimensions of the T_p, $p \in T$. We suppose in the rest of the paper $d = \dim T$.

We look for a good representation of T. We start with:

LEMMA 2.1. Let T be a semialgebraic set and Λ a decomposition of T. Then there is a refinement Λ' of Λ such that

(a) Γ'_c is irreducible for every Γ' of Λ'.

(b) If $x \in \Gamma'$, $\Gamma' \in \Lambda'$, x is regular of Γ'_c of dimension equal to $\dim \Gamma'_c$.

(c) For every $\Gamma' \in \Lambda'$, $VI(\Gamma') = \Gamma'_c$.

PROOF. Let $\Gamma \in \Lambda$ and $E = VI(\Gamma)$. We have $\Gamma = \Gamma_o \cap \Gamma_c = \Gamma_o \cap E$. If $E = \bigcup_k E_k$ is the decomposition of E in irreducible components then $\Gamma = \Gamma_o \cap E = \bigcup_k (\Gamma_o \cap E_k)$ and it is immediate that $VI(\Gamma \cap E_k) = E_k$. Replacing Γ by $\Gamma_o \cap E_k$ we may assume $\Gamma_c = VI(\Gamma)$ and irreducible for every $\Gamma \in \Lambda$. So, if $\Gamma \in \Lambda$ there exists $x \in \Gamma$ regular of Γ_c of dimension $\dim \Gamma_c$ (if not, from 1.5 $VI(\Gamma) \subsetneq \Gamma_c$). From 1.5, there exists $\Gamma^1_c \subsetneq \Gamma_c$, algebraic with $\dim \Gamma^1_c < \dim \Gamma = s$, such that $\Gamma_c \setminus \Gamma^1_c = R_s(\Gamma_c)$. Then $\Gamma^1_o = \Gamma_o \setminus \Gamma^1_c$ is open and semialgebraic and

$$\Gamma = (\Gamma^1_o \cap \Gamma_c) \cup (\Gamma_o \cap \Gamma^1_c).$$

Γ_c being irreducible, is $\Gamma_c = VI(\Gamma^1_o \cap \Gamma_c)$ and $\Gamma^1 = \Gamma^1_o \cap \Gamma_c$ verifies all the conditions of the lemma. Let, now, $E = VI(\Gamma_o \cap \Gamma^1_c)$.

Since $\dim E < \dim \Gamma_c$, repeating the process, the proof can be finished in a finite number of steps.

DEFINITION 2.2. A decomposition Λ of T verifying conditions (a), (b) and (c) of proposition 2.1 and irredundant, i.e., $\Gamma^i \not\subset \bigcup_{j \neq i} \Gamma^j$, will be called a <u>normal decomposition</u> of T. Furthermore, if Λ is a partition of T, Λ will be called a <u>normal partition</u> of T.

Note that in the proof of 2.1 if the initial decomposition is a partition we obtain a normal partition just replacing the decomposition in irreducible components:

$$\Gamma_0 \cap E = \bigcup_k (\Gamma_0 \cap E_k)$$

by

$$\Gamma_0 \cap E = (\Gamma_0 \cap E_1) \cup ((\Gamma_0 \setminus E_1) \cap E_2) \cup ((\Gamma_0 \setminus (E_1 \cup E_2)) \cap E_3) \cup \ldots\ldots$$

So, since evidently each decomposition Γ of T can be refined to a partition, we obtain:

COROLLARY 2.3. Let T be a semialgebraic set and Λ a decomposition of T. There is a refinement of Λ which is a normal partition.

In what follows, given $p \in T$, X_p represents the smallest algebraic germ which contains the germ T_p. It is immediate to check that $X_p = VJ_p(T)_p$ and that $VJ_p(T)$ is the smallest representative of X_p. So, all the assertions made about X_p can be said globally just replacing X_p by $VJ_p(T)$. The following definition is of fundamental importance:

DEFINITION 2.4. Let T be a semialgebraic set, Λ a decomposition of T and $p \in T$. We call index of p in T along Λ, and we write $\mathrm{ind}_\Lambda^T(p)$, $\mathrm{ind}_\Lambda(p)$ or simply $\mathrm{ind}(p)$ if there is no confusion, the number card $\{\Gamma \in \Lambda \mid p \in \overline{\Gamma} \setminus \Gamma\}$.

So, given a decomposition, we have defined a function from T to $\mathbb{N}$, namely the function ind_Λ which is upper semicontinuous.

One can also check that a point p of T is regular, if and only if there is a normal partition of T along which $\mathrm{ind}(p) = 0$. This shows the close relationship between the value zero for the index function and the regularity. It will be pointed out more specifically in section 3.

The concept of index is useful to prove some results as the following proposition, which shows how normal decompositions give information about the local behaviour of T.

PROPOSITION 2.5. Let T be a semialgebraic set, Λ a normal decomposition of T and p a point of T. Let $\Lambda_1 = \{\Gamma \in \Lambda \mid p \in \overline{\Gamma}\}$. Then:

(a) $X_p = \bigcup_{\Gamma \in \Lambda_1} \Gamma_{c,p}$ (or $VJ_p(T) = \bigcup_{\Gamma \in \Lambda_1} \Gamma_c$)

(b) If $X = VI(T)$ then $X = \bigcup_{\Gamma \in \Lambda} \Gamma_c$

(c) If $\Gamma \in \Lambda_1$ then $\dim \Gamma_{c,p} \le \dim T_p$.

PROOF. We have $T_p = \bigcup_{\Gamma\in\Lambda_1} \Gamma_p \subset X_p$. So, if $\Gamma \in \Lambda_1$, $\Gamma_p \subset X_p$ and there is an open neighborhood U of p such that $\Gamma \cap U \subset \tilde{X} \cap U$, where $\tilde{X}$ is an algebraic set with germ X_p at p. $\Gamma \in \Lambda_1$, so $\Gamma_o \cap U \neq \emptyset$ and there is $q \in \Gamma_c \cap (U \cap \Gamma_o) \subset \tilde{X} \cap (U \cap \Gamma_o)$. But q is regular of Γ_c of dimension $\dim \Gamma_c$ and $\Gamma_{c,q} \subset \tilde{X}_q$. Γ_c being irreducible we have (cor. 1.4 (c)) $J_q(\Gamma_c) = I(\Gamma_c)$. Then:

$$I(\Gamma_c) = J_q(\Gamma_c) \supset J_q(\tilde{X}) \supset I(\tilde{X}).$$

So, $\Gamma_c \subset \tilde{X}$ and $\Gamma_{c,p} \subset X_p$. Trivially $X_p \subset \bigcup_{\Gamma\in\Lambda_1} \Gamma_{c,p}$ because of the minimality of X_p and the proof is finished.

(b) is trivial. To prove (c) we will use induction on the index of p along Λ. Let $\Lambda_2 = \{\Gamma \in \Lambda \mid p \in \overline{\Gamma} \setminus \Gamma\}$. If $\mathrm{ind}_\Lambda^T(p) = 0$ then $T_p = X_p = \bigcup_{\Gamma\in\Lambda_1} \Gamma_{c,p}$ and by prop. 1.8 the result holds. Suppose the result is known if $y \in T$ and $\mathrm{ind}(y) < s$. Let, now, $p \in T$ with $\mathrm{ind}(p) = s$. For each Γ, $\overline{\Gamma} \setminus \Gamma$ is closed and so, we can choose an open neighborhood U of p such that:

(i) $U \cap \overline{\Gamma} = \emptyset$ if $\Gamma \notin \Lambda_1$;

(ii) $U \cap (\overline{\Gamma} \setminus \Gamma) = \emptyset$ if $\Gamma \in \Lambda_2$; and

(iii) if $q \in U \cap T$ then $\dim T_q \leq \dim T_p$.

Thus, if $\Gamma \in \Lambda_2$ there exists $q \in \Gamma \cap U$ such that $\dim \Gamma_{c,q} = \dim \Gamma_c$ and $\mathrm{ind}(p) > \mathrm{ind}(q)$. Applying the induction hypothesis we have:

$$\dim \Gamma_{c,p} = \dim \Gamma_{c,q} \leq \dim T_q \leq \dim T_p.$$

On the other hand, suppose $\Gamma \in \Lambda_1 \setminus \Lambda_2$. If there is $q \in \Gamma \cap U$ with $\mathrm{ind}(q) < \mathrm{ind}(p)$ we get, as above, $\dim \Gamma_{c,p} \leq \dim T_p$. In any case, for every $q \in \Gamma \cap U$, $\mathrm{ind}(q) = \mathrm{ind}(p)$ and then $q \in \overline{\Gamma'} \setminus \Gamma'$ for every $\Gamma' \in \Lambda_2$, in particular $q \in \Gamma'_c$. Similarly to (a) we obtain $\Gamma_{c,q} \subset \Gamma'_{c,q}$, q regular of Γ_c of dimension $\dim \Gamma_c$ and Γ_c irreducible, and we conclude $\Gamma_c \subset \Gamma'_c$. Thus $\Gamma_{c,p} \subset \Gamma'_{c,p}$ with $\Gamma' \in \Lambda_2$ and so

$$\dim \Gamma_{c,p} \leq \dim \Gamma'_{c,p} \leq \dim T_p.$$

From this proposition a lot of classical results are obtained [cf. S.G.A.R.M. 1980]:

PROPOSITION 2.6. Let S, T be semialgebraic sets, $S \subset T$, $p \in S$. Then, $\dim S_p \leq \dim T_p$. As a consequence $\dim S \leq \dim T$.

PROPOSITION 2.7. Let S, T be semialgebraic sets. Then, $\dim (S_1 \cup S_2) = \max \{\dim S_1, \dim S_2\}$.

COROLLARY 2.8. Let T be a semialgebraic set and Λ a normal decomposition of T. Using the notation of 2.5 we have:

$$\dim T_p = \max_{\Gamma \in \Lambda_1} \{\dim \Gamma_{c,p}\} = \max_{\Gamma \in \Lambda_1} \{\dim \Gamma_c\}$$

Therefore, $\dim T_p = \dim X_p$. Moreover, $\dim T = \dim X$.

COROLLARY 2.9. [Brumfiel 1979] Let T be a semialgebraic set. The number of ideals of degeneracy of T is finite.

PROOF. Let Λ be a normal decomposition of T and let Λ_1 be as in 2.6 by 2.6(a) we have:

$$J_p(T) = J_p(VIJ_p(T)) = \bigcap_{\Gamma\in\Lambda_1} J_p(\Gamma_c)$$

But $J_p(\Gamma_c) = I(\Gamma_c \cap W)$ for some neighborhood W of p. Let $q \in \Gamma \cap W$. Then:

$$I(\Gamma_c) \subset J_p(\Gamma_c) = I(\Gamma_c \cap W) \subset J_q(\Gamma_c) = I(\Gamma_c),$$

so $J_p(\Gamma_c) = I(\Gamma_c)$ and $J_p(T) = \bigcap_{\Gamma\in\Lambda_1} I(\Gamma_c)$.

The number of Γ being finite the proof is complete.

We study now the problem of uniqueness in normal decompositions. Of course, under present conditions the decomposition of a semialgebraic set, even in normal partitions, need not be unique. For example: let $T = V(Z-X^2) \cap \mathbb{R}^3$ and the normal partitions:

(1) $T = \Gamma = \mathbb{R}^3 \cap V(z-x^2)$ (i.e., $\Gamma_o = \mathbb{R}^3$ and $\Gamma_c = V(z-x^2)$)

(2) $T = \Gamma^1 \cup \Gamma^2$, where $\Gamma^1 = \{z > 0\} \cap V(z-x^2)$ and $\Gamma^2 = \mathbb{R}^3 \cap V(x,z)$.

The difference between both decompositions is given by the index function: the former is constantly zero though the second gets the value one. The first fact about uniqueness arises from looking at the points with index zero:

PROPOSITION 2.10. Let Λ, Λ' be normal partitions of T. If $\Gamma \in \Lambda$ is such that there is $x \in \Gamma$ with $\mathrm{ind}_\Lambda(x) = 0$ then there is $\Gamma' \in \Lambda'$ such that $\Gamma_c = \Gamma'_c$. (That means that the algebraic sets of the members Γ with points with index zero are univoquely determined——in particular the algebraic sets of the highest dimension.)

PROOF. If $x \in \Gamma$ verifies that $\mathrm{ind}_\Lambda(x) = 0$ then $T_p = \Gamma_p = \Gamma_{c,p}$ and p is regular of T. Let Λ'_1 as in 2.6. $T_p = X_p$ and then $T_p = \bigcup_{\Gamma'\in\Lambda'_1} \Gamma'_c$. But $T_p = \Gamma_{c,p}$, Γ_c irreducible implies $\Gamma_{c,p} = \Gamma'_{c,p}$ for some

$\Gamma' \in \Lambda_1'$. But Γ_c and Γ_c' are irreducible, and p is regular of both of them of dimension $s = \dim \Gamma_c = \dim \Gamma_c' = \dim T_p$, so $\Gamma_c = \Gamma_c'$.

In this way there is a natural relationship between the minimality of the index function and a possible uniqueness in the normal partitions. That will be the goal of the next section. We finish this one with the following results similar to [Lojasiewicz 1965], theorem 4, page 77.

PROPOSITION 2.11. Let T be a semialgebraic set of dimension d. The set of the regular points of dimension d, $R_d(T)$, is semialgebraic. Moreover its complementary has dimension less than d.

THEOREM 2.12. Let T be a semialgebraic set of dimension d. The set $R_k(T)$ is semialgebraic, $0 \le k \le d$.

PROOF. Let Λ be a normal partition of T. Let Λ^i be the set of elements of Λ of dimension less than or equal to i and let $T^k = \bigcap_{\Gamma \in \Lambda^k} \Gamma$. We have:

$$R_k(T) = R_k(T^k) \setminus \overline{(T \setminus T^k)}.$$

From 2.11, $R_k(T^k)$ is semialgebraic and this ends the proof.

PROOF of 2.11. Let Λ be a normal partition of T. We may assume that $\Gamma_c \neq \Gamma_c'$ if $\Gamma \neq \Gamma'$. With the notation of 2.12 for every $\Gamma \in \Lambda \setminus \Lambda^{d-1}$ we define:

$$M_\Gamma = \operatorname{Reg} \Gamma_c \cap \overline{\Gamma} \cap (\widehat{T \cap \Gamma_c}^{\circ}),$$

where $\operatorname{Reg} \Gamma_c$ represents the set of regular points of Γ_c of dimension equal to $\dim \Gamma_c$, i.e., the set of regular points in the sense of the algebraic geometry, the adherence is taken in $\mathbb{R}^n$ and $\widehat{T \cap \Gamma_c}^{\circ}$ is the interior of $T \cap \Gamma_c$ in Γ_c. M_Γ is semialgebraic and $\Gamma \subset \overline{M}_\Gamma$. First of all, note that in general $\Gamma \not\subset M_\Gamma$. Let $p \in \Gamma$ and U an open neighborhood of p. We have $U \cap \Gamma_c \not\subset \bigcup_{\Gamma' \subset \Gamma_c} \overline{\Gamma}'$ (in other case, as p is regular of Γ_c and Γ_c is irreducible, we would have $\Gamma_c \subset \Gamma_c'$ for some Γ', contradiction). Then, there is $q \in U \cap \Gamma_c$ such that $q \in \operatorname{Reg} \Gamma_c$, $q \in \overline{\Gamma}$ and $q \notin \overline{\Gamma}'$ if $\Gamma' \not\subset \Gamma_c$. But then, if W is a neighborhood of p such that $W \subset U$ and $W \cap \overline{\Gamma}' = \emptyset$ we obtain:

$$W \cap T = \bigcup_{\Gamma' \subset \Gamma_c} (W \cap \Gamma') \subset W \cap \Gamma_c \subset \Gamma_c$$

and $q \in \widehat{T \cap \Gamma_c}^{\circ}$. This proves that $q \in \overline{M}_\Gamma$.

Let, now $\Gamma_1' = \Gamma' \setminus \bigcup_{\Gamma \in \Lambda^{d-1}} M_\Gamma$ for every $\Gamma' \in \Lambda^{d-1}$ and we consider for each $\Gamma \notin \Lambda^{d-1}$ $Y_\Gamma = \bigcup_{\Gamma' \neq \Gamma} Y_{\Gamma\Gamma'}$ where $Y_{\Gamma\Gamma'} = M_\Gamma \cap \overline{\Gamma}_1'$ if

$\Gamma' \in \Lambda^{d-1}$ and $Y_{\Gamma\Gamma'} = M_\Gamma \cap \overline{M}_{\Gamma'}$ if $\Gamma' \notin \Lambda^{d-1}$. Let, finally, $M'_\Gamma = M_\Gamma \setminus Y_\Gamma$. We claim that $R_d(T) = \bigcup_{\Gamma \notin \Lambda^{d-1}} M'_\Gamma$.

To check it, note first that $R_d(T) \cap \Gamma_c \subset M_\Gamma$ for every $\Gamma \notin \Lambda^{d-1}$ and so $R_d(T) \subset \bigcup_{\Gamma \notin \Lambda^{d-1}} M_\Gamma$. We will see, now, that $R_d(T) \cap Y_\Gamma = \emptyset$ if $\Gamma \notin \Lambda^{d-1}$. If $x \in R_d(T) \cap Y_\Gamma$ there is $\Gamma' \neq \Gamma$ such that $x \in R_d(T) \cap Y_{\Gamma\Gamma'} \subset R_d(T) \cap M_\Gamma$. Then $x \in \widehat{T \cap \Gamma_c}$ and there exists a neighborhood U of x such that $U \cap \Gamma_c = U \cap T \subset R_d(T) \cap \Gamma_c$. Suppose $\Gamma' \in \Lambda^{d-1}$. Then there is $z \in U \cap \Gamma'_1 \subset U \cap T \subset R_d(T) \cap \Gamma_c \subset M_\Gamma$, in contradiction with $\Gamma'_1 \cap M_\Gamma = \emptyset$. But if $\Gamma' \notin \Lambda^{d-1}$ there is $z \in U \cap M_{\Gamma'}$ and then $X_z \supset \Gamma_c \cap \Gamma'_c$ both of maximal dimension and z would not be regular. Thus $R_d(T) \cap Y_\Gamma = \emptyset$ and so $R_d(T) \subset \bigcup_{\Gamma \notin \Lambda^{d-1}} M'_\Gamma$.

Conversely, if $p \in M'_\Gamma$ then $p \in M_\Gamma$ and there is a neighborhood U of p such that $U \cap \Gamma_c \subset T$. So $\Gamma_{c,p} \subset T_p$. Since $p \notin Y_\Gamma$ we have $p \notin \overline{M}_{\Gamma'}$ and $p \notin \overline{\Gamma}'_1$ if $\Gamma' \neq \Gamma$. But, $T = (\bigcup_{\Gamma \notin \Lambda^{d-1}} \overline{M}_\Gamma) \cup (\bigcup_{\Gamma \in \Lambda^{d-1}} \Gamma_1)$ and then $T_p = (\overline{M}_\Gamma)_p \subset \overline{\Gamma}_p \subset \Gamma_{c,p}$. Thus $\Gamma_{c,p} = T_p$. Since $p \in M_\Gamma$, $p \in \operatorname{Reg} \Gamma_c$ and so p is regular of T of dimension d.

To prove the assertion about the dimension, it is enough to see that the sets Y_Γ are of dimension less than d. If $\Gamma' \in \Lambda^{d-1}$, trivially $\dim Y_{\Gamma\Gamma'} < d$. If $\Gamma' \in \Lambda^{d-1}$, $Y_{\Gamma\Gamma'} \subset \Gamma_c \cap \Gamma'_c$ and then $\dim Y_{\Gamma\Gamma'} \leq \dim (\Gamma_c \cap \Gamma'_c) < d$. So, $\dim Y_\Gamma = \dim (\bigcup_{\Gamma' \neq \Gamma} Y_{\Gamma\Gamma'}) < d$ which completes the proof.

REMARK 2.13. If $p_1, \ldots, p_s$ are the ideals of degeneracy of $R_k(T)$ all of them are prime and there is a normal partition of $R_k(T)$, $\Lambda = \{\Gamma^1, \ldots, \Gamma^s\}$, such that $\Gamma^i = V(p_i)$ $1 \leq i \leq s$. To do this, note that $R_k(T) \cap V(p_i)$ is open in $V(p_i)$ and then

$$A^i = \mathbb{R}^n \setminus [V(p_i) \setminus (R_k(T) \setminus V(p_i))]$$

is open and semialgebraic. It is enough to set $\Gamma^i = A^i \cap V(p_i)$.

An analogous result holds for $R(T)$. Furthermore, in the above partition we may assume that the Γ^i are connected, just taking connected components. As these are connected components of $R(T)$ too, we obtain a canonical normal partition Λ of $R_d(T)$ by considering its connected components. It is such that if $x \in \Gamma$, $\Gamma \in \Lambda$, $T_x = R(T)_x = \Gamma_x$.

3. NORMAL STRATIFICATIONS. ORDERS BETWEEN NORMAL STRATIFICATIONS.

In this section we study the question pointed out in section 2, namely the uniqueness of the normal partitions related with the index function that they define. The stratifications arise in a natural way as the appropriate frame to study this problem. By a stratification of T we mean a finite partition where each member is connected and verifying the Whitney frontier condition. We begin with the following definition:

DEFINITION 3.1. Let Λ, Λ' be normal partitions of T. We will say that Λ' is thinner than Λ and we will write $\Lambda < \Lambda'$ if for every $x \in T$, $\mathrm{ind}^T_{\Lambda'}(X) \leq \mathrm{ind}^T_\Lambda(x)$.

The relation $<$ does not define an order in general, but it does if we restrict ourselves to consider normal stratifications, that is, normal partitions which are stratifications. We will use the following notation: given Λ normal strafification of T, $\dim T = d$, let $\Lambda^k = \{\Gamma \in \Lambda \mid \dim \Gamma \leq k\}$, $T^k_\Lambda = \bigcup_{\Gamma\in\Lambda^k} \Gamma$, $Q_k = T^k - T^{k-1}$, $Q_o = T^o$ and $\Lambda_k = \Lambda^k - \Lambda^{k-1}$, $\Lambda^o = \Lambda_o$.

LEMMA 3.2. Let Λ, Λ' normal stratifications of T. Then:

(a) If $x \in Q_d$ then $\mathrm{ind}^T_\Lambda(x) = 0$

(b) If $Q_k = Q'_k$, $r + 1 \leq k \leq d$, $0 \leq r \leq d$, then:

(1) $T^r_\Lambda = T^r_{\Lambda'}$ and

(2) the strata of dimension higher than r are the same in Λ and Λ', i.e., $\Lambda \setminus \Lambda^r = \Lambda' \setminus \Lambda'^r$.

(c) If $\Lambda < \Lambda'$ and $\Lambda' < \Lambda$ then $Q_k = Q'_k$ $0 \leq k \leq d$.

(d) The relation $<$ defines an order in the set of the normal stratifications of T, i.e., the index function characterises the stratification.

PROOF. (a) is trivial. We will prove (b) for the case $r = d-1$. The rest of the cases are immediate by recurrence. Suppose, then, $Q_d = Q'_d$. Trivially $T^{d-1}_\Lambda = T^{d-1}_\Lambda$. Let $Q_d = \bigcup_{\Gamma\in\Lambda_d} \Gamma$ and $Q'_d = \bigcup_{\Gamma'\in\Lambda'_d} \Gamma'$. For each $\Gamma \in \Lambda_d$ we have $\Gamma = \bigcup_{\Gamma'\in\Lambda'_d} (\Gamma \cap \Gamma')$. But if $\Gamma' \cap \Gamma \neq \emptyset$ then, from (a), $\Gamma_c = \Gamma'_c$. Thus, for each $\Gamma' \in \Lambda'_d$, $\Gamma' \cap \Gamma$ is open in Γ. Since the Γ' are disjoint and Γ is connected there is unique $\Gamma' \in \Lambda'_d$ such that $\Gamma' \cap \Gamma \neq \emptyset$. Then $\Gamma = \Gamma \cap \Gamma'$ and $\Gamma \subset \Gamma'$. Symmetrically we obtain the equality.

To prove (c) note first of all that if $x \in Q_d$ then x is a regular point of T of dimension d and $\mathrm{ind}^T_\Lambda(x) = 0$. Then $\mathrm{ind}^T_{\Lambda'}(x) = 0$ and so $x \in \Gamma' \in \Lambda'_d$. Thus $x \in Q'_d$. Again by symmetry we conclude

$Q_d = Q'_d$. Assume, now, that $Q_k = Q'_k$, $r < k \le d$. Then by (b) $T^r = T^r_\Lambda$ and $\Lambda \setminus \Lambda^r = \Lambda' \setminus \Lambda'^r$. So, for every $x \in T^r_\Lambda$,

$$\operatorname{ind}^{T^r}_{\Lambda^r \big| T^r}(x) = \operatorname{ind}^{T^{r}{}'}_{\Lambda'^r \big| T^r}(x)$$

and repeating the same argument as above replacing T and Q_d by T^r and Q_r respectively we get $Q_r = Q'_r$.

(d) is an immediate consequence of (b) and (c).

After Lemma 3.2 another order in the set of normal stratifications of T can be considered:

DEFINITION 3.3. Let Λ, Λ' normal stratifications of T. We shall write $\Lambda \lhd \Lambda'$ if $Q_k = Q'_k$ for $r < k \le d+1$ and $Q_r \subset Q'_r$ for some r, where $Q_{d+1} = Q'_{d+1} = \emptyset$. (Lexicographic order in the Q_r.)

The relation between $<$ and $\lhd$ is shown in Lemma 3.2.

COROLLARY 3.4. The order $<$ is stronger than the order $\lhd$, i.e., if $\Lambda < \Lambda'$ then $\Lambda \lhd \Lambda'$.

We try to find a maximum for the order $<$. We shall look for, first of all, a maximum for $\lhd$. Then, if $<$ has a maximum, this must be this last one.

CONSTRUCTION 3.5. We shall construct a sequence of semialgebraic closed sets in T, $Z_d = T \supset \dots \supset Z_n \supset Z_{n-1} \supset \dots Z_0 \supset Z_{-1}$ such that for each n, $\dim Z_n \le n$ and $T \setminus Z_n$ is stratified, $T \setminus Z = \bigcup_{\alpha \in A_n} S_\alpha$ in the following way:

(i) $\{S_\alpha\}_{\alpha \in A_n}$ is a normal partition of $T \setminus Z_n$.

(ii) Each S_α is a semialgebraic connected set of dimension higher than n, which is a smooth analytical variety.

(iii) $\{S_\alpha\}_{\alpha \in A_n}$ verifies the frontier Whitney condition, i.e., if $S_\alpha \cap \overline{S}_\beta \ne \emptyset$ then $S_\alpha \subset \overline{S}_\beta$ and $\dim S_\alpha < \dim S_\beta$.

It is evident that $Z_{-1} = \emptyset$ and that in $d+1$ steps we obtain a normal stratification of T. All the adherences to appear are considered in T.

Suppose that Z_{n+1} is constructed under the conditions above. We construct Z_{n+1} as follows: for each $\alpha \in A_{n+1}$ let $X_\alpha = \overline{(Z_{n+1} \setminus \overline{S}_\alpha)} \cap \overline{S}_\alpha$. Then $X_\alpha \subset \overline{(Z_{n+1} \setminus \overline{S}_\alpha)} \setminus (Z_{n+1} \setminus \overline{S})$ and so $\dim X_\alpha < \dim X_{n+1}$. Now $\operatorname{Sing}(Z_{n+1}) = \{X \in Z_{n+1} \mid x$ is not regular on $Z_{n+1}\}$ is semialgebraic and closed in Z_{n+1} and a fortiori in T of dimension less than Z_{n+1} (prop. 2.11). Let

$$Z_n^* = \text{Sing } Z_{n+1} \cup (\bigcup_\alpha X_\alpha)$$

and consider the connected components of $Z_{n+1} \setminus Z_n^*$. We take, finally:

$$Z_n = Z_n^* \cup \{\text{connected components of } Z_{n+1} \setminus Z_n^* \text{ of dimension less than } n+1\}.$$

Then Z_n is closed in Z_{n+1} and $\dim Z_n \leq n$. Moreover,

$$Z_{n+1} \setminus Z_n = \{\text{conn. components of } Z_{n+1} \setminus Z_n^* \text{ of dimension } n+1\}.$$

Let these components be $\{S'_{\alpha'}\}_{\alpha' \in A'}$. We claim that $\{S_\alpha\}_{\alpha \in A_n} \cup \{S'_{\alpha'}\}_{\alpha' \in A}$ is a partition of $T \setminus Z_n$ in the required conditions.

Indeed, $\{S'_{\alpha'}\}_{\alpha' \in A'}$ is a normal partition of $Z_{n+1} \setminus Z_n$ since $Z_{n+1} \setminus Z_n$ is open in $R_{n+1}(Z_{n+1})$ and remark 2.13 applies. To check the frontier property, it is enough to see the case $S'_{\alpha'} \cap \overline{S}_\alpha \neq \emptyset$, the others being trivial. But $X_\alpha = (Z_{n+1} \setminus \overline{S}_\alpha) \cap \overline{S}_\alpha \subset Z_n$ and by Lemma 3.6 $\overline{S}_\alpha \cap (Z_{n+1} \setminus Z_n)$ is open and closed in $Z_{n+1} \setminus Z_n$ and thus it contains all components it intersects. So, $S'_{\alpha'} \subset \overline{S}_\alpha$ and $\dim S'_{\alpha'} = n+1 < \dim S_\alpha$, which completes the construction.

LEMMA 3.6. Let $T \subset \mathbf{R}^n$, $E \subset T$, Z' and Z closed in T, $Z' \subset Z$. If $(Z - \overline{E}) \cap \overline{E} \subset Z'$ then $\overline{E} \cap (Z \setminus Z')$ is open and closed in $Z \setminus Z'$.

PROOF. Straightforward.

REMARK 3.7. (a) In 3.5, in the construction of Z_n, only strata of dimension $n+1$ are introduced. So, the strata are introduced step by step in decreasing dimension order. Thus, if we call Λ the normal stratification obtained in 3.5, we have $Z_n = T^n_\Lambda$.

(b) Z_n is the complementary of $R_{n+1}(T^{n+1}_\Lambda)$, plus the elements of $R_{n+1}(T^{n+1}_\Lambda)$ that lie in some X_α. So if $x \in R_{n+1}(T^{n+1}_\Lambda)$ and x is not in any strata of dimension $n+1$ of Λ, then $x \in X_\alpha$ for some S_α with $\dim S > n+1$.

(c) $Q_d = R_d(T)$ and $Q_n \subset R_n(T^n)$, $0 \leq n \leq d$.

We characterise now the stratification obtained:

PROPOSITION 3.8. Let Λ be the stratification of 3.5. We have:

(a) Λ is a maximum for the order $\lhd$.

(b) If $Y = \{x \in T \mid \text{there is a normal partition } \beta \text{ of } T \text{ with } \operatorname{ind}^T_\beta(x) = 0\}$, and $Y_\Lambda = \{x \in T \mid \operatorname{ind}^T(x) = 0\}$ then $Y = Y_\Lambda$ (i.e., the set of points with index zero is the biggest possible in Λ).

PROOF. (a) Let Λ' be another normal stratification of T. We shall see that $\Lambda' \lhd \Lambda$. Let $x \in T$, $x \in Q'_d$. Then x is regular of T of dimension d and by 3.7.(c) $x \in Q_d$. Thus $Q'_d \subset Q_d$. If $Q'_d \subsetneq Q_d$ the result holds directly. Otherwise, assume $Q'_s = Q_s$ if $k+1 \le s \le d$. Then by 3.2.(b) $T^k_\Lambda = T^k_{\Lambda'}$ and $\Lambda \setminus \Lambda^k = \Lambda' \setminus \Lambda'^k$. Let $x \in Q'_k$ and $x \notin Q_k$. Then $x \in R_k(T^k_\Lambda)$ and x does not belong to any strata of dimension k of Λ; hence by 3.7.(b) there is $S_\alpha \in \Lambda$ with $\dim S_\alpha \ge k+1$ and $x \in X_\alpha$. But $S_\alpha \in \Lambda'$ and, if Γ' is the strata of Λ' to which x belongs, we have $\Gamma' \subset \overline{S}_\alpha$ as a consequence of the frontier condition. But since x is regular on $T^k_\Lambda = Z_k$ there is an open neighborhood U of x such that $U \cap T^k_\Lambda = \overline{U \cap \Gamma'}$. Thus $U \cap T^k_\Lambda = U \cap Z_k \subset \Gamma' \subset \overline{S}_\alpha$ and then $U \cap (Z_k \setminus \overline{S}_\alpha) = \emptyset$ so $x \notin \overline{(Z_k \setminus \overline{S}_\alpha)}$ and $x \notin X_\alpha$. Contradiction!

To prove (b) we shall prove that the set of regular points of T of dimension k is a union of strata of dimension k. So they have index zero in Λ. Since, if x verifies that ind $(x) = 0$ for some partition β then x is a regular point of T, we obtain $Y \subset Y_\Lambda$. The other inclusion is trivial just taking $B = \Lambda$. So, the proof is complete after the two following lemmas:

LEMMA 3.9. Let Γ be a connected component of $X_{n+1} \setminus Z^*_n$ such that $\dim \Gamma = k \le n$ and for every $x \in \Gamma$, x is regular on T of dimension k Then Γ is a connected component of $Z_n \setminus Z^*_{n-1}$.

PROOF. Γ is open in $Z_{n+1} \setminus Z^*_n$ that is open in Z_{n+1}. So Γ is open in Z_n. Let us see that Γ is also closed. Since Γ is a connected component of $Z_{n+1} \setminus Z^*_n$, $\overline{\Gamma} \setminus \Gamma \subset Z^*_n$. Let $x \in \overline{\Gamma} \setminus \Gamma$. Then x must belong to Sing Z_{n+1} or to some X_α with $\dim S_\alpha > n+1$. If $x \in$ Sing Z_{n+1} and $x \notin \overline{S}'_\alpha$ for any $S'_{\alpha'}$ with $\dim S'_{\alpha'} = n+1$ (i.e., $S'_{\alpha'}$ is another component of $Z_{n+1} \setminus Z^*_n$) the germs of Z_{n+1} and Z_n at p are the same and $x \in$ Sing Z_n.

On the other hand, if $x \in \overline{S}'_{\alpha'}$ then $x \in X_{\alpha'} = \overline{(Z_n \setminus \overline{S}'_{\alpha'})} \cap \overline{S}'_{\alpha'}$. (If not it would be an open neighborhood U of x such that $U^x \cap (Z_n \setminus \overline{S}'_{\alpha'}) = \emptyset$, i.e., $U^x \cap Z_n \subset \overline{S}_\alpha$ and then $\Gamma \alpha \overline{S}'_{\alpha'}$ which contradicts the hypothesis about the regularity of the points of Γ.) Therefore $x \in X_{\alpha'}$ and $x \in Z^*_{n-1}$.

Finally if $x \in X_\alpha$ for some S_α with $\dim S_\alpha > n+1$ then $x \in \tilde{X}_\alpha$ where $\tilde{X}_\alpha = \overline{(Z_n \setminus \overline{S}_\alpha)} \cap \overline{S}_\alpha$. If not, we would have $\Gamma \subset \overline{S}_\alpha$ and again we get a contradiction.

So, in any case we have $x \in Z^*_{n-1}$ and then $\overline{\Gamma} \setminus \Gamma \subset Z^*_{n-1}$. But then Γ is closed in $Z_n \setminus Z^*_{n-1}$ providing that $Z^*_{n-1} \subset Z^*_n$.

LEMMA 3.10. $Z^*_n \subset Z^*_{n+1}$ for $0 \leq n \leq d-2$.

PROOF. We have:

$$Z^*_{n+1} = \text{Sing } Z_{n+2} \cup (\cup X_\alpha)$$

and

$$X_{n+1} = Z^*_{n+1} \cup A_1 \cup \ldots \cup A_t$$

where $A_1,\ldots,A_t$ are the connected components of $Z_{n+2} \setminus Z^*_{n+1}$ of dimension less or equal than $n+1$. We have $A_i \subset R(Z_{n+2})$ and $A_i \subset R(Z_{n+1})$, $1 \leq i \leq t$, because Z^*_{n+1} is closed. Therefore $\text{Sing } Z_{n+1} \subset Z^*_{n+1}$. Moreover, if $S'_{\alpha'}$ is a component of $Z_{n+2} \setminus Z^*_{n+1}$ of dimension $n+2$ then $S'_{\alpha'} \cap A_i = \emptyset$, $1 \leq i \leq t$, and so $\overline{S'_{\alpha'}} \cap Z_{n+1} = \overline{S'_{\alpha'}} \cap Z^*_{n+1}$ and $X_{\alpha'} = \overline{(Z_{n+1} \setminus \overline{S'_{\alpha'}})} \cap \overline{S'_{\alpha'}} \subset Z_{n+1} \cap \overline{S'_{\alpha'}} \subset Z^*_{n+1}$.

Finally, let S_α with $\dim S_\alpha > n+2$. Then

$$\tilde{X}_\alpha = \overline{(Z_{n+1} \setminus \overline{S}_\alpha)} \cap \overline{S}_\alpha \subset \overline{(Z_{n+2} \setminus \overline{S}_\alpha)} \cap \overline{S}_\alpha = X_\alpha \subset Z^*_{n+1}.$$

Thus, $Z^*_n = \text{Sing } Z_{n+1} \cup (\cup X_{\alpha'}) \cup (\cup \tilde{X}_\alpha) \subset Z^*_{n+1}$.

So, we obtain in the end:

COROLLARY 3.12. The set $R_k(T)$ is a union of strata of dimension k of Λ.

PROOF. Clearly $R_k(T)$ is the union of the components of dimension k of $Z_d \setminus Z^*_{d-1}$. But from lemma 3.9 all of them are components of $Z_k \setminus Z^*_k$ and strata of Λ of dimension k.

This completes the proof of 3.8.(b).

DEFINITION 3.13. The normal stratification obtained in 3.5 is called <u>the standard decomposition of</u> T. So, as a summary of this section we can establish the following:

THEOREM 3.14. Given a semialgebraic set T we can associate to it families $\{A_1,\ldots,A_t\}$ and $\{C_1,\ldots,C_t\}$ of open semialgebraic and algebraic sets respectively, depending only on T such that: $\Lambda = \{A_i \cap C_i\}$ is a normal stratification of T.

PROOF. The standard decomposition is uniquely determined by T. Let it be $\Lambda = \{\Gamma^i\}_{i=1,\ldots,t}$. Then, take $C_i = \Gamma^i_c$ and $A_i = \mathbb{R}^n \setminus (C_i \setminus \Gamma_i)$.

Coming back to the question about the existence of a maximum for the order $<$, it is clear that if it does exist it must be the standard decomposition. Unfortunately it is easy to check that it is not the case

EXAMPLE 3.15. Let $C = V(z^3 + xz + y^3$ (see picture below) and $W = \{y < 0\} \cap \{x \geq 0\}$. We consider the semialgebraic $T = C \setminus W$.

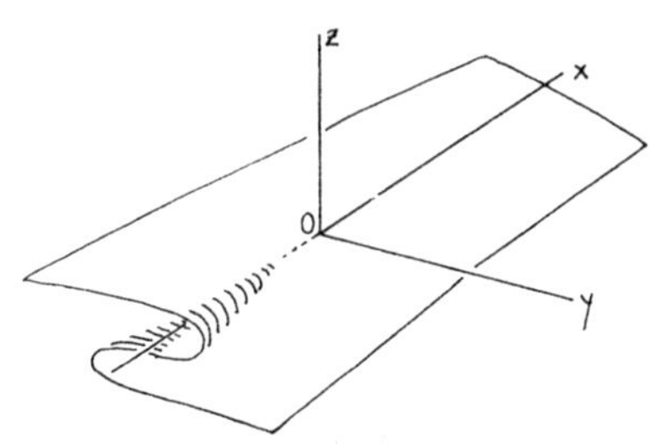

The standard decomposition of T is $T = \Gamma_1 \cup \Gamma_2 \cup \Gamma_3$ with

$\Gamma_1 = R(T)$, $\Gamma_2 = \{(x,0,0) \mid x > 0\}$ and $\Gamma_3 = \{0 = (0,0,0)\}$.

Then $\mathrm{ind}_\Lambda^T(0) = 2$.

On the other hand, the normal stratification $\Lambda' = \{\Gamma_1', \Gamma_2', \Gamma_3'\}$, where

$\Gamma_3' = \{B = (-1,0,0)\}$, $\Gamma_2' = \{(x,0,0) / x > -1\}$ and $\Gamma_1' = T \setminus (\Gamma_2' \cup \Gamma_3')$

is such that $\mathrm{ind}_{\Lambda'}^T(0) = 1$.

4. APPLICATIONS

Now, using the standard decomposition of T (or more generally normal stratifications) we prove some facts on semialgebraic sets. The most important is the result about coherence analogous to Galbiati (1976) for semianalytic sets.

PROPOSITION 4.1. Let T be a semialgebraic set and I an ideal of degeneracy of T. The set $T(I) = \{x \in T \mid J_x(T) = I\}$ is semialgebraic.

PROOF. It is clear that $T(I) = \{x \in T \mid V(J_x(T)) = V(I)\}$. Let Λ be the standard decomposition of T. We shall prove that $T(I)$ is compatible with Λ, i.e., if $\Gamma \cap T(J) \neq \emptyset$, then $\Gamma \subset T(I)$. So $T(I)$ is a union of strata and semialgebraic. It is enough to prove that each $\Gamma \in \Lambda$ has the following property: if $x,y \in \Gamma$ then $V(J_x(T)) = V(J_y(T))$. But, by 2.5, $V(J_x(T)) = \cup \Gamma_c'$ for those $\Gamma' \in \Lambda$ such that $p \in \overline{\Gamma}'$. Since Λ is a stratification, $\Gamma \subset \overline{\Gamma}'$ for those Γ', and then $q \in \overline{\Gamma}'$. Again by 2.5, $V(J_y(T)) \subset V(J_x(T))$. The symmetry proves the equality.

REMARK 4.2. Note that if p, q are points of the same strata, $J_p(T) = J_q(T) = \cap p_i$ where the p_i are the primes $J(V(\Gamma_i))$ for those $\Gamma_i \notin \Lambda$ such that $\Gamma \subset \overline{\Gamma}_i$, and that this happens for every normal stratification, not merely the standard one.

DEFINITION 4.3. Let T be a semialgebraic set. We say that $x \in T$ is coherent if there is an open neighborhood U of x and polynomials $f_1,\ldots,f_r$ such that for each $y \in U \cap T$ one has

$$(f_1,\dots,f_r)\ \mathbb{R}[\underline{X}]_y = J_y(T)\ \mathbb{R}[\underline{X}]_y$$

REMARK 4.4. If we consider the sheaf $\mathfrak{J}_T$ of ideals of T defined for each open set U of $\mathbb{R}^n$ as

$$\mathfrak{J}_T(U) = \{f \in \mathbb{R}(\underline{X}) \mid f \text{ is regular on } U \cap T \text{ and } f(U \cap T) = 0\},$$

the above definition coincides with the usual one, namely that the sheaf $\mathfrak{J}_T$ is coherent at x.

PROPOSITION 4.5. A point x of T is coherent if and only if there is an open neighborhood U of x such that $I(U \cap T) = J_x(T)$ and for each $y \in U$ one has $J_x(T)\mathbb{R}[\underline{X}]_y = J_y(T)\mathbb{R}[\underline{X}]_y$.

PROOF. Suppose x is coherent. Shrinking, if necessary, the open U of the definition 4.3, we may assume that $I(U \cap T) = J_x(T)$ and that if $y \in U$ then $J_y(T) \supset J_x(T)$. Taking $y = x$ in 4.3 we obtain:

$$(f_1,\dots,f_r)\ \mathbb{R}[\underline{X}]_x = J_x(T)\ \mathbb{R}[\underline{X}]_x$$

and then $(f_1,\dots,f_r) \subset J_x(T)$. But then, for each $y \in U$ we have

$$(f_1,\dots,f_r)\ \mathbb{R}[\underline{X}]_y \subset J_x(T)\ \mathbb{R}[\underline{X}]_y \subset J_y(T)\ \mathbb{R}[\underline{X}]_y$$

and by the coherence of x, $J_x(T)\ \mathbb{R}[\underline{X}]_y = J_y(T)\ \mathbb{R}[\underline{X}]_y$.

The converse is immediate.

Before we prove the proposition about the semialgebricity of the non-coherent points of T we need some preliminary considerations. Let Λ be the standard decomposition of T. The family of prime ideals $\{I(\Gamma)\}_{\Gamma\in\Lambda}$ (respectively algebraic sets $\{\Gamma_c\}_{\Gamma\in\Lambda}$), will be called the associated primes (resp. algebraic sets) of T. Then, given $x \in T$, the ideal of degeneracy $J_x(T)$ is a real radical ideal which has a decomposition in minimal primes: $J_x(T) = p_1 \cap \dots \cap p_r$, where all the p_i are associated primes of T.

LEMMA 4.6. Let x be a coherent point of T, and suppose $J_x(T) = p_1 \cap \dots \cap p_r$ to be the decomposition of $J_x(T)$ in minimal primes. Then, there is an open neighborhood U of x such that if $y \in U$ the minimal primes of $J_y(T)$ are among $p_1,\dots,p_r$. That means $J_y(T) = p_{i_1} \cap \dots \cap p_{i_j}$ with $p_{i_k} \in \{p_1,\dots,p_r\}$, $1 \le k \le j$. Moreover, if $p_s \notin \{p_{i_1},\dots,p_{i_p}\}$, $1 \le s \le r$, then $p_s\ \mathbb{R}[\underline{X}]_y = (1)$.

PROOF. Let us take U such that if $y \in U$, $J_x(T) \subset J_y(T)$ and that for each $y \in U$, $J_x(T)\ \mathbb{R}[\underline{X}]_y = J_y(T)\ \mathbb{R}[\underline{X}]_y$. Then, if p_{i_j} is a minimal prime of $J_y(T)$ which is not among $p_1,\dots,p_r$, let $\{p_1,\dots,p_j\}$ be the subset of $\{p_1,\dots,p_r\}$ such that $p_k \subsetneq p_{i_j}$, $1 \le k \le j$. Then, let $z \in T$ be a point of U such that $J_z(T) = p_{i_j}$. (It is clear that such a point exists as a consequence of the minimality of p_{i_j}.) Thus, we have:

$$J_x(T)\,\mathbb{R}[\underline{X}]_z = (p_1 \cap \cdots \cap p_r)\,\mathbb{R}[\underline{X}]_z \subset (p_1 \cap \cdots \cap p_j)\,\mathbb{R}[\underline{X}]_z \subsetneq p_{i_j}\,\mathbb{R}[\underline{X}]_z,$$

contradiction. Furthermore, if $p_s \notin \{p_{i_1},\ldots,p_{i_j}\}$ and $p_s\,\mathbb{R}[\underline{X}]_y \neq (1)$ then $(p_{i_1} \cap \cdots \cap p_{i_j})\,\mathbb{R}[\underline{X}]_y \subset (p_{i_1} \cap \cdots \cap p_{i_j} \cap p_s)\,\mathbb{R}[\underline{X}]_y$ and we obtain $p_{i_1} \cap \cdots \cap p_{i_j} \subset p_s$ which contradicts the minimality of p_s. So, $p_s\,\mathbb{R}[\underline{X}]_y = (1)$.

Consider now a normal stratification of T that is compatible with the family of associated algebraic sets of T. (This can be obtained defining for each associated algebraic set of T, C_k, the set $Y_k = \overline{(Z_{n+1} \setminus C_k)} \cap C_k$ and repeating again the construction 3.5 with $Z_n^* = \mathrm{Sing}(Z_{n+1}) \cup (\bigcup_\alpha X_\alpha) \cup (\bigcup_k Y_k)$.

PROPOSITION 4.7. Let T be a semialgebraic set. The set of non-coherent points of T is semialgebraic. Moreover, its dimension is less than or equal to $d-2$.

PROOF. Denote by Λ a normal stratification compatible with the family $C_1,\ldots,C_t$ of associated algebraic sets of T. We shall see that set of coherent points, $\mathrm{Coh}(T)$, is compatible with Λ.

Let $x \in \Gamma \cap \mathrm{Coh}(T)$, $\Gamma \in \Lambda$. Then there is an open neighborhood U of x such that: (1) lemma 4.6 holds, and (2) if $\Gamma \not\subset \overline{\Gamma'}$ then $U \cap \Gamma' = \emptyset$. Let x' be another point of Γ and choose U', open neighborhood of x', such that (1) if $\Gamma \not\subset \overline{\Gamma'}$ then $U' \cap \Gamma' = \emptyset$, and (2) $I(U' \cap T) = J_{x'}(T)$. Let, finally, $y' \in U' \cap T$ and suppose $y' \in \Gamma'' \in \Lambda$. Then there exist $y \in U \cap \Gamma''$ and by remark 4.2, $J_x(T) = J_{x'}(T)$ and $J_y(T) = J_{y'}(T)$. Thus, with the notation of 4.6, we have

$$J_{x'}(T)\,\mathbb{R}[\underline{X}]_{y'} = J_x(T)\,\mathbb{R}[\underline{X}]_{y'} = (p_1 \cap \cdots \cap p_s)\,\mathbb{R}[\underline{X}]_{y'}.$$

But, by lemma 4.6, if $p_s \notin \{p_{i_1},\ldots,p_{i_j}\}$ then $p_s\,\mathbb{R}[\underline{X}]_y = (1)$, i.e., $y \notin V(p_s) = C_s$ and by the compatibility of Λ with C_s, $y' \notin V(p_s)$ and $p_s\,\mathbb{R}[\underline{X}]_{y'} = (1)$. Thus,

$$J_{x'}(T)\,\mathbb{R}[\underline{X}]_{y'} = (p_{i_1} \cap \cdots \cap p_{i_j})\,\mathbb{R}[\underline{X}]_{y'} = J_y(T)\,\mathbb{R}[\underline{X}]_{y'} = J_{y'}(T)\,\mathbb{R}[\underline{X}]_{y'}$$

which proves that x' is coherent.

Suppose now that $\Gamma \in \Lambda$ is of dimension greater or equal than $d-1$. If $\dim \Gamma = d$ the points of Γ are regular and hence coherent. If

$\dim \Gamma = d-1$ then $\Gamma \subset \overline{\Gamma}_1 \cap \dots \cap \overline{\Gamma}_r$ where $\dim \Gamma_i = d$, $1 \le i \le r$. (This intersection can be empty but then the case is trivial.) Then, if $x \in \Gamma$, $J_x(T) = p_1 \cap \dots \cap p_r$ with p_i prime of dimension d, $1 \le i \le r$. But if we take U such that $I(U \cap T) = J_x(T)$, and $U \cap \Gamma' = \emptyset$ if $\Gamma' \in \Lambda - \{\Gamma, \Gamma_1, \dots, \Gamma_r\}$, we have

(i) for each $y \in U - \Gamma$, $J_y(T) = p_i$ for some i, $1 \le i \le r$.

(ii) if $p_j \ne p_i$ then $p_j \mathbb{R}[\underline{X}]_y = (1)$. In other case $y \in V(p_j)$ and by the compatibility of Λ with $V(p_i)$ and $V(p_j)$ one would have $y \in \Gamma_i \subset V(p_i) \cap V(p_j)$, which implies $\dim \Gamma_i < d$, contradiction.

But then,

$$J_x(T)\, \mathbb{R}[\underline{X}]_y = (p_1 \cap \dots \cap p_r)\, \mathbb{R}[\underline{X}]_y = p_i\, \mathbb{R}[X]_y = J_y(T)\, \mathbb{R}[\underline{X}]_y$$

and therefore x is coherent.

So, the set of non-coherent points of T is contained in the union of the strata of dimension less or equal than $d-2$, which completes the proof.

REMARK 4.8. A direct proof, without the use of stratifications, that the set of non-coherent points is semialgebraic has been given by M. Alonso and J. M. Gamboa, who have proved that if $T = \cup \Gamma$ is a normal decomposition of T then

$$\mathrm{Coh}(T) = [T \setminus \overline{(T \setminus M)}] \cup \mathrm{Coh}\,(T \setminus M),$$

where $M = \bigcup_i (\Gamma \setminus \bigcup_{\Gamma'_c \ne \Gamma_c} \Gamma'_c)$. Nevertheless this way doesn't prove the result about the dimension.

REFERENCES

BRUMFIEL, 1979: "Partially ordered rings and semi-algebraic geometry". Lect. Notes of the London Math. Soc. 37.

GALBIATI, 1976: "Stratifications et ensemble de non coherénce d'un espace analytique réel". Inv. Math. 34, pp. 113-133.

LOJASIEWICZ, 1965: "Ensembles semi-analytiques". Number A66765, Ecole Polytechnique, Paris.

RECIO, 1977a: "Conjuntos preanalíticos, prenáshicos y prealgebraicos". Mem. del Inst. Jorge Juan, Madrid.

RECIO, 1977b: "Una descomposición de un conjunto semialgebraico". Actas V Congreso de Matemáticos de Expresión Latina, Mallorca, Spain.

S.G.A.R.M., 1980: Notas del Seminario de Geometría Algebraica Real, Madrid.

WHITNEY, 1957: "Elementary structure of real algebraic varieties". Ann. of Math. 66, pp. 545-556.

DEPARTAMENTO DE ALGEBRA Y FUNDAMENTOS
UNIVERSIDAD (COMPLUTENSE) DE MADRID
MADRID

and

DEPARTMENT OF MATHEMATICS AND STATISTICS
UNIVERSITY OF NEW MEXICO
ALBUQUERQUE

Contemporary Mathematics
Volume 8, 1982

SOME OPEN PROBLEMS

Collected by Gregory Brumfiel

PREFACE. We have Professor Brumfiel to thank for the problems collected here, for it was his idea to include them, and in San Francisco he furnished the persuasion which resulted in the contributions. Moreover, he teamed with Carlos Andradas and Victor Espino to put the problems together. There were some corrections mailed in by contributors and I have made a few others. I take full responsibility for any errors that may still appear.

Problems posed by Hans Delfs and Manfred Knebusch (Regensburg)

For the basic definitions and some results in the theory of semi-algebraic spaces we refer the reader to our paper "Semialgebraic topology over a real closed field" in these proceedings, marked by [*].

1) Let M be a separated semialgebraic space over a real closed field R. Is M an affine semialgebraic space?

An affirmative answer to this question would, e.g., imply that every separated semialgebraic space can be triangulated and that Poincaré duality is valid for the set $V(R)$ of real points of every smooth complete variety V over R.

2) Let M be an affine semialgebraic space.

DEFINITION. A singular q-simplex in M is semialgebraic map $\sigma : \Delta_q \longrightarrow M$ from the q-dimensional standard simplex

$$\Delta_q = \{(x_0,\dots,x_n) \in R^{n+1} \;\Big|\; \text{all } x_i \geq 0, \; \sum_{i=0} x_i = 1\}$$

into M.

Using this definition one can form in the usual way singular homology and cohomology groups of M. Do these groups coincide with the semialgebraic homology and cohomology groups of M, as they were defined in [D] (cf. [*])?

1980 Mathematics Subject Classification 12D15, 12J15, 14G30, 58A07, 10C04, 13K05.

0271-4132/81/0000-0672/$02.75

The methods used to prove the corresponding fact in the theory of topological spaces, relying on barycentric subdivision and simplicial approximation, do not apply in the case of a non-archimedian real closed field R.

3) Let V be a smooth algebraic variety over an algebraically closed field C of characteristic 0. Choose a real closed subfield R of C with $C = R(\sqrt{-1})$. Identifying C with R^2 we can regard the set $V(C)$ of C-rational points of V as a semialgebraic space over R. We can then form the semialgebraic cohomology groups $H^q_{sa}(V(C),\mathbb{Z}/n)$ of $V(C)$ with coefficients in $\mathbb{Z}/n (n > 0)$. Do these groups coincide with the etale cohomology groups $H^q_{et}(V,\mathbb{Z}/n)$?

In the case $(C,R) = (\mathbb{C},\mathbb{R})$ this is true as is well known and due to M. Artin ([SGA4, 3]).

4) <u>The extended signature problem</u>.

Let V be a quasicompact separated scheme. Consider the real spectrum $X(V)$ of V as it is defined by M. F. Coste-Roy and M. Coste ([C]) and by L. Bröcker ([B]). The points of $X(V)$ are the pairs (P,p) consisting of a point p of V and an ordering P of the residue class field $\kappa(p) = \mathcal{O}_{V,p}/m_{V,p}$ of V in p. A subbasis of the topology of $X(V)$ is given by the sets

$$D_{U,f} = \left\{ (P,p) \;\middle|\; \begin{array}{l} p \in U;\ \text{the image of } f \text{ in } \kappa(p) \text{ is not} \\ \text{no zero or positive in the ordering } P \end{array} \right\},$$

where U runs through the open affine subsets of V and f through the coordinate ring $\Gamma(U,\mathcal{O}_U)$ of U. $X(V)$ is quasicompact.

Let $W(V)$ denote the Wittring of V, cf. [K]. Every point $P = (P,p)$ of $X(V)$ yields a signature $\tau_p : W(V) \longrightarrow \mathbb{Z}$ in the obvious way:

$$\tau_p : W(V) \xrightarrow[\text{canonical}]{} W(\kappa(p)) \xrightarrow[\text{Sign}_p]{} \mathbb{Z}\,.$$

It is easily seen that the map $P \longrightarrow \tau_p$ from the space $X(V)$ to the profinite space $\mathrm{Sign}(V)$ of signatures on V is continuous. Assume now that V is also divisorial (a very weak assumption, cf. [K, III §1]). Then this map $\pi : X(V) \longrightarrow \mathrm{Sign}(V)$ is also <u>surjective</u>. This is just the theorem, explained in [K, Chapter III] and going back to Dress and Kanzaki-Kitmura, that every signature of V factors through the residue class field $\kappa(p)$ of some point p of V.

LEMMA. If $Q = (Q,q)$ is a specialization of $P = (P,p)$ then $\tau_Q = \tau_P$.

PROOF. q lies in the closure $\overline{\{p\}}$ of the one point set $\{p\} \subset V$. Let $\xi \in W(V)$ be given. We choose an affine open neighbourhood U of q such that $\xi|U$ can be represented by a diagonal form:

$$\xi|U \sim \langle g_1,\ldots,g_n\rangle ,$$

with $g_i \in \mathcal{O}_V^*(U)$. There exist signs $\varepsilon_i = \pm 1$ such that $\varepsilon_i g_i$ is positive with respect to Q, hence also positive with respect to P. Thus

$$\tau_P(\xi) = \tau_Q(\xi) = \sum_{i=1}^{n} \varepsilon_i .$$

The subspace $X^{max}(V)$ of the maximal (= closed) points of $X(V)$ is Hausdorff and compact. By the lemma already the restriction

$$\pi^{max} : X^{max}(V) \longrightarrow \mathrm{Sign}(V)$$

of π is surjective and thus identifying.

PROBLEM. For which divisorial schemes V are the fibres of π^{max} just the connected components of $X^{max}(V)$ and hence the fibres of π the connected components of $X(V)$? (Recall that $\mathrm{Sign}(V)$ is totally disconnected.)

If $V = \mathrm{Spec}(A)$ with A a semilocal ring then it follows easily from the theory of signatures of A and their associated prime ideals [K_1, §4 and App. B] that π^{max} is a bijection, hence a homeomorphism from $X^{max}(V)$ onto $\mathrm{Sign}(V)$. Thus the problem has an affirmative answer in the semilocal case

Assume now that V is a divisorial variety over a real closed field. The set $V(R)$ of rational points of V can be identified with the subset $X_{max}(V)$ of $X^{max}(V)$ consisting of all points (P,p) of $X(V)$ with p maximal (= closed) in V. This subset is dense in $X^{max}(V)$ ([C],[B]). Every connected component Δ of $X^{max}(V)$ yields by intersection a semi-algebraic component $\Delta \cap V(R)$, and in this way the connected components of $X^{max}(V)$ correspond uniquely with the semialgebraic components of $V(R)$ ([C]). Recall that $V(R)$ has finitely many components $\Gamma_1,\ldots,\Gamma_r$. Now it has been proved in [DK, §5] that already the restriction of π to $V(R) = X_{max}(V)$ is surjective. Thus our problem boils down to the following slight generalization of Problem 16 in [K_2]: Every point P of Γ_j yields a signature $\tau_P = \tau_j$ which is only dependent on Γ_j, and $\tau_1,\ldots,\tau_r$ are all signatures of V. Is $\tau_i \neq \tau_j$ for $i \neq j$?

This "special signature problem" has been solved affirmatively in the following three cases:

a) V a (possibly singular) curve over an arbitrary real closed base field [Di];

b) V a smooth projective surface over $\mathbb{R}$ [CT-S];

c) V an abelian variety over $\mathbb{R}$ [CT-S].

Already trying to solve the problem for a smooth projective surface over a real closed nonarchimedian base field R seems to be an interesting task. The considerations of Colliot-Thélène and Sansuc for $R = \mathbb{R}$ use the Stone-Weierstraß theorem in an essential way, and this theorem is known to be wrong if we replace $\mathbb{R}$ by a nonarchimedian real closed field.

Despite our apparent lack of knowledge on the special signature problem we pose the general problem above to attract attention to a possibly very interesting connection between the theory of real spectra and the theory of quadratic forms.

References

[SGA 4,3] M. Artin, A. Grothendieck, et al., "Théorie des Topos et Cohomologie étale des Schemas", Tome 3, Sém. Géom. alg. du Bois Marie 1963/64, Lecture Notes Math. 305, Springer (1973).

[Br] L. Bröcker, manuscript on spectra of algebraic varieties, Munster 1980.

[CT-S] J. L. Colliot-Thélène, J. J. Sansuc, Fibrés quadratiques et composantes connexes reelles, Math. Ann. 244, 105-134 (1979).

[C] M. F. Coste-Roy, Spectre réel d'un anneau et topos etale réel, Thèse Paris 1980.

[D] H. Delfs, Kohomologie affiner semialgebraischer Räume, thesis Regensburg 1980.

[DK] Delfs-Knebusch, Semialgebraic topology I, Paths and components in the set of rational points of an algebraic variety, Math. Z., to appear.

[Di] G. Dietel, thesis, Universität Regensburg, in preparation.

[K] M. Knebusch, Symmetric bilinear forms over algebraic varieties, In: Conference on quadratic forms (Kingston 1976), pp. 103-283. Queen's papers in Pure Appl. Math. 46 (1977).

[K_1] M. Knebusch, Real closures of commutative rings I, J. reine angew. Math. 274/275, 61-89 (1975).

[K_2] M. Knebusch, Some open problems, In: Conference on quadratic forms (Kingston 1976), pp. 361-370. Queen's papers in Pure Appl. Math. 46 (1977).

Problems posed by Tomás Recio (Málaga)

5) (Suggested by Gondard-Ribenboim's paper: "Fonctions defines positives sur les variétes réeles". Bull. Sci. Math. 2^3serie, 98 (1974) 39-47.) Let $V \subset \mathbb{R}^n$ be a real variety, $u \in \mathbb{R}[X_1,\ldots,X_n] = \mathbb{R}[X]$ a polynomial function. It is known that if $u(p) > 0$ for all $p \in V$, then there is a $u' \in \mathbb{R}[X]$ with $u' - u \in I(V)$ and $u'(q) > 0$ for all $q \in \mathbb{R}^n$. Here $I(V) \subset \mathbb{R}[X]$ is the ideal of functions which vanish on V. If it is only assumed that $u(p) \geq 0$ for all $p \in V$, the analogous statement is false.

PROBLEM. Give conditions on V which imply that if $u(p) \geq 0$ for all $p \in V$, then there is a $u' \in \mathbb{R}[X]$ with $u' - u \in I(V)$ and $u'(q) \geq 0$ for all $q \geq \mathbb{R}^n$. A variation of this problem asks only for a u' with $u'(q) \geq 0$ for all q in a neighborhood of V.

6) Let V be a real variety in $\mathbb{R}^n$ and let V_c be the strong closure of the set of simple points of V. It is known that V_c is a semialgebraic set of the type

$$V_c = \bigcup_{i \in I} \{p \in \mathbb{R}^n;\ f_{ij}(p) \geq 0 \ \text{ for all } \ j \in J_i\},$$

where I, J_i are finite sets and the f_{ij} belong to $\mathbb{R}[\underline{X}]$.

PROBLEM. Give conditions on V which imply that V_c is a simple closed semialgebraic set, i.e.,

$$V_c = \{p \in \mathbb{R}^n;\ f_1(p) \geq 0, \ldots, f_r(p) \geq 0\}$$

for some polynomials $f_1,\ldots,f_r$ in $\mathbb{R}[\underline{X}]$.

7) Let us call $D = V - V_c$ the <u>degenerate set</u> of V. Given two semialgebraic sets S_1, S_2, in $\mathbb{R}^n$, when do we have an algebraic variety V such that $V_c = S_1$, $D = S_2$?

8) Let S_1, S_2 be disjoint semialgebraic sets in $\mathbb{R}^n$ (more generally in an open semialgebraic set Ω in $\mathbb{R}^n$).

PROBLEM. Under what conditions can a polynomial g be found which separates S_1 and S_2, i.e., $g(S_1) > 0$ and $g(S_2) < 0$? A variation of this problem asks for separation of S_1 and S_2 by a simple semialgebraic set T, i.e., $T \supset S_1$ and $T \cap S_2$ is empty.

9) We know that the image of a semialgebraic set under a polynomial (more generally, semialgebraic) mapping is semialgebraic.

PROBLEM. Characterize the semialgebraic sets S in $\mathbb{R}^n$ such that $S = p(\mathbb{R}^m)$, where $p : \mathbb{R}^m \longrightarrow \mathbb{R}^n$ is a polynomial mapping (cf. Bochnak-Efroymson's paper "Real Algebraic Geometry and the 17th Hilbert Problem", Math. Ann. 251 (1980) pp. 213-241).

10) Hilbert's Basis Theorem reduces any system of polynomial equations to a finite one. For inequalities there is no such a tool.

PROBLEM. Given the system $\{f_i \geq 0;\ i \in I\}$ of polynomial inequalities, I being an arbitrary set. When do we have a semialgebraic solution set?

11) Several algebraic theories associating an algebraic object to a semialgebraic set havebeen developed recently, e.g., Brumfiel's RHJ algebras or Coste-coste-Roy's real spectra.

PROBLEM. How can we handle problems about unions of semialgebraic sets in this algebraic fashion? For example, how can one characterize algebraically sets such as $\{x \geq 0\} \cup \{y \geq 0\}$ in $\mathbb{R}^2$ which are not simple.

12) A representation of a semialgebraic set $S = \bigcup_{i \in I} \{f_{ij} > 0, g_{ij} = 0, j \in J_i\}$ is said to be trim if $\overline{S} = \bigcup_{i \in I} \{f_{ij} \geq 0, g_{ij} = 0, j \in J_i\}$, where $\overline{S}$ denotes the strong closure of S (cf. McEnerney's paper "Trim stratification of semianalytic sets", Man. Math. 25 (1978).

PROBLEM. Can we obtain an algebraic criterion for trimness? And, more generally, can we obtain some basic inequalities describing a given semi-algebraic set in the same way as we obtain a basis of the ideal of a given algebraic variety?

Problems posed by T. M. Viswanathan and A. J. Engler (Campinas)

13) The following problem naturally arises in the study of intersections of two real closed fields:

Let R be a real closed field and $\Omega = R(\sqrt{-1})$. Let F be a subfield of Ω containing $\sqrt{-1}$. When can we say that the degree $(F : F \cap R) = 2$?

14) The following problem is suggested by Ware's paper (Proposition 3.2, J. Alg. 58, 227-237 (1979)).

Let k be an ordered field with algebraic closure Ω. Assume that the Galois group $G(\sqrt{-1}) = \mathrm{Gal}(\Omega | k(\sqrt{-1}))$ is a free pro-2-group on n generators. Let σ be a k-automorphism of Ω of order 2. Describe the action of σ on $G(\sqrt{-1})$. (for the case $n = 1$ and 2, see the author's paper in these Proceedings).

15) Question 8 may throw further light on the classification of real pythagorean fields. What are the real pythagorean fields k for which the absolute Galois group $\mathrm{Gal}(\overline{k} | k(\sqrt{-1}))$ is a free pro-2-group? In particular, does there exist one of these having $|\dot{k}/\dot{k}^2| = 8$?

16) How does one construct examples of ordered fields dense in their real closures and having any number of archimedean and non-archimedean orderings?

17) Is it possible to describe in a concrete way all the hereditarily euclidean subfields of the algebraic closure $\overline{Q}$ of the rationals?

Contemporary Mathematics
Volume 8, 1982

LA TOPOLOGIE DU SPECTRE REEL

Michel Coste and Marie-Françoise Roy

1°) INTRODUCTION

Nous voulons dégager ici un outil algébrique qui permette d'appréhender les particularités topologiques de l'ensemble des points réels d'une variété algébrique; il s'agit bien sur ici de la topologie euclidienne (ou usuelle) et pas de la topologie de Zariski. Nous nous attacherons plus particulièrement a deux phénomènes:

Le premier est qu'une variété, même irréductible, peut avoir plusieurs composantes connexes, comme la cubique $y^2 - x^3 + x = 0$ qui en a deux.

Le second est que l'on peut avoir une "chute de dimension" sur une partie de la variété, comme le point isolé de la cubique $y^2 - x^3 + x^2 = 0$; ou le "manche" du "parapluie de Whitney" d'équation $(x^2 + y^2)z - y^3 = 0$.

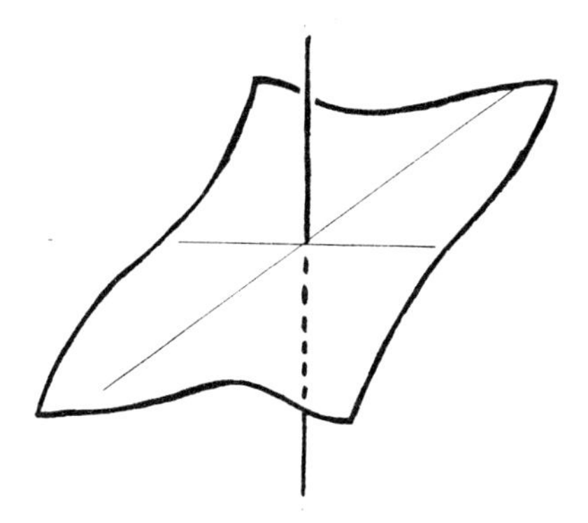

Nous associons à chaque anneau (toujours commutatif) A un espace topologique, le spectre réel de A, note $\mathrm{Spec}_R A$. Ce spectre réel a les mêmes propriétés formelles et topologiques générales que le spectre premier avec la topologie de Zariski. Mais dans le cas où $A = \mathbb{R}[V]$ est l'anneau de coordonnées d'une variété algébrique affine V définie sur $\mathbb{R}$, $\mathrm{Spec}_R A$

1980 Mathematics Subject Classification 12D15, 12J15, 13J25, 14G30, 58A07.

est suffisamment proche de l'ensemble $V(\mathbb{R})$ des points réels de V avec la topologie euclidienne pour rendre compte des phénomènes signalés ci-dessus.

Il nous faut introduire les points du spectre réel, qui joueront le rôle des idéaux premiers. Les idéaux premiers de A sont liés aux morphismes de A dans des corps algébriquement clos de la façon suivante: tout morphisme de A dans un corps algébriquement clos factorise à travers un plus petit corps algébriquement clos, la clôture algébrique du corps résiduel en l'idéal premier noyau du morphisme. Pour la géométrie algebrique réelle il est naturel de s'intéresser aux corps réels clos et non aux corps algébriquement clos. Ici aussi, tout morphisme de A dans un corps réel clos K factorise à travers un plus petit corps réel clos, à savoir la clôture réelle du corps résiduel en l'idéal premier noyau du morphisme $A \longrightarrow K$ pour l'ordre induit par celui de K; la donnée de l'ordre sur le corps résiduel est indispensable. Un point du spectre réel sera un couple formé d'un idéal premier $\mathfrak{p}$ de A et d'un ordre sur le corps résiduel $k(\mathfrak{p})$. Remarquons que le fait que $k(\mathfrak{p})$ soit ordonnable équivaut à dire que l'idéal $\mathfrak{p}$ est réel, c.a.d. vérifie

$$\sum_{i=1}^{n} x_i^2 \in \mathfrak{p} \longrightarrow x_i \in \mathfrak{p} \quad i=1, \ldots, n$$

Nous allons transformer la donnée d'un tel couple en celle d'une partie de A, à savoir celle formée des éléments qui deviennent négatifs ou nuls dans le corps résiduel (ce choix vise à donner aux axiomes une forme semblable à ceux des idéaux premiers).

DEFINITION 1.1. Un <u>précone premier</u> α de A est une partie α de A qui vérifie:

1°) $1 \notin \alpha$

2°) $\forall x \in A \qquad -x^2 \in \alpha$

3°) $\forall x, y \in \alpha \qquad x+y \in \alpha$

4°) $\forall x, y \in A$

$$xy \in \alpha \longleftrightarrow (x \in \alpha \text{ et } -y \in \alpha) \text{ ou } (-x \in \alpha \text{ et } y \in \alpha)$$

Les éléments de A qui deviennent négatifs ou nuls dans un ordre sur un corps résiduel vérifient bien sur les propriétes ci-dessus. Réciproquement, si α est un précone premier, on vérifie sans peine que

$\mathfrak{p} = \alpha \cap -\alpha$ est un idéal premier réel et qu'il y a sur $k(\mathfrak{p})$ un unique ordre pour lequel les images d'éléments de α deviennent négatives ou nulles.

NOTATION. Si $\mathfrak{p} = \alpha \cap -\alpha$, $k(\alpha)$ désignera la clôture réelle du corps résiduel $k(\mathfrak{p})$ pour l'ordre déterminé par α.

Nous allons maintenant faire de l'ensemble des précones premiers un espace topologique sur le modèle du spectre de Zariski.

DEFINITION 1.2. Le spectre réel de l'anneau A (noté $\mathrm{Spec}_R A$) est l'ensemble des précones premiers de A muni de la topologie donnée par la base d'ouverts

$$D(a_1, \ldots, a_n) = \{\alpha \in \mathrm{Spec}_R A \mid a_1 \notin \alpha \text{ et } \ldots \text{ et } a_n \notin \alpha\} \quad a_1, \ldots, a_n \in A.$$

$D(a_1, \ldots, a_n)$ est donc l'ensemble des précones premiers α tels que les éléments $a_1, \ldots, a_n$ deviennent simultanément strictement positifs dans $k(\alpha)$. On n'a pas ici $D(a_1) \cap D(a_2) = D(a_1 a_2)$.

Il est clair que Spec_R est en fait un foncteur contravariant de la catégorie des anneaux dans la catégorie des espaces topologiques. Si $f : A \longrightarrow B$ est un morphisme d'anneaux et si β est un précone premier de B, on pose $\mathrm{Spec}_R f(\beta) = f^{-1}(\beta)$; $\mathrm{Spec}_R f$ est bien continue.

Donnons un exemple de spectre réel: celui de l'anneau $\mathbb{R}[X]$. Les idéaux premiers réels de $\mathbb{R}[X]$ sont d'une part les idéaux maximaux réels qui correspondent aux points réels de la droite, d'autre part l'idéal (0) point générique de la droite. Le corps résiduel en un idéal maximal réel est $\mathbb{R}$, qui a un ordre unique. Un réel r détermine donc un précone premier que l'on encore écrire $r = \{p \in \mathbb{R}[X] \mid P(r) \leqslant 0\}$. Le corps résiduel au point generique est $\mathbb{R}(X)$; pour ordonner $\mathbb{R}(X)$, il faut situer X par rapport aux constantes réelles. On peut le mettre à $+\infty$, à $-\infty$, en r_- (resp. en r_+) c.a.d. juste à gauche (resp. à droite) d'un réel r. Un polynôme sera strictement positif en r_- s'il l'est sur un intervalle $]r-\varepsilon, r[$.

En termes de précones premiers:

$$r_- = \{P \in \mathbb{R}[X] \mid \exists\, \varepsilon > 0 \quad \forall x \in]r-\varepsilon, r[\;\; P(x) \leqslant 0\}$$

On peut dessiner $\mathrm{Spec}_R \mathbb{R}[X]$ en representant les precones prémiers autres que les points réels de la droite par de petites flèches (on verra que pour une courbe réele quelconque de tels points doivent être vus comme les points génériques de "demi branches" reelles de la courbe).

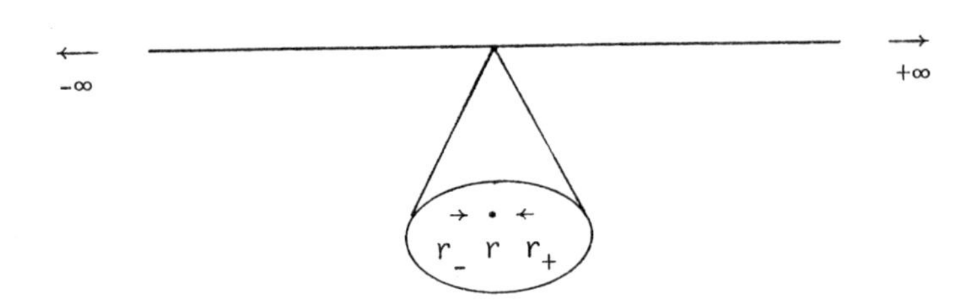

Une base d'ouverts de $\mathrm{Spec}_R\mathbb{R}[X]$ est donnée par les "intervalles" du genre $[-\infty, r_-] = D(r-X)$, $[s_+, +\infty] = D(X-s)$, $[r_+, s_-] = D(X-r, s-X)$ pour $r < s$.

Nous montrerons dans la partie 2 que $\mathrm{Spec}_R A$ est toujours un espace spectral, c.a.d. que les $D(a_1, \ldots, a_n)$ sont quasi-compacts et que les fermés irréductibles sont l'adhérence d'un point et d'un seul. On peut s'en rendre compte sur l'exemple ci-dessus. On utilisera la terminologie habituelle pour le spectre premier.

DEFINITION 1.3. Soient α et β deux precones premiers de A. On dira que α est une <u>spécialisation</u> de β (ou β une <u>générisation</u> de α) si on a $\alpha \supset \beta$. Ceci veut dire aussi que α est dans l'adhérence de β dans $\mathrm{Spec}_R A$.

L'exemple plus haut montre que l'on n'a plus de "point generique" de la droite: celui-ci a explosé en donnant les points génériques des demi-branches réelles $-\infty$, r_-, r_+, $+\infty$. Les points r_- ou r_+ se spécialisent en un seul point réel de la droite, à savoir r.

Dans la partie 3 nous montrons que si $A = \mathbb{R}[V]$ est l'anneau de coordonnées d'une variété algébrique affine définie sur $\mathbb{R}$, $\mathrm{Spec}_R\mathbb{R}[V]$ contient naturellement $V(\mathbb{R})$ avec la topologie euclidienne (comme pour l'exemple de la droite). La restriction à $V(\mathbb{R})$ fournit une bijection entre les ouverts quasi-compacts de $\mathrm{Spec}_R\mathbb{R}[V]$ et les ouverts semi-algébriques de $V(\mathbb{R})$*. Il faut remarquer ici que deux ouverts distincts de $\mathrm{Spec}_R\mathbb{R}[X]$ comme $[r_+, s_-]$ et $]r_+, s_-[= \bigcup_{r<r'<s'<s} [r'_+, s'_-]$ peuvent avoir la même trace sur les points reels de la droite; le premier est quasi-compact, pas le second.

Dans la partie 4 nous réexprimons des résultats de la partie précédente sous la forme d'un principe qui donnera d'utiles renseignements topologiques sur le spectre réel; c'est la combinaison de l'élimination des quantificateurs pour les corps réels clos (plus connue sous le nom de principe de Tarski-Seidenberg) et du fait qu'un ouvert semi-algébrique peut toujours être donné par une disjonction finie de conjonctions d'inégalités strictes.

---

*ceci montre que $\mathrm{Spec}_R\mathbb{R}[V]$ et $V(\mathbb{R})$ ont les mêmes composantes connexes.

Dans la partie 5 nous étendons les résultats de la partie 3 au cas d'un corps réel clos K quelconque. Bien sur V(K) n'a pas en général de bonnes propriétés topologiques: par exemple il n'est pas localement connexe. Cependant $\mathrm{Spec}_R K$ [V] l'est toujours et a un nombre fini de composantes connexes qui ne dépendent que des équations et pas du corps réel clos K choisi.

Dans la partie 6 nous établissons les propriétés topologiques de certains morphismes. Si f est entier, $\mathrm{Spec}_R f$ est fermé (ceci est l'analogue du théorème de montée de Cohen-Seidenberg). Si f est étale, $\mathrm{Spec}_R f$ est un homéomorphisme local; ceci permet de raisonner au voisinage d'un point réel regulier comme dans un espace affine.

Dans la partie 7 nous précisons les liens entre précones premiers et places réelles en renvoyant pour certains points au livre de Brumfiel [4]. Nous définissons les variétés réellement complètes.

Dans la partie 8 nous nous intéressons à la dimension de l'espace $\mathrm{Spec}_R A$, calculée grâce aux chaînes de précones premiers. Nous introduisons la dimension réelle d'une variété V en un point x de V(K) (K réel clos) et nous montrons quelques résultats sur cette dimension locale: elle correspond à ce que l'on voit sur les dessins, comme par exemple ceux du début de cette introduction.

Enfin dans la partie 9 nous montrons l'invariance birationnelle des composantes connexes des variétés affines non singulières réellement complètes sur un corps réel clos (en fait celles du spectre réel correspondant).

Il nous faut faire quelques remarques

1) $\mathrm{Spec}_R A$ est muni d'un faisceau structural; ceci est très important. On a ainsi un espace annelé en anneaux locaux henséliens de corps résiduels réel clos qui est l'analogue réel du topos étale du schéma Spec A (voir [6]). Dans le cas de l'anneau de coordonnées d'une variété V affine non singulière sur $\mathbb{R}$, ce faisceau structural est l'image directe du faisceau des fonctions de Nash (ou analytiques - algébriques) [1] sur $V(\mathbb{R})$. Nous n'aborderons pas cet aspect important ici. Nous avons choisi de nous limiter aux propriétés topologiques du spectre réel.

2) Nous ne sortons pas du cadre affine. On pourrait recoller des spectres réels, avec leurs faisceaux structuraux, pour avoir des "schémas réels". La limitation au cadre affine n'est cependant pas très génante en géométrie algébrique réelle (on peut plonger un espace projectif réel dans un espace affine réel de dimension suffisamment grande). On aura par exemple des variétés réellement complètes bien qu'affines, comme le cercle.

3) Si A est un corps k, $\mathrm{Spec}_R k$ est l'espace booléen des ordres sur k, qui intervient dans la classification des formes

quadratiques sur k. Si $W(k)$ est l'anneau de Witt de k on a un morphisme (signature globale) de $W(k)$ dans l'anneau $\mathrm{Cont}(\mathrm{Spec}_R k, \mathbb{Z})$ des fonctions continues de $\mathrm{Spec}_R k$ dans $\mathbb{Z}$. Le principe local-global dit que le noyau et le conoyau de ce morphisme sont de 2 torsion [12]. Dans le cas d'un anneau A on peut encore construire un morphisme de $W(A)$ (comme defini par Knebusch [10]) dans $\mathrm{Cont}(\mathrm{Spec}_R A, \mathbb{Z})$. Que peut-on dire ce de morphisme?

4) Nous avons quelques résultats analogues à ceux de Brumfiel [4]: l'existence d'un nombre fini de "composantes connexes" pour les variétés sur un corps réel clos (son théorème 8.13.14) ou le "place perturbation theorem" (8.4.9) semblable à notre proposition 8.7. Pour ce dernier résultat nous avons d'ailleurs modifié notre preuve originale pour la rapprocher de celle de Brumfiel. D'un autre coté H. Delfs a aussi montré la finitude du nombre de composantes connexes en utilisant une notion de connexité par arcs. Notre originalité essentielle est l'introduction du spectre réel. Nous sommes convaincus que cet outil facilite la compréhension des problèmes et permet de donner des résultats plus complets.

5) Le lecteur (la lectrice) sera peut être surpris du rôle de la logique dans le nombreuses démonstrations. On pourrait donner un exposé du même type pour certains résultats sur le spectre premier (quasi compacité, théorème de Chevalley, théorème de montée, ...). Ce serait une redémonstration de résultats obtenus par des techniques algébriques. Ici au contraire la logique nous permet d'obtenir des résultats nouveaux dans un domaine où les outils algébriques sont insuffisamment développés.

2°) PROPRIETES TOPOLOGIQUES GENERALES DU SPECTRE REEL

PROPOSITION 2.1. Soit A un anneau commutatif

i) les ouverts $D(a_1, \ldots, a_n)$ de $\mathrm{Spec}_R A$ sont quasi-compacts. En particulier $\mathrm{Spec}_R A$ est quasi-compact.

ii) Les fermés irréductibles de $\mathrm{Spec}_R A$ sont l'adhérence d'un point et d'un seul.
On a $\alpha \in \mathrm{adh}\,\{\beta\}$ ssi $\beta \subset \alpha$.

iii) Si $\alpha \subset \beta$ et $\alpha \subset \gamma$, on a $\beta \subset \gamma$ ou $\gamma \subset \beta$.

Les propriétés i et ii veulent dire que $\mathrm{Spec}_R A$ est un espace spectral au sens de Hochster (c.a.d. homéomorphe au spectre premier d'un anneau). La propriété iii montre que les espaces spectraux obtenus ont la propriété que les spécialisations de n'importe quel point forment toujours une chaîne (totalement ordonnée). Nous ne savons pas si cette propriété est caractéristique des spectres réels.

PREUVE. i) Considérons la théorie propositionelle dont les variables sont les $N(a)$ (pour "a négatif ou nul") où a décrit l'anneau A et dont les axiomes sont:

$$\neg N(1) \qquad\qquad N(-a^2)$$
$$N(a) \wedge N(b) \longrightarrow N(a+b)$$
$$N(ab) \longleftrightarrow (N(a) \wedge N(-b)) \vee (N(-a) \wedge N(b)).$$

Il est clair qu'un modèle de cette théorie propositionelle revient exactement à un précone premier de A. Donc si on a $D(a_1, \ldots, a_n) = = \bigcup_{i\in I} D(b_{i,1}, \ldots, b_{i,p_i})$, c'est que l'ensemble de formules

$$\{\neg(N(a_1) \vee \ldots \vee N(a_n))\} \cup \{N(b_{i,1}) \vee \ldots \vee N(b_{i,p_i}) \mid i \in I\}$$

est inconsistant avec la théorie propositionelle. Un sous ensemble fini est déjà inconsistant, ce qui montre que $D(a_1, \ldots, a_n)$ est l'union d'un nombre fini des $D(b_{i,1}, b_{i,p_i})$.

ii) La deuxième partie de l'assertion est claire. Si F est un fermé irréductible, on vérifie sans peine que

$\alpha = \{a \in A \mid D(a) \cap F = \emptyset$ est un précone premier de A et que $F = \mathrm{adh}(\{\alpha\})$.

iii) Supposons l'assertion fausse. On a alors $b \in \beta$, $b \notin \gamma$ et $c \in \gamma$, $c \notin \beta$. On sait que $b-c$ ou $c-b$ est dans α; dans le premier cas, on a $b = (b-c) + c \in \gamma$ et dans la deuxième $c = (c-b) + b \in \beta$.

Puisque le spectre réel est un espace spectral, on peut parler de ses constructibles:

DEFINITION et PROPOSITION 2.2. Un <u>constructible</u> de $\mathrm{Spec}_R A$ est une combinaison booléenne d'ouverts $D(a_1, \ldots, a_n)$ (c.a.d. obtenue par intersections et réunions finies et passages au complémentaire). Une partie C de $\mathrm{Spec}_R A$ est constructible ssi il y a une formule close Φ_C de la théorie des corps ordonnés à paramètres dans A telle que

$$C = \{\alpha \in \mathrm{Spec}_R A \mid k(\alpha) \text{ satisfait } \Phi_C\}.$$

Les constructibles de $\mathrm{Spec}_R A$ sont les ouverts fermés d'une topologie compacte et totalement discontinue.

PREUVE. Si on prend pour $\Phi_{D(a_1, \ldots, a_n)}$ la formule $a_1 > 0 \wedge \ldots \wedge a_n > 0$, on a bien ce qui est annoncé. En faisant des combinaisons booléennes on associe bien à tout constructible C une formule Φ_C avec la propriété voulue. Réciproquement si on part d'une formule Φ on peut toujours la supposer sans quantificateur (par l'élimination des quantificateurs pour les corps réels clos) et on peut alors l'écrire comme combinaison booléenne de formules du genre $a_1 > 0 \wedge \ldots \wedge a_n > 0$. Ceci montre que $\{\alpha \in \mathrm{Spec}_R A \mid k(\alpha)$ satisfait $\Phi\}$ est un constructible de $\mathrm{Spec}_R A$. La dernière partie de la proposition est bien connue pour les constructibles de n'importe quel espace spectral.

PROPOSITION 2.3. ("Théorème de Chevalley réel"): Soit $f : A \longrightarrow B$ une A-algèbre de présentation finie et C un constructible de $\mathrm{Spec}_R B$. Alors $\mathrm{Spec}_R f(C)$ est un constructible de $\mathrm{Spec}_R A$.

PREUVE. Posons $B = A[X_1, \ldots, X_n]/_{(P_1, \ldots, P_q)}$ et $C = \{\beta \in \mathrm{Spec}_R B \mid k(\beta)$ satisfait $\Phi(Q_1(\underline{X}), \ldots, Q_r(\underline{X})\}$ où Φ est une formule dont les paramètres sont les polynômes $Q_1(\underline{X}), \ldots, Q_r(\underline{X})$ a coefficients dans A. Il suffit de vérifier que l'on a

$$\mathrm{Spec}_R f(C) = \{\alpha \in \mathrm{Spec}_R A \mid k(\alpha) \text{ satisfait } \exists x_1 \ldots \exists x_n (\bigwedge_{i=1}^{q} P_i(\underline{x}) = 0 \wedge \Phi(\underline{Q}(\underline{x})))\}.$$

Si $\alpha = \mathrm{Spec}_R f(\beta)$ avec $\beta \in C$, $k(\beta)$ satisfait la formule ci-dessus (prendre pour $x_1, \ldots, x_n$ les images de $X_1, \ldots, X_n$) et donc $k(\alpha)$ aussi puisque l'on a un carré commutatif:

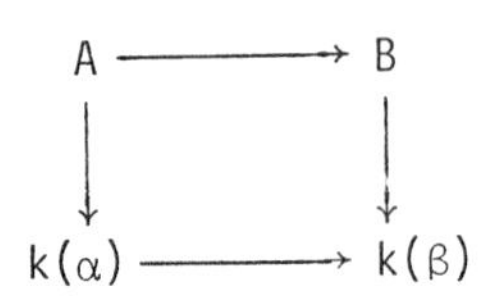

Réciproquement supposons que $k(\alpha)$ satisfait cette formule. On peut alors envoyer B dans $k(\alpha)$ en choisissant pour images de $X_1, \ldots, X_n$ des $x_1, \ldots, x_n$ qui satisfont la formule. Ce morphisme de B dans $k(\alpha)$ donne un précone premier β de B (formé des éléments qui deviennent négatifs ou nuls dans $k(\alpha)$) qui est au-dessus de α et qui est bien dans C.

PROPOSITION 2.4. Soit $(A_i)_{i \in I}$ un système filtrant d'anneaux. Alors $\mathrm{Spec}_R(\varinjlim_i A_i)$ est homéomorphe à $\varprojlim_i (\mathrm{Spec}_R A_i)$.

PREUVE. Indiquons seulement comment passer de $\varprojlim_i (\mathrm{Spec}_R A_i)$ à $\mathrm{Spec}_R(\varinjlim_i A_i)$. Les corps $k(\alpha_i)$ forment un système filtrant et donc

$k = \varinjlim_i k(\alpha_i)$ est un corps réel clos. $A = \varinjlim_i A_i$ s'envoie canoniquement dans k, ce qui donne un précone premier α de A.

PROPOSITION 2.5. Soit $f : A \longrightarrow B$ un morphisme d'anneaux, α un précone premier de A. On a une bijection entre $\mathrm{Spec}_R(k(\alpha) \otimes_A B)$ et $(\mathrm{Spec}_R f)^{-1}(\alpha)$. (Nous montrerons plus tard que c'est un homéomorphisme).

PREUVE. Si $\mathrm{Spec}_R f(\beta) = \alpha$, le carré commutatif

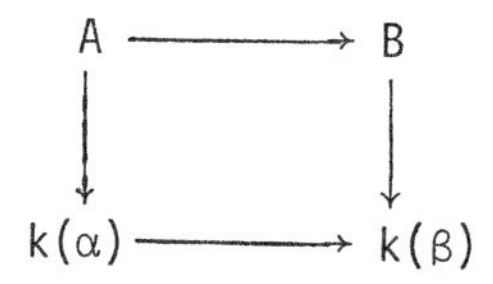

donne un morphisme de $k(\alpha) \otimes_A B$ dans $k(\beta)$ et donc un précone premier de $k(\alpha) \otimes_A B$ au-dessus de β. Supposons que β_1 et β_2 soient deux précones premiers de $k(\alpha) \otimes_A B$ au-dessus de β. D'après l'amalgamation des corps réels clos, on peut trouver un corps réel clos k qui complète le carré commutatif

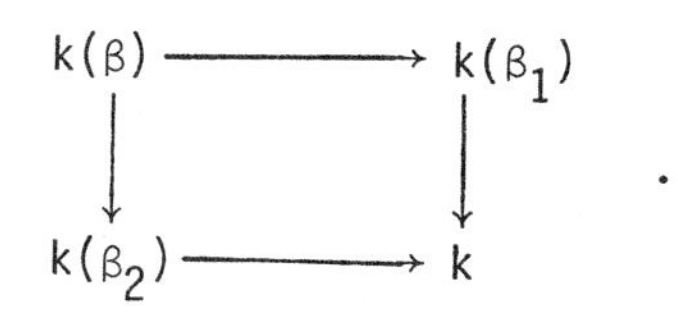

Mais alors le carré

$$\begin{array}{ccc} k(\alpha) \otimes_A B & \longrightarrow & k(\beta_1) \\ \downarrow & & \downarrow \\ k(\beta_2) & \longrightarrow & k \end{array}$$

est lui aussi commutatif ce qui montre que l'on a $\beta_1 = \beta_2$.

PROPOSITION 2.6. Soient $f : A \longrightarrow B$ un morphisme d'anneaux et $(C_i)_{i\in I}$ une famille filtrante (à gauche) de constructibles de $\mathrm{Spec}_R B$. Alors $\mathrm{Spec}_R f(\bigcap_{i\in I} C_i) = \bigcap_{i\in I} \mathrm{Spec}_R f(C_i)$.

PREUVE. L'inclusion non évidente est $\bigcap_{i\in I} \mathrm{Spec}_R f(C_i) \subset \mathrm{Spec}_R f(\bigcap_{i\in I} C_i)$.
Soit $\alpha \in \bigcap_{i\in I} \mathrm{Spec}_R f(C_i)$. On a pour tout i

$$(\mathrm{Spec}_R f)^{-1}(\alpha) \cap C_i \neq \emptyset.$$

On peut d'après 2.5. identifier $C_i \cap (\mathrm{Spec}_R f)^{-1}(\alpha)$ à un constructible de $\mathrm{Spec}_R(k(\alpha) \otimes_A B)$. On a ainsi une famille de constructibles avec la propriété d'intersection finie, d'où $\bigcap_{i\in I} C_i \cap (\mathrm{Spec}_R f)^{-1}(\alpha) \neq \emptyset$, ce qui veut dire $\alpha \in \mathrm{Spec}_R f(\bigcap_{i\in I} C_i)$.

REMARQUE. On peut montrer [7] que les fermés pour la topologie constructible de $\mathrm{Spec}_R A$ (ces pro-constructibles) sont exactement les images de spectres réels de A-algèbres quelconques.

3°) VARIETES ALGEBRIQUES AFFINES SUR LE CORPS DES NOMBRES REELS

Soit V une variété algébrique affine définie sur $\mathbb{R}$. On notera $\mathbb{R}[V]$ son anneau de coordonnées. On pose $\mathbb{R}[V] = \mathbb{R}[X_1, \dots, X_n]/_{(P_1, \dots, P_q)}$.

PROPOSITION 3.1. $V(\mathbb{R})$ muni de la topologie euclidienne s'identifie à un sous-espace dense de $\mathrm{Spec}_R \mathbb{R}[V]$.

PREUVE. Soit x un point de $V(\mathbb{R})$. Le corps résiduel en x est $\mathbb{R}$, ordonné de manière unique. Le point x détermine donc un précone premier de $\mathbb{R}[V]$, à savoir

$$\{P \in \mathbb{R}[V] \mid P(x) \leq 0\}.$$

Dans la suite nous identifierons toujours un point réel de la variété avec le précone premier correspondant. La trace de $D(Q_1, \dots, Q_m)$ sur $V(\mathbb{R})$ est

$$\{x \in V(\mathbb{R}) \mid Q_1(x) > 0 \text{ et } \dots \text{ et } Q_m(x) > 0\}.$$

Il est clair que ceci donne une base de la topologie euclidienne sur $V(\mathbb{R})$. La densité de $V(\mathbb{R})$ dans $\mathrm{Spec}_R \mathbb{R}[V]$ sera une conséquence de la proposition suivante.

Rappelons qu'un sous ensemble semi-algébrique de $V(\mathbb{R})$ est une combinaison booléenne de sous ensembles de $V(\mathbb{R})$ donnes par des inégalités polynomiales strictes.

PROPOSITION 3.2. La restriction à $V(\mathbb{R})$ fournit une bijection entre les constructibles de $\mathrm{Spec}_R \mathbb{R}[V]$ et les sous-ensembles semi-algébriques de $V(\mathbb{R})$.

PREUVE. Soit C un constructible de $\mathrm{Spec}_R \mathbb{R}[V]$ donné par la formule $\Phi_C(Q_1, \dots, Q_m)$ a pàramètres dan $\mathbb{R}[V]$ que l'on peut supposer sans quantificateur (cf. 2.2). La trace de C sur $V(\mathbb{R})$ est

$$\{x \in V(\mathbb{R}) \mid \Phi_C(Q_1(x), \dots, Q_m(x))\}.$$

Il est clair que tout semi-algébrique de $V(\mathbb{R})$ est ainsi atteint. Soit maintenant C' un autre constructible de $\mathrm{Spec}_R \mathbb{R}[V]$, donné par $\Phi_{C'}(Q'_1, \ldots, Q'_{m'})$, tel que

$$C \cap V(\mathbb{R}) = C' \cap V(\mathbb{R}).$$

Ceci veut dire que $\mathbb{R}$ satisfait la formule

$$\forall x_1 \ldots \forall x_n, \quad P_1(\underline{x}) = \ldots = P_q(\underline{x}) = 0 \rightarrow$$
$$(\Phi_C(Q_1(\underline{x}), \ldots, Q_m(\underline{x})) \leftrightarrow \Phi_{C'}(Q'_1(\underline{x}), \ldots, Q'_{m'}(\underline{x}))).$$

Cette formule est aussi satisfaite par les corps réels clos $k(\alpha)$ pour tout précone premier α de $\mathbb{R}[V]$. En prenant pour $x_1, \ldots, x_n$ les images de $X_1, \ldots, X_n$ dans $k(\alpha)$ on trouve que $k(\alpha)$ satisfait $\Phi_C(Q_1, \ldots, Q_m)$ ssi il satisfait $\Phi_{C'}(Q'_1, \ldots, Q'_m)$. On a donc $C = C'$.

NOTATION. Si S est un sous ensemble semi-algébrique de $V(\mathbb{R})$ on notera $\tilde{S}$ le constructible correspondant de $\mathrm{Spec}_R(\mathbb{R}[V])$.

THEOREME 3.3. S est ouvert dans $V(\mathbb{R})$ ssi $\tilde{S}$ est ouvert dans $\mathrm{Spec}_R \mathbb{R}[V]$. Autrement dit, la restriction à $V(\mathbb{R})$ fournit une bijection entre les ouverts constructibles (c.a.d. quasi-compacts) de $\mathrm{Spec}_R \mathbb{R}[V]$ et les ouverts semi-algébriques de $V(\mathbb{R})$.

PREUVE. Il suffit de montrer que si S est ouvert, $\tilde{S}$ l'est aussi. Pour ceci il faut voir que S est une union finie de semi-algébriques de $V(\mathbb{R})$ de la forme

$$\{\underline{x} \in V(\mathbb{R}) \mid Q_1(\underline{x}) > 0 \text{ et } \ldots \text{ et } Q_m(x) > 0\}.$$

Nous le déduirons du résultat suivant de G. Efroymson [9] (on peut trouver une autre preuve dans [2]; le résultat est annoncé sans démonstration dans [4]):

DEFINITION 3.4. Soient $Q_1, \ldots, Q_m$ des polynômes de $\mathbb{R}[X_1, \ldots, X_n]$. La famille $(Q_1, \ldots, Q_m)$ induit une partition de l'espace affine $A^n(\mathbb{R})$ en ensembles semi-algébriques qui sont les composantes connexes des parties de $A^n(\mathbb{R})$ du genre

$$\{\underline{x} \in A^n(R) \mid Q_1(\underline{x}) \ ?_1\ 0 \text{ et } \ldots \text{ et } Q_m(x) \ ?_m\ 0\}$$

où $?_i$ est $>$, $<$ ou $=$. On dira que la famille $(Q_1, \ldots, Q_m)$ est <u>séparante</u> si pour tous morceaux S et T de cette partition tels que toute condition de signe stricte (> 0 ou < 0) portant sur les Q_i vraie dans S est vraie dans T, on a $S \subset \mathrm{adh}(T)$.

PROPOSITION 3.5. i) Pour toute famille $Q_1, \ldots, Q_r$ dans $\mathbb{R}[X_1, \ldots, X_n]$ on peut trouver des polynômes $Q_{r+1}, \ldots, Q_{r+s}$ tels que la famille $Q_1, \ldots, Q_{r+s}$ soit séparante.

ii) Etant donnés n, r et le degré maximum des polynômes $Q_1, \ldots, Q_r$ on peut borner s et le degré maximum des polynômes $Q_{r+1}, \ldots, Q_{r+s}$.

iii) On peut choisir $Q_{r+1}, \ldots, Q_{r+s}$ de façon que leurs coefficients soient des expressions polynomiales à coefficients entiers en les coefficients de $Q_1, \ldots, Q_r$.

La partie i est le résultat d'Efroymson. Les précisions ii et iii viennent par exemple de la preuve donnée dans [5]. Elles seront utiles par la suite.

Nous allons terminer la preuve de 3.3. Pour cela nous montrerons un peu plus que ce qui nous est nécessaire; le surplus nous servira par la suite.

PROPOSITION 3.6. Soit S un sous-ensemble semi-algébrique de l'espace affine $A^n(\mathbb{R})$ et U un semi-algébrique ouvert dans S. Alors U est une union finie de semi-algébriques du genre

$$\{x \in S \mid Q_1(x) > 0 \text{ et } \ldots \text{ et } Q_m(x) > 0 .$$

PREUVE. S et U sont donnés par une combinaison booleenne d'inégalités portant sur des polynômes $Q_1, \ldots, Q_r$. On complete en une famille séparante $Q_1, \ldots, Q_{r+s}$. On a $U = \bigcup_{i=1}^{t} S_i$ où les S_i sont des morceaux de la partition de $A^n(\mathbb{R})$ en semi-algébriques connexes induite par $Q_1, \ldots, Q_{r+s}$ comme en 3.4. Soit U_i l'ensemble des points de S où sont vérifiées toutes les conditions de signe strictes sur $Q_1, \ldots, Q_{r+s}$ vérifées dans S_i. U_i est bien sur un semi-algébrique ouvert dans S. La proposition sera établie si l'on montre $U = \bigcup_{i=1}^{t} U_i$. Il suffit de voir que U_i est inclus dans U. U_i lui-même est réunion de morceaux de la partition. Soit T un de ceux-ci. On a $S_i \subset \mathrm{adh}(T)$. Puisque U est ouvert dans S et que T est inclus dans S on a $U \cap T \neq \emptyset$ et donc $T \subset U$.

COROLLAIRE 3.7. $V(\mathbb{R})$ et $\mathrm{Spec}_R(\mathbb{R}[V])$ ont les mêmes composantes connexes. Si S est un semi-algébrique de $V(\mathbb{R})$, S et $\tilde{S}$ ont les mêmes composantes connexes.

PREUVE. On said que S a un nombre fini de composantes connexes $U_1, \ldots, U_s$ qui sont des semi-algébriques ouverts dans S. Par conséquent $\tilde{U}_1, \ldots, \tilde{U}_s$ font une partition de $\tilde{S}$ d'après 3.2 et les $\tilde{U}_i$ sont ouverts dans $\tilde{S}$ d'après 3.6. Si U_i n'est pas connexe, on a $U_i = C \cup C'$ où C et C' sont disjoints ouverts dans S, non vides et constructibles; comme $U_i = (C \cup V(\mathbb{R}) \cup (C' \cup V(\mathbb{R}))$, U_i ne serait pas connexe.

REMARQUE. Une autre preuve de 3.6 se trouve dans la thèse de Delzell [16].

4°) COMMENT MONTRER QU'UN CONSTRUCTIBLE EST OUVERT

La proposition 3.6 peut se reénoncer sous un habillage logique qui ajoute une précision topologique à l'élimination des quantificateurs pour les corps réels clos:

PROPOSITION 4.1. Soit $\Phi(x_1, \ldots, x_n)$ une formule du langage des corps ordonnés, telle que l'extension de Φ dans $\mathbb{R}^n$ soit ouverte. Alors Φ est équivalente dans la théorie des corps réels clos à une disjonction finie de conjonctions d'inégalités strictes.

PREUVE. La proposition 3.6 entraîne que, dans $\mathbb{R}$, Φ est équivalente à une formule

$$\Psi : \bigvee_{i=1}^{m} \bigwedge_{j=1}^{n_i} P_{i,j}(\underline{x}) > 0$$

où les $P_{i,j}$ sont à coefficients entiers d'après 3.5 iii. Φ et Ψ sont alors équivalentes dans tout corps réel clos.

THEOREME 4.2. Soient A un anneau commutatif et C un constructible de $\mathrm{Spec}_R A$, donné par

$$C = \{\alpha \in \mathrm{Spec}_R A \mid k(\alpha) \text{ satisfait } \Phi(a_1, \ldots, a_n)\}$$

où les $a_i \in A$ sont les paramètres de Φ. C est ouvert (resp. fermé) si l'extension de la formule $\Phi(x_1, \ldots, x_n)$ dans $\mathbb{R}^n$ est ouverte (resp. fermée).

PREUVE. D'après 4.1 si l'extension de $\Phi(\underline{x})$ est ouverte $\Phi(\underline{x})$ est équivalente dans tout corps réel clos à une formule

$$\bigvee_{i=1}^{m} \bigwedge_{j=1}^{n_i} P_{i,j}(\underline{x}) > 0.$$

On a donc $C = \bigcup_{i=1}^{m} D(P_{i,1}(\underline{a}), \ldots, P_{i,n_i}(a))$.

Ce théorème est un outil fondamental. Il permet, si l'on veut, d'utiliser des méthodes "transcendantes" pour avoir des résultats de caractère algébrique. Comme exemple d'application, nous allons compléter la proposition 2.5.

PROPOSITION 4.3. Soit $f : A \longrightarrow B$ un morphisme d'anneaux, α un précone premier de A. $(\mathrm{Spec}_R f)^{-1}(\alpha)$ est homéomorphe à $\mathrm{Spec}_R k(\alpha) \underset{A}{\otimes} B)$.

PREUVE. Il reste à montrer qu'un ouvert de $\mathrm{Spec}_R(k(\alpha) \underset{A}{\otimes} B)$ est la trace d'un ouvert de $\mathrm{Spec}_R B$. On peut se contenter d'un ouvert $D(\sum_{i=1}^{n} \lambda_i \otimes b_i)$. Soit $\mathfrak{p} = \alpha \cap -\alpha$. λ_i a un polynôme minimal P_i sur le corps résiduel $k(\mathfrak{p})$, que l'on peut choisir à coefficients dans A; soit a_i le coefficient dominant de P_i. λ_i est, disons, la k_i^{eme} racine de P_i dans $k(\alpha)$. Considérons la formule $\Phi(\underline{a}, \underline{b})$ dont les paramètres sont les coefficients des P_i (il faudrait en fait considérer leurs images par f ...) et les b_i: "$\exists t_1 \ldots \exists t_n \bigwedge_{i=1}^{n}$ (a_i inversible et discriminant (P_i) inversible et t_i est la k_i^{eme} racine de P_i) et $\sum_{i=1}^{n} t_i b_i > 0$". On vérifie d'une part que la trace de

$$\{\beta \in \mathrm{Spec}_R B \mid k(\beta) \text{ satisfait } \Phi(\underline{a}, \underline{b})\}$$

sur $\mathrm{Spec}_R(k(\alpha) \underset{A}{\otimes} B)$ est bien $D(\sum_{i=1}^{n} \lambda_i \otimes b_i)$ et d'autre part que l'extension de $\Phi(\underline{x}, \underline{y})$ dans la puissance convenable de $\mathbb{R}$ est ouverte. Le théorème 4.2 permet alors de conclure.

5°) VARIETES ALGEBRIQUES AFFINES SUR UN CORPS REEL CLOS QUELCONQUE

Nous allons maintenant examiner ce qui se passe quand on remplace $\mathbb{R}$ par un corps réel clos quelconque K.

THEOREME 5.1. Soit V une variété affine sur K réel clos.

i) $V(K)$ muni de la topologie euclidienne s'identifie à un sous-espace dense de $\mathrm{Spec}_R(K[V])$.

ii) La restriction à $V(K)$ donne une bijection entre les constructibles de $\mathrm{Spec}_R(K[V])$ et les sous-ensembles semi-algébriques de $V(K)$.

iii) Cette bijection se spécialise en une bijection entre les ouverts constructibles de $\mathrm{Spec}_R K[V]$ et les ouverts semi-algébriques de $V(K)$.

PREUVE. i et ii se montrent de la même façon que pour $\mathbb{R}$. Pour iii, nous montrons la généralisation de 3.6:

PROPOSITION 5.2. Soit S un sous-ensemble semi-algébrique de l'espace affine $A^n(K)$ et U un semi-algébrique ouvert dans S. Alors U est une union finie de semi-algébriques du genre

$$\{\underline{x} \in S \mid Q_1(\underline{x}) > 0 \text{ et } \ldots \text{ et } Q_m(\underline{x}) > 0\}.$$

PREUVE. On a $S = \{\underline{x} \in A^n(K) \mid \Phi(\underline{x})\}$ et $U = \{\underline{x} \in A^n(K) \mid \Psi(x)\}$ où Φ et Ψ sont des combinaisons booléennes d'inégalités portant sur les polynômes $Q_1(\underline{x}), \ldots, Q_r(\underline{x})$.

Soit $\mathrm{Ouv}_{\Psi,\Phi}$ la formule suivante, dont les paramètres sont les coefficients de $Q_1, \ldots, Q_r$:

$$\forall \underline{x}(\Psi(\underline{x}) \longrightarrow [\exists\, \varepsilon > 0\ \forall \underline{y}(\Phi(y) \wedge \sum_{i=1}^{n} (x_i - y_i)^2 < \varepsilon) \longrightarrow \Psi(y)]).$$

$\mathrm{Ouv}_{\Psi,\Phi}$ est faite pour que U soit ouvert dans S ssi K satisfait $\mathrm{Ouv}_{\Psi,\Phi}$. Replaçons-nous maintenant dans $\mathbb{R}$. La proposition 3.6 et sa démonstration montrent que $\mathbb{R}$ satisfait

$$(*) \qquad \forall Q_1 \ldots \forall Q_r(\mathrm{Ouv}_{\Psi,\Phi} \longrightarrow [\exists Q_{r+1} \ldots \exists Q_{r+s} \bigvee_{k\in I} (\Psi \leftrightarrow \Phi \wedge \Theta_k)])$$

où $\{\Theta_k\}_{k\in I}$ est l'ensemble fini des combinaisons positives (c.a.d. obtenues uniquement avec $\wedge$ et $\vee$) d'inégalités strictes sur $Q_1, \ldots, Q_{r+s}$. Il faut entendre les quantifications sur les polynômes comme des quantificateurs sur la suite de leurs coefficients. Telle quelle, la formule (*) n'a pas de sens puisqu'il y a une quantification existentielle sur une suite de polynômes; cependant ceci est justifié par le ii de la proposition 3.5 qui permet de borner s et le degré des $Q_{r+1}, \ldots, Q_{r+s}$.

Puisque $\mathbb{R}$ satisfait (*), K la satisfait aussi, ce qui donne le résultat cherché.

REMARQUE 1. On aurait pu aussi généraliser au cas des corps réels clos quelconques le lemme d'Efroymson (3.5). La preuve qui en est donnée en [5] est en effet élémentaire à ceci près qu'on utilise le fait que $\mathbb{R}^n$ est localement connexe. On peut tourner cette difficulté en utilisant la notion de connexité par arcs développée par M. Knebusch et H. Delfs. Mais pourquoi se priver du principe de Tarski-Seidenberg?

REMARQUE 2. La propriété iii de 3.7 entraîne qu'il revient au même de se donner un faisceau sur $\mathrm{Spec}_R K[V]$ ou de s'en donner un sur le site des ouverts semi-algébriques de $V(K)$ avec la topologie des recouvrements finis. Brumfiel suggère de s'intéresser à ce site ([4] 8.13, p. 248) sans remarquer qu'on a là en fait un espace topologique.

NOTATION. On utilisera cella déjà introduite pour $\mathbb{R}$. Le constructible de $\mathrm{Spec}_R K[V]$ correspondant au semi-algébrique S de $V(K)$ sera noté $\tilde{S}$. On utilisera aussi $\widetilde{V(K)}$ (ou $\tilde{V}$ si cela ne prête pas à confusion) au lieu de $\mathrm{Spec}_R K[V]$. Si $f : V \longrightarrow W$ est un morphisme de variétés algébriques affines sur K et $f^* : K[W] \longrightarrow K[V]$ le morphisme correspondant sur les anneaux de coordonnées, on écrira $\tilde{f}$ plutôt que $\mathrm{Spec}_R f^*$.

PROPOSITION 5.3. L'opération $S \longrightarrow \tilde{S}$ commute

i) aux intersections et réunions finies, au complémentaire,

ii) à l'intérieur et à l'adhérence,

iii) à l'image et à l'image réciproque par un morphisme de variétés algébriques défini sur K.

PREUVE. i) est clair.

ii) On sait que l'intérieur d'un semi-algébrique est un semi-algébrique. On a surement $\widetilde{\mathrm{int}(s)} \subset \mathrm{int}(\tilde{S})$ et ces deux ouverts ont même trace sur $V(K)$. On conclut grâce au lemme suivant:

LEMME 5.4. Soit U un ouvert semi-algébrique de $V(K)$. $\tilde{U}$ est le plus grand ouvert de $\tilde{V}$ de trace U.

PREUVE. Soit W un ouvert de $\tilde{V}$ de trace U. On a $W = \bigcup_{i\in I} W_i$ où les W_i sont des ouverts constructibles. Puisque l'on a $W_i \cap V(K) \subset U$, on a bien $W = \bigcup_{i\in I} \widetilde{(W_i \cap V(K))} \subset \tilde{U}$.

iii) Soit $f : V \longrightarrow W$ un morphisme. $\tilde{f}(\tilde{S})$ est un constructible de $\tilde{W}$ d'après 2.3, et sa trace sur $W(K)$ est bien $f(S)$. Même chose pour l'image réciproque.

Nous en venons maintenant au problèmes des composantes connexes. Quand le corps réel clos K n'est pas $\mathbb{R}$, $V(K)$ n'a pas de bonnes propriétés topologiques; il n'est pas localement connexe. $\mathrm{Spec}_R K[V]$ se comporte beaucoup mieux, et il permet d'avoir une notion algébrique satisfaisante de composante connexe.

THEOREME 5.5. V est une variété affine sur K réel clos.

i) $\mathrm{Spec}_R K[V]$ est localement connexe et a un nombre fini de composantes connexes qui sont des ouverts constructibles. Ceci est vrai également de tout constructible de $\mathrm{Spec}_R K[V]$.

ii) Si L est une extension réelle close de K, les images réciproques de ces composantes connexes par $\mathrm{Spec}_R L[V] \longrightarrow \mathrm{Spec}_R K[V]$ sont les composantes connexes de $\mathrm{Spec}_R L[V]$.

PREUVE. i) Il suffit de montrer qu'un constructible C de $\mathrm{Spec}_R K[X_1, \ldots, X_n]$ a un nombre fini de composantes connexes. On suppose C donné par $C = \{\alpha \mid k(\alpha) \text{ satisfait } \Phi\}$ où Φ est une combinaison booléenne d'inégalités portant sur les polynômes $Q_1, \ldots, Q_r$ (cf. 2.2). On fabrique une formule Conn_Φ, à paramètres les coefficients de $Q_1, \ldots, Q_r$:

$$\Phi \neq \emptyset \wedge [\forall Q_{r+1} \ldots \forall Q_{r+s} \bigwedge_{(k,k')\in I\times I} (\Phi \subset \Theta_k \cup \Theta_{k'} \wedge \Phi \cap \Theta_k \cap \Theta_{k'} = \emptyset$$

$$\longrightarrow \Phi \cap \Theta_k = \emptyset \vee \Phi \cap \Theta_{k'} = \emptyset)|.$$

Le I est le même que celui de la preuve de 5.2, la quantification sur la suite de polynômes est justifiée par les bornes de 3.5. ii et les autres abus de notation ont une signification claire.

LEMME 5.6. C est connexe ssi K satisfait Conn_Φ.

PREUVE. Il est clair que si C est connexe, K satisfait Conn_Φ. Si C n'est pas connexe et n'est pas vide, on a une formule Γ:

$$(\Phi \subset \Lambda_1 \cup \Lambda_2) \wedge (\Phi \cap \Lambda_1 \cap \Lambda_2 = \emptyset) \wedge (\Phi \cap \Lambda_1 \neq \emptyset) \wedge (\Phi \cap \Lambda_2 \neq \emptyset)$$

où Λ_1 et Λ_2 sont des combinaisons positives d'inégalités strictes sur des polynômes $R_1, \ldots, R_t$ qui est satisfaite par K. Puisque d'après 3.5 $\mathbb{R}$ satisfait

$$\forall Q_1 \ldots \forall Q_r \forall R_1 \ldots \forall R_t (\Gamma \longrightarrow \neg \mathrm{Conn}\, \Phi)$$

K satisfait aussi cette formule et donc K ne satisfait pas Conn_Φ. Revenons à la démonstration du théorème. Toujours d'après 3.5, $\mathbb{R}$ satisfait

$$\forall Q_1 \ldots \forall Q_r \exists Q_{r+1} \ldots \exists Q_{r+s} \bigvee_{J\in\Phi(I)} (\bigwedge_{k\in J} \mathrm{Conn}_{\Phi\wedge\Theta_k} \ldots$$

$$\ldots \Phi \subset \bigcup_{k\in J} \Theta_k \wedge \bigwedge_{\substack{k,k'\in J\\ k\neq k'}} \Phi \cap \Theta_k \cap \Theta_{k'} = \emptyset).$$

K satisfait aussi cette formule, et vu le lemme ceci donne le résultat.

ii) est clair puisque le fait d'être connexe de dit par une formule.

REMARQUE 1. On peut définir la notion de connexité d'un ensemble semi-algébrique C sans parler de spectre réel: C est connexe s'il n'est pas réunion de deux semi-algébriques disjoints ouverts dans C (ce qui revient à dire aue $\tilde{C}$ est connexe au sens usuel). C'est ce que fait Brumfiel et il montre la partie i de notre théorème ([4] p. 262). H. Delfs* a montré que cette notion de connexité est équivalente à une notion de "connexité par arc", et il obtient un résultat équivalent à notre théorème. Il est tout de même agréable d'avoir un espace où ces notions de connexité coincident avec la notion usuelle.

REMARQUE 2. Ce que l'on a dit jusqu'ici est valable sans aucune hypothèse sur la variété V. Par la suite il sera quelquefois plus commode de se restreindre à des variétés qui ont "suffisamment de points réels" (en écartant par exemple la variété d'équation $x^2 + y^2 + z^2 = 0$).

DEFINITION 5.7. Une variété affine V sur un corps réel clos K sera dite réelle si son anneau de coordonnées K[V] est réel, c.a.d. s'il vérifie

$$\sum_{i=1}^{n} x_i^2 = 0 \Longrightarrow x_i = 0, \qquad i = 1, \ldots, n.$$

On passe d'une variété quelconque à une variété réelle en quotientant l'anneau de coordonnées par l'idéal des x tels qu'une expression $x^{2m} + \sum_{i=1}^{n} y_i^2$ soit nulle. Ceci ne change pas l'ensemble des points réels, ni le spectre réel.

6°) PROPRIETES TOPOLOGIQUES DES MORPHISMES

Le premier résultat que nous voulons montrer est l'analogue du théorème de montée de Cohen-Seidenberg. Ce dernier n'est pas valable pour les idéaux premiers réels: Considérons l'exemple de la projection de la parabole $x = y^2$ sur l'axe des x.

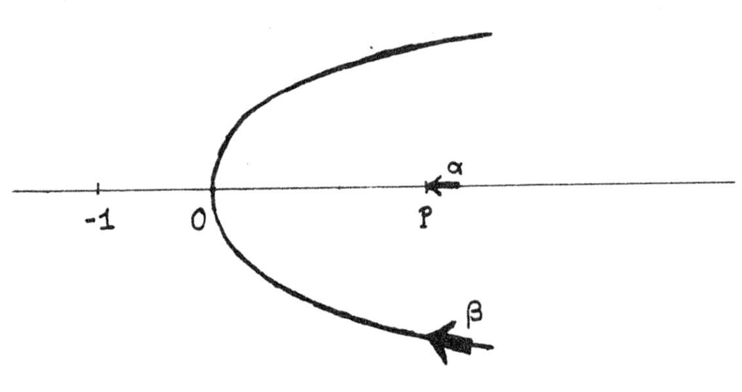

*et M. Knebusch, voir [15].

Le point générique de la parabole s'envoie sur le point générique de la droite. Si on spécialise ce dernier au point d'abscisse -1, il n'y a aucun point réel de la parabole au-dessus. Reprenons maintenant cet exemple avec les précones premiers. Soit β le point générique d'une demi-branche réelle de la parabole et α son image. Si α se spécialise en un point réel P, il y a bien un point réel de la parabole au-dessus de P qui est une spécialisation de β. On voit bien ici comment l'explosion du point générique est nécessaire pour avoir des résultats dans le cas réel.

THEOREME 6.1. Soient $f : A \longrightarrow B$ un morphisme entier d'anneaux, β un précone premier de B, $\alpha = \mathrm{Spec}_R f(\beta)$, α' une spécialisation de α. Il y a une spécialisation β' de β au-dessus de α' $(\alpha' = \mathrm{Spec}_R f(\beta'))$.

Ce théorème de montée est un corollaire du résultat suivant:

THEOREME 6.2. Soit $f : A \longrightarrow B$ un morphisme entier d'anneaux. L'application $\mathrm{Spec}_R f : \mathrm{Spec}_R B \longrightarrow \mathrm{Spec}_R A$ est fermée.

PREUVE. Soit F un fermé de $\mathrm{Spec}_R B$. F est intersection filtrante de fermés constructibles F_i et on a $\mathrm{Spec}_R f(F) = \bigcap_i \mathrm{Spec}_R f(F_i)$ d'après 2.6. On peut donc se ramener au cas où F est constructible, et même au cas

$$F = \{\beta \in \mathrm{Spec}_R B \mid b_1 \in \beta \text{ et } \dots \text{ et } b_n \in \beta\}$$

puisqu'un fermé constructible est une union finie de tels fermés. Posons $g : B \longrightarrow C = B[X_1, \dots, X_n]/(X_1^2 + b_1, \dots, X_n^2 + b_n)$.

g est entier et l'image de $\mathrm{Spec}_R g$ est F. Ainsi il suffit de montrer que si f est entier l'image de $\mathrm{Spec}_R f$ est fermée. B est limite inductive filtrante de A-algèbres entières de présentation finie B_i:

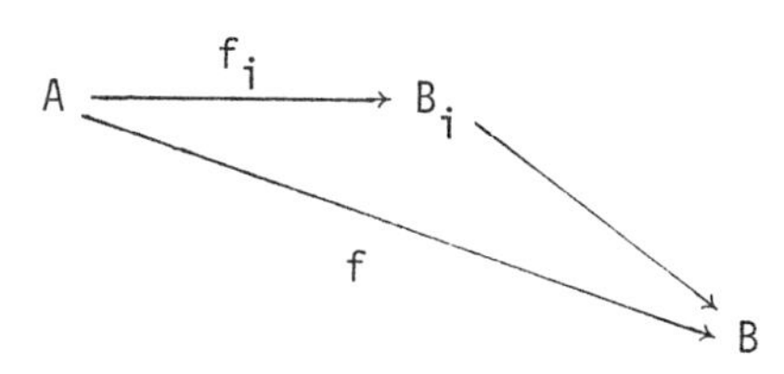

Montrons que l'image de $\mathrm{Spec}_R f$ est l'intersection des images des $\mathrm{Spec}_R f_i$. Il est clair qu'elle est contenue dans cette intersection. Soit alors α dans $\bigcap_i \mathrm{Im}(\mathrm{Spec}_R f_i)$.

Tensorisons le tout par $k(\alpha)$:

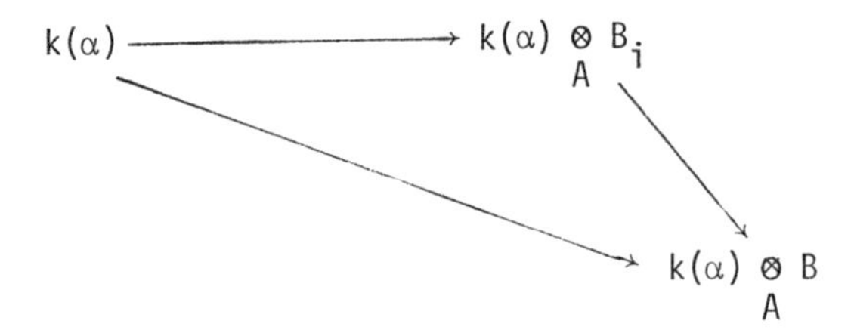

$(\mathrm{Spec}_R f_i)^{-1}(\alpha) \simeq \mathrm{Spec}_R(k(\alpha) \underset{A}{\otimes} B_i)$ est le spectre réel d'une $k(\alpha)$-algèbre finie donc il est fini discret et non vide par hypothèse. $(\mathrm{Spec}_R f)^{-1}(\alpha) \simeq \mathrm{Spec}_R(k(\alpha) \otimes B)$ est la limite projective filtrante des $\mathrm{Spec}_R(k(\alpha) \otimes B_i)$ (2.4) et est donc non vide. Donc on a $\alpha \in \mathrm{Im}(\mathrm{Spec}_R f)$.

Finalement il suffit de montrer que si B est une A-algèbre entière de présentation finie, l'image de $\mathrm{Spec}_R B$ dans $\mathrm{Spec}_R A$ est fermée. On peut écrire:

$$B = A[X_1, \ldots, X_m] / (F_1(X_1), \ldots, F_m(X_m), G_1(\underline{X}), \ldots, G_p(\underline{X}))$$

où F_i est un polynôme unitaire en la seule indéterminée X_i et G_j un polynôme quelconque en $\underline{X} = X_1, \ldots, X_m$. L'image de $\mathrm{Spec}_R B$ est l'emsemble des α de $\mathrm{Spec}_R A$ tels que le corps $k(\alpha)$ satisfait la formule

$$\exists \underline{x} \quad (F_1(x_1) = \ldots = F_m(x_m) = G_1(\underline{x}) = \ldots = G_p(\underline{x}) = 0).$$

Cette formule a pour paramètres les coefficients des F_i (sauf le coefficient dominant qui est 1) et ceux des G_j. Si l'on fait varier ces paramètres dans $\mathbb{R}$, l'extension de la formule est fermée. D'après le théorème 4.2 l'image de $\mathrm{Spec}_R B$ est fermée. Ceci achève la preuve du théorème.

Remarquons que l'on n'a pas l'analogue du théorème de descente de Cohen-seidenberg, comme le montre l'exemple du début de cette partie: Le point 0_- de $\mathrm{Spec}_R \mathbb{R}[X]$ n'est pas l'image d'un point générique d'une demi-branche réelle de la parabole et pourtant il se spécialise en l'origine de l'axe des x qui se relève en un point réel de la parabole. Cet exemple montre aussi qu'un morphisme plat n'est pas ouvert pour le spectre réel. Pour ce qui est des propriétés d'ouverture, on a les résultats suivants:

THEOREME 6.3. Soit $f : A \longrightarrow B$ un morphisme d'anneaux.

i) Si f est étale, $\mathrm{Spec}_R f$ est un homéomorphisme local.

ii) Si f est lisse, $\mathrm{Spec}_R f$ est ouvert.

PREUVE. i) C'est le corollaire 1 de [6].

ii) Vu i), il suffit de s'intéresser au cas d'un morphisme $A \longrightarrow A[X_1, \ldots, X_n]$ qui se traite facilement en utilisant 4.2.

COROLLAIRE 6.4. Soit V une variété affine sur K réel clos, x un point non singulier de $V(K)$, $T_x(V)$ l'espace tangent à V en x. Un voisinage de x dans $\tilde{V}$ est homéomorphe à un voisinage de x dans $\widetilde{T_x(V)}$.

PREUVE. On peut projeter V sur $T_x(V)$ de façon étale au voisinage de x (pour la topologie de Zariski).

7°) PRECONES PREMIERS ET PLACES REELES

Soient V une variété affine sur un corps K réel clos, α un precone premier de $K[V]$, $\mathfrak{p} = \alpha \cap -\alpha$. α induit un ordre sur $K(\mathfrak{p})$. On peut construire une place réelle [13] ϕ_α de $K(\mathfrak{p})$ selon un procédé bien connu depuis Krull [11]: l'anneau de valuation est formé des éléments de $K(\mathfrak{p})$ qui ne sont pas infiniment grands par rapport aux constantes de K. Le corps résiduel K' est canoniquement ordonné: ses éléments strictement positifs sont ceux qui proviennent d'éléments strictement positifs de $K(\mathfrak{p})$. K' est une extension archimédienne de K. Supposons ϕ_α finie sur $K[V]/_{\mathfrak{p}}$ on dira que α est à <u>distance finie</u>. ϕ_α induit alors un morphisme de $K[V]$ dans le corps ordonné K' et un tel morphisme donne un précone premier de $K[V]$: il suffit bien sûr de prendre les éléments dont l'image est négative ou nulle. Il est clair que ce précone premier est une spécialisation de α.

DEFINITION 7.1. Si α est à distance finie nous appellerons centre de α, noté $c(\alpha)$, le précone premier construit ci-dessus.

Si $K = \mathbb{R}$, on a toujours $K' = \mathbb{R}$ et donc le centre d'un précone premier à distance finie est un point réel de la variété.

Supposons maintenant que la place $\phi_\alpha : K(\mathfrak{p}) \longrightarrow K' ,\infty$ se décompose en $K(\mathfrak{p}) \longrightarrow K_1,\infty \longrightarrow K',\infty$. De la même façon que ci-dessus, K_1 est canoniquement ordonné et si la place $K(\mathfrak{p}) \longrightarrow K_1,\infty$ est finie sur $K[V]/_{\mathfrak{p}}$, le morphisme $K[V] \longrightarrow K_1$ donne une spécialisation α_1 de α. Si on a une décomposition $\phi_\alpha : K(\mathfrak{p}) \longrightarrow K_1,\infty \longrightarrow K_2,\infty \longrightarrow \ldots \longrightarrow K_r,\infty \longrightarrow K',\infty$ telle que la place $K(\mathfrak{p}) \longrightarrow K_r,\infty$ est finie sur $K[V]/_{\mathfrak{p}}$ on obtient des précones premiers $\alpha_o = \alpha \subset \alpha_1 \subset \ldots \subset \alpha_r$. Il faut faire attention: l'inclusion $\alpha_i \subset \alpha_{i+1}$ n'est pas forcément stricte pour $i > 1$ même si $K_i \longrightarrow K_{i+1},\infty$ n'est pas triviale. Voici un contre-example: Prenons $K[V] = \mathbb{R}[X,Y]$. Soit

α l'ordre sur $\mathbb{R}(X,Y) = \mathbb{R}(X,T)$, $T = \frac{Y}{X}$, dans lequel sont positifs les polynômes $P(X,T)$ tels que: $\exists M > 0\ \forall\ \varepsilon \in]0,M[\ \exists N > 0\ \forall\ \eta \in]0,N[P(\eta,\varepsilon) > 0$. La place $\phi_\alpha : \mathbb{R}(X,Y) \longrightarrow \mathbb{R},\infty$ se décompose en $\mathbb{R}(X,Y) \simeq \mathbb{R}(X,T) \longrightarrow \mathbb{R}(T),\infty \longrightarrow \mathbb{R},\infty$ et est finie sur $\mathbb{R}[X,Y]$. La chaîne (non strict !) de précones premiers correspondants est $\alpha \subset 0 \subset 0$ (où 0 est l'origine du plan). Ce contre-exemple vient de Zariski.

Nous allons maintenant voir comment à partir d'une chaîne de spécialisations de α on peut récupérer une décomposition de la place ϕ_α:

PROPOSITION 7.2. Supposons donnée une chaîne de précones premiers $\alpha = \alpha_0 \subset \alpha_1 \subset \ldots \subset \alpha_r$ de $K[V]$. On peut décomposer la place ϕ_α en $K(\mathfrak{p}) \longrightarrow K_1,\infty \longrightarrow K_2,\infty \longrightarrow \ldots \longrightarrow K_r,\infty \longrightarrow K'$ de telle sorte que la place $K(\mathfrak{p}) \longrightarrow K_r,\infty$ soit finie sur $K[V]$ et que la chaîne de précones premiers induite par cette décomposition soit $\alpha \subset \alpha_1 \subset \ldots \subset \alpha_r$.

PREUVE. Considérons l'anneau $K[V]/_{\mathfrak{p}}$ ordonné par l'ordre induit par α. La donnée d'une spécialisation β de α revient alors à celle d'un idéal premier convexe $\mathfrak{q}$ de $K[V]$ (à savoir l'image de $\beta \cap -\beta$). On peut utiliser la proposition 7.7.4 de Brumfiel [4], qui donne un anneau de valuation A' * convexe pour l'ordre sur $K(\mathfrak{p})$, d'idéal maximal $\mathfrak{m}_{A'}$ avec $\mathfrak{m}_{A'} \cap K[V]/_{\mathfrak{p}} = \mathfrak{q}$. Comme A' est convexe et continent K, il contient aussi l'anneau de valuation de la place ϕ_α et donc correspond à une décomposition $K(\mathfrak{p}) \longrightarrow L,\infty \longrightarrow K',\infty$ de ϕ_α. La place $K(\mathfrak{p}) \longrightarrow L,\infty$ est finie sur $K[V]/_{\mathfrak{p}}$ et la spécialisation correspondante de α est bien sûr β.

COROLLAIRE 7.3. La longueur de la chaîne des spécialisations de α est inférieure ou égale au rang de la place ϕ_α. L'inégalité stricte peut avoir lieu même si α a un centre.

Donnons maintenant un idéal premier réel $\mathfrak{p}$ de $K[V]$ et ϕ une place réelle de $K(\mathfrak{p})$ au-dessus de K, de corps résiduel L. Un ordre sur $K(\mathfrak{p})$ sera dit compatible avec ϕ si l'anneau de valuation de ϕ (ou, ce qui revient au même, l'idéal maximal de cet anneau) est convexe pour cet ordre. On sait qu'il y a sur $K(\mathfrak{p})$ au moins un ordre compatible avec ϕ [13]. Précisément, si on choisit un ordre sur L, le nombre d'ordres sur $K(\mathfrak{p})$ compatibles avec ϕ et qui induisent l'ordre donné sur L est égal à l'ordre de $\Gamma/_{2\Gamma}$ ou Γ est le groupe de valuation de ϕ [3].

Ceci montre qu'une description des précones premiers de $K[V]$ doit reposer sur la connaissance des places rélles des corps résiduels $K(\mathfrak{p})$.

---

*contenant $K[V]$.

Considérons par exemple le cas d'une courbe affine irréductible réelle C sur $\mathbb{R}$. Les points de $\tilde{C}$ sont d'une part les points de $C(\mathbb{R})$, correspondant aux idéaux maximaux réels de $\mathbb{R}[C]$, d'autre part les points correspondant aux ordres sur $\mathbb{R}(C)$. Un tel ordre induit une valuation (discrète de rang 1) de corps résiduel $\mathbb{R}$. Réciproquement à chaque valuation sur $\mathbb{R}(C)$ de corps résiduel $\mathbb{R}$ correspondent deux ordres (pour déterminer l'ordre, il suffit de choisir le signe d'un paramètre uniformisant). Les ordres sur $\mathbb{R}(C)$ correspondent donc aux demi-branches réelles de la courbe que l'on peut définir comme classes d'équivalences de paramétrisations réelles pour des changements de paramètres croissants. Voici justifiée l'assertion faite dans l'introduction:

PROPOSITION 7.4. Soit C une courbe affine irréductible réelle sur $\mathbb{R}$. Les points de $\tilde{C}$ sont, outre les points de $C(\mathbb{R})$, les points génériques des demi-branches réelles de C.

Quand la dimension de la variété est plus grande que 1, la description est plus délicate. Pour ce qui est du plan affine sur $\mathbb{R}$ par exemple, on a les points réels du plan, les points génériques de demi-branches réelles de courbes algébriques du plan et les points correspondants aux ordres sur le corps $\mathbb{R}(X,Y)$. On peut pour avoir une idée de ce que sont ces ordres se reporter au paragraphe 8.12 de Brumfiel [4]; signalons simplement que la donnée d'une demi-branche réelle d'une courbe algébrique C et d'une orientation du plan détermine un ordre sur $\mathbb{R}(X,Y)$, générisation du point générique de la demi-branche en question. La place associée à cet ordre est discrète de rang 2 et se décompose en $\mathbb{R}(X,Y) \longrightarrow \mathbb{R}(C),\infty \longrightarrow \mathbb{R},\infty$. Il faut faire attention à ce que tous les ordres correspondant à des places de rang 2 ne sont pas obtenues ainsi, comme le montre le contre-exemple plus haut; il y a aussi bien sûr des ordres correspondant à des places de rang 1. Dans le cas d'un espace affine de dimension n, on peut construire des ordres sur le corps de fractions rationnelles en se donnant une suite de variétés linéaires emboitées, chacune avec une orientation. Ceci permet de montrer le résultat suivant:

PROPOSITION 7.5. Soit $x \in A^n(K)$, K réel clos. Il y a dans $K[X_1, \ldots, X_n]$ une chaîne de précones premiers $\alpha_0 \subsetneq \alpha_1 \subsetneq \ldots \subsetneq \alpha_n = x$.

PREUVE. On peut supposer que x est l'origine de K^n, et on prend

$$\alpha_i = \{P \in K[X_1, \ldots, X_n] \mid \exists M_n > 0 \quad \forall \varepsilon_n \in]0,M_n[\quad \exists M_{n-1} > 0 \quad \forall \varepsilon_{n-1} \in]0,M_{n-1}[\ \ldots$$

$$\ldots \exists M_{i+1} > 0 \quad \forall \varepsilon_{i+1} \in]0,M_{i+1}[\quad P(0, \ldots, 0, \varepsilon_{i+1}, \ldots, \varepsilon_n) \leq 0 .$$

La place ϕ_{α_o} se décompose en $K(X_1, \ldots, Xn) \longrightarrow K(X_1, \ldots, X_{n-1}), \infty \longrightarrow \ldots \longrightarrow K$. Nous en venons maintenant à la définition des variétés réellement complètes, qui est du type "critère valuatif de complétude":

DEFINITION 7.6. Une variété affine V sur un corps K réel clos sera dite réellement complète si tout précone premier de $K[V]$ a un centre (autrement dit si pour tout précone premier α la place réelle ϕ_α est finie sur l'image de $K[V]$).

PROPOSITION 7.7. V est réellement complète ssi tout élément de $K[V]$ est borné sur $V(K)$ (autrement dit, $V(K)$ est borné).

PREUVE. Soit f un élément de $K[V]$. f est un morphisme de V dans la droite affine sur K, et on a donc $\tilde{f} : \tilde{V} \longrightarrow \widetilde{A^1(K)}$. On sait que $\widetilde{f(V)} = \tilde{f}(\tilde{V})$ est un constructible. Il est alors clair que $f(V)$ est borné ssi $\widetilde{f(V)}$ ne contient ni $+\infty$, ni $-\infty$. Or $\tilde{f}(\alpha) \neq \pm\infty$ ssi f est fini à la place ϕ_α.

REMARQUE 1. On pourrait penser que si V est irréductible et réelle (5.7) il suffit de dire que toute place réelle sur $K(V)$ est finie sur $K[V]$. Ce n'est pas le cas, comme le montre le contre-exemple de la surface $(1-z^2)x^2 = x^4 + y^4$.

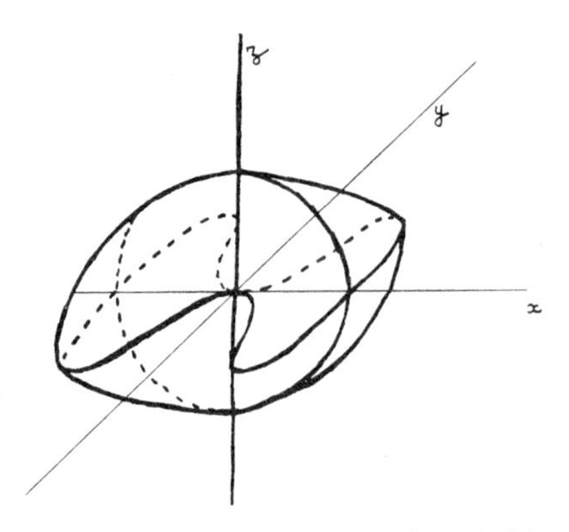

Cette surface n'est pas réellement complète puisqu'elle n'est pas bornée. Cependant sa "partie de dimension 2" est bornée (ces problèmes de dimension seront précisés au 8°).

REMARQUE 2. On peut montrer que si V est réellement complète, l'application "centre" est continue. Si $K = \mathbb{R}$, c'est alors une rétractation continue de l'inclusion $V(\mathbb{R}) \longrightarrow \tilde{V}$ [7].

8°) DIMENSION REELLE

DEFINITION 8.1. Soit A un anneau commutatif. La <u>dimension réelle</u> de A, notée $\dim_R A$, est le sup les longueurs de chaînes de précones premiers de A.

D'après 2.1.ii la dimension réelle de A est la dimension combinatoire de $\text{Spec}_R A$, c.a.d. le sup des longueurs de chaînes de fermés irréductibles de cet espace.

D'autrepart puisqu'une chaîne de précones premiers de A induit une chaîne d'idéaux premiers (réels) de A on a toujours $\dim_R A \leqslant \dim A$.

PROPOSITION 8.2. Soit V un variété algébrique affine sur K réel clos. $\dim_R K[V]$ est égal au sup des longueurs de chaînes de sous-ensembles algébriques irréductibles de $V(K)$, qui est la dimension de $V(K)$ comme ensemble algébrique. On notera $\dim_R V$ ce nombre.

PREUVE. Le théorème des zéros réels [14] nous dit que l'on a une bijection entre les sous-ensembles algébriques irréductibles de $V(K)$ et les idéaux premiers réels de $K[V]$. La dimension de $V(K)$ est donc surement au moins égale à la dimension réelle de $K[V]$. Pour montrer l'égalité, on utilisera le résultat suivant:

PROPOSITION 8.3. Soit V une variété affine irréductible réelle (5.7) sur K. $\dim_R K[V]$ est égal au degré de transcendance de $K(V)$ sur K.

PREUVE. Soit d ce degré de transcendance. Le lemme de normalisation de E. Noether donne un morphisme entier injectif $\Pi^* : K[X_1, \ldots, X_d] \longrightarrow K[V]$ correspondant au morphisme de variétés $\Pi : V \longrightarrow A^d$. D'après 6.2 l'application $\tilde{\Pi} : \tilde{V} \longrightarrow \tilde{A}^d$ est fermée. Puisque $K[V]$ est réel, $K(V)$ est ordonnable. Soit α un ordre sur $K(V)$ et β son image par $\tilde{\Pi}$; β est l'ordre sur $K(X_1, \ldots, X_d)$ induit par α. $\tilde{\Pi}(\tilde{V})$ est un constructible de $\tilde{A}^d$ d'apres 2.3.

LEMME 8.4. Soit C un constructible de $\tilde{A}^d$ qui contient un ordre β sur $K(X_1, \ldots, X_d)$. C contient un voisinage ouvert de β. (On peut remplacer A^d par n'importe quelle variété affine irréductible réelle sur K).

PREUVE. On a $C = \{\alpha \mid k(\alpha) \text{ satisfait } \Phi\}$ où Φ peut s'écrire comme disjonction de conjonctions d'inégalités polynomiales strictes ou larges. Soit Ψ la formule obtenue en remplaçant les inégalités larges par des inégalités strictes, et $D = \{\alpha \mid k(\alpha) \text{ satisfait } \Psi\}$. D est un ouvert contenu dans C, et il contient β.

Revenons à la preuve de 8.3. Le lemme ci-dessus entraîne que $\tilde{\Pi}(\tilde{V})$ contient un ouvert constructible non vide $\tilde{U}$. Soit y un point de U. D'après 7.5 on peut trouver une chaîne $\beta_o \subsetneqq \beta_1 \subsetneqq \ldots \subsetneqq \beta_d = y$ dans $\tilde{U}$. Choisissons $\alpha_o \in \tilde{V}$ tel que $\tilde{\Pi}(\alpha_o) = \beta_o$. Le théorème de montée nous donne alors une chaîne $\alpha_o \subsetneqq \alpha_1 \subsetneqq \ldots \subsetneqq \alpha_d$ dans $\tilde{V}$.

La dimension de l'ensemble algébrique $V(K)$ est égale au sup des longueurs de chaînes d'idéaux premiers réels de $K[V]$. La proposition 8.2 ne suffit donc pas à établir l'utilité de la notion de dimension réelle calculée avec les precones premiers. L'intérêt de cette dimension reelle est qu'elle permet d'obtenir algébriquement la dimension d'une variété en un point réel conformément à ce que l'on voit sur les dessins (comme ceux de l' intro duction).

DEFINITION 8.5. Soit V une variété affine sur K réel clos, x un point de $V(K)$. La <u>dimension réelle de V en x</u>, notée $\dim_R(V,x)$, est le sup des longueurs de chaînes de précones premiers de $K[V]$ se terminant en x.

REMARQUE. Le sous espace de $\mathrm{Spec}_R K[V]$ formé des générisations de x est homéomorphe au spectre réel de $(K[V]_x)^h$, le hensélisé de l'anneau local en x; ceci est une conséquence de la démonstration de la spatialité du topos étale réel [6]. On a donc:

$$\dim_R(V,x) = \dim_R(K[V]_x)^h.$$

Nous allons maintenant étudier les propriétés de la demension réelle en un point. Pour ceci nous avons besoin d'un outil qui permette d'éliminer les précones premiers qui ont une chaîne de spécialisation trop courte. Précisons ce point.

DEFINITION 8.6. Soit α un précone premier de $K[V]$. La <u>profondeur</u> de α est le degré de transcendance de $k(\alpha)$ sur K. Si $\mathfrak{p} = \alpha \cap -\alpha$, la profondeur de α est aussi la dimension réelle (ou la dimension de Krull) de $K[V]/_{\mathfrak{p}}$ d'après 8.3. Les "bons" précones premiers sont ceux dont la profondeur est égale à la longueur de la chaîne des spécialisations. (En général elle est supérieure ou égale.) Le résultat suivant dit qu'il y a suffisamment de "bons" precones premiers:

PROPOSITION 8.7. Soit x un point de $V(K)$, α un précone premier de $K[V]$ centré en x et de profondeur d. Il existe un précone premier β de $K[V]$ qui a même idéal premier réel associé que α $(\alpha \cap -\alpha = \beta \cap -\beta)$ et qui a une chaîne de spécialisations $\beta = \beta_0 \subsetneqq \beta_1 \subsetneqq \cdots \subsetneqq \beta_d = x$. Si de plus S est un semi-algébrique de $V(K)$ tel que S contient α, on peut choisir β dans S.

PREUVE. On peut se ramener au cas $\alpha \cap -\alpha = (0)$; α est alors un ordre sur $K(V)$. On a un morphisme entier injectif:

$$\Pi^* : K[X_1, \ldots, X_d] \longrightarrow K[V].$$

Soit $y = \Pi(x)$. $\Pi^{-1}(y)$ est fini discret et donc il y a un ouvert semi-algébrique U de $V(K)$ contenant x tel que $\Pi^{-1}(y) \cap U = \{x\}$. On a $\alpha \in \tilde{U} \cap \tilde{S}$. Le lemme 8.4 montre que $\Pi(U \cap S)$ contient un ouvert semi-algébrique W de $A^d(K)$ avec $\tilde{\Pi}(\alpha) \in \tilde{W}$. y est donc adhérent à W.

LEMME 8.8. Soit W un ouvert semi-algébrique de $A^d(K)$, y un point adhérent à W. Il existe une chaîne de précones premiers

$$\gamma_o \subsetneqq \gamma_1 \subsetneqq \ldots \subsetneqq \gamma_d = y$$

avec $\gamma_o \in \tilde{W}$.

Finissons d'abord la preuve de la proposition. On a $\beta_o \in \tilde{U} \cap \tilde{S}$ tel que $\tilde{\Pi}(\beta_o) = \gamma_o$ et le théorème de montée donne alors une chaîne $\beta_o \subsetneqq \beta_1 \subsetneqq \ldots \subsetneqq \beta_d = x$.

PREUVE DU LEMME. Le lemme est clair pour $d = 1$ (W est une union finie d'intervalles ouverts). Supposons le montré pour $d-1$. On peut se ramener au cas

$$W = \{\underline{z} \in K^d \mid \bigwedge_{i=1}^{n} P_i(\underline{z}) > o\}, \quad \underline{y} = 0$$

où les P_i sont des polynômes en X unitaires en X_d. On peut de plus supposer $\underline{z} \in W \Longrightarrow |z_d| < M$ pour un $M \in K$ fixé avec M suffisamment petit pour qu'il n'y ait sur l'axe des X_d aucun zéro d'un des P_i entre $-M$ et $+M$ autre que l'origine (éventuellement). Il y a deux cas de figure:

i) Les points $(0, \ldots, 0, \varepsilon)$ pour $0<\varepsilon<M$ (rep. $-M<\varepsilon<0$) sont dans W. Alors on fabrique sans peine sur le modèle de 7.5 une chaîne de précones premiers $\gamma_o \subsetneqq \gamma_1 \subsetneqq \ldots \subsetneqq \gamma_d = y$ où γ_{d-1} est la demi-branche de l'axe des X_d centrée à l'origine et dirigée vers le haut (resp. vers le bas) qui est bien dans $\tilde{W}$.

ii) Dans le cas contraire, l'intersection de l'adhérence de W avec l'axe des X_d est réduite à l'origine. Soit $p : A^d(K) \longrightarrow A^{d-1}(K)$ la projection parallèlement à l'axe des X_d. $p(y)$ est adhérent à l'ouvert semi-algébrique $p(w)$: on a donc une chaîne $\delta = \delta_1 \subsetneqq \ldots \subsetneqq \delta_d = p(y)$ avec $\delta \in \widetilde{p(W)}$. $\tilde{p}^{-1}(\delta)$ est homéomorphe à $\mathrm{Spec}_R(k(\delta)[X_d]) \simeq \widetilde{A^1(k(\delta))}$ d'après 4.3. Comme $\tilde{p}^{-1}(\delta) \cap \tilde{W}$ est non vide, moyennant l'identification de $\tilde{p}^{-1}(\delta)$ avec $\widetilde{A^1(k(\delta))}$, on a un point γ de $\widetilde{A^1(k(\delta))}$ qui est dans $\tilde{W}$ puisque $A^1(k(\delta))$ est dense dans $\widetilde{A^1(k(\delta))}$ (5.1); $k(\gamma)$ est isomorphe à $k(\delta)$.

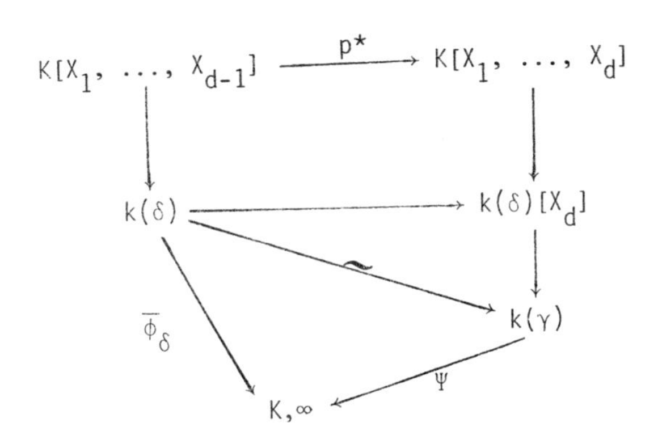

La place ϕ_δ s'étend en une place $\overline{\Phi}_\delta$ sur $k(\delta)$ discrète de rang $d-1$ et de corps résiduel K (7.2). Par isomorphisme on a une place Ψ sur $k(\gamma)$ avec les mêmes propriétés. Ψ est finie sur $K[X_1, \ldots, X_d]$ car $X_1, \ldots, X_{d-1}$ s'envoient sur 0 (comme pour ϕ_δ) et X_d reste fini, majorée en valeur absolue par M, puisque $\gamma \in \tilde{W}$. On a ainsi d'après le début du 7° une chaîne de précones premiers $\gamma = \gamma_1 \subsetneq \cdots \subsetneq \gamma_d$. Ces précones premiers sont bien tous distincts car leurs images par p sont $\delta_1 \subsetneq \cdots \subsetneq \delta_d$. γ_d est un point de l'axe des X_d dans l'adhérence de W; c'est donc y. Prenons maintenant pour γ_0 une générisation γ_+ ou γ_- de γ dans $\widetilde{A^1(k(\delta))}$ (même description que celle donnée dans l'introduction pour le corps des nombres réels). On a bien finalement une chaîne $\gamma_0 \subsetneq \gamma_1 \subsetneq \cdots \subsetneq \gamma_d = y$ avec $\gamma_0 \in \tilde{W}$.

Nous pouvons maintenant passer au résultat essentiel pour l'étude de la dimension réelle en un point.

THEOREME 8.9. Soient V une variété affine irréductible réelle sur K réel clos, x un point de $V(K)$. On a $\dim_R(V,x) = \dim_R V$ ssi x est limite pour la topologie euclidienne de points non singuliers de $V(K)$.

PREUVE. Soit $d = \dim_R V$; d est aussi la dimension de Krull de $K[V]$. Soit x un point non singulier de $V(K)$. D'après 6.4 un voisinage de x dans V est homéomorphe à un voisinage de x dans $\widetilde{T_x(V)} \simeq \widetilde{A^d(K)}$. La proposition 7.5 montre alors que l'on a $\dim_R(V,x) = \dim_R V$. Supposons maintenant que x est limite de points non singuliers. Soit $(U_i)_{i \in I}$ une base de voisinages ouverts semi-algébriques de x dans $V(K)$. Tout U_i contient un point où la dimension réelle est d; un tel point est le centre d'un précone premier qui est un ordre sur $K(V)$. Ceci montre que l'on a pour tout i, $\tilde{U}_i \cap \mathrm{Spec}_R K(V) \neq \emptyset$. Puisque $\mathrm{Spec}_R K(V)$ est compact on en déduit $(\bigcap_{i \in I} \tilde{U}_i) \cap \mathrm{Spec}_R K(V) \neq \emptyset$. On vérifie facilement que $\bigcap_{i \in I} \tilde{U}_i$ est l'ensemble des générisations de x. x est donc le centre d'un

ordre sur $K(V)$, c'est-à-dire d'un précone premier de profondeur d. D'après 8.7 on a $\dim_R(V,x) = d$.
Réciproquement soit x un point qui n'est pas limite de points non singuliers de V. Il y a donc un voisinage ouvert semi-algébrique U de x contenu dans $\mathrm{Sing}(V)$, le lieu singulier de V. $\mathrm{Sing}(V)$ est une sous variété de V de dimension (ordinaire, et donc à fortiori réelle) strictement plus petite que d. Soit α un précone premier de $K[V]$ de centre x. On a $\alpha \in \tilde{U} \subset \widetilde{\mathrm{Sing}(V)}$ et donc la profondeur de α est strictement inférieure à d. Ceci montre $\dim_R(V,x) < d$.
REMARQUE. On a vu plus haut que $\dim_R(V,x) = \dim_R(K[V]_x)^h$. Sous les hypothèses du théorème 8.9. On a donc immédiatement le résultat suivant: $\dim_R(V,x) = \dim_R V$ ssi $(K[V]_x)^h$ a un idéal premier minimal qui est réel. On peut trouver l'équivalence de cette dernière propriété avec le fait que x est limite de points non singuliers réels dans [8]. Efroymson annonce aussi l'équivalence avec une autre propriété: le normalise de $K[V]_x$ a un idéal maximal réel. Ceci est faux; la surface d'équation $x^2 + y^2 + z^2 - x^3 = 0$ est normale, de dimension réelle 2, et pourtant l'origine est un point réel isolé, c'est-à-dire que la dimension réelle y est 0 (ce contre exemple a été suggéré par J. J. Risler). De fait, dans la bijection entre idéaux premiers minimaux du hensélisé et idéaux maximaux du normalisé, un idéal premier minimal réel du premier donne toujours un idéal maximal réel du second, mais pas réciproquement.
La notion de dimension en un point s'étend naturellement aux ensembles semi-algébriques:

DEFINITION 8.10. Soient S un semi-algébrique de $A^n(K)$, x un point de S. La dimension réelle de S en x, notée $\dim_R(S,x)$ est le sup des longueurs de chaînes de précones premiers dans $\tilde{S}$ se terminant en x.

PROPOSITION 8.11. Soit d la dimension de la clôture de Zariski de S dans $A^n(K)$. On a $d = \sup\{\dim_R(S,x) \mid x \in S\}$.

PREUVE. S est réunion finie de semi-algébriques localement fermés du genre

$$(*) \quad T = \{x \in A^n(K) \mid P_1(x) = \dots = P_m(x) = 0 \text{ et } Q_1(x) > 0 \text{ et } \dots \text{ et } Q_p(x) > 0\}$$

où l'idéal $(P_1, \dots, P_m)$ est celui des polynômes qui s'annulent sur T; il est donc réel. On peut supposer en plus qu'il est premier. On a ainsi

$$S = T_1 \cup \dots \cup T_n$$

où les T_i sont comme ci-dessus. Un au moins de ces T_i a une clôture de Zariski de dimension d; supposons le donné par $(*)$. Soit V la variété affine réelle irréductible $P_1(x) = \ldots = P_m(x) = 0$. Puisque T est dense pour la topologie de Zariski dans $V(K)$, il n'est pas contenu dans le lieu singulier de V et il y a un point x de T tel que $\dim_R(V,x) = d$ (8.9). Comme T est ouvert dans $V(K)$ on a aussi $\dim_R(T,x) = d$ et donc $\dim_R(S,x) \geqslant d$. Ceci prouve $d \leqslant \sup\{\dim_R(T,x) \mid x \in S\}$. L'inégalité inverse est claire.

PROPOSITION 8.12. Soit S un ensemble semi-algébrique. La fonction $x \longmapsto \dim_R(S,u)$ est semi continue supérieurement sur S. Pour k fixé, $\{x \in S \mid \dim_R(S,x) < k\}$ est un semi-algébrique ouvert dans S.

PREUVE. Soit encore d la dimension de la clôture de Zariski de S. Il suffit de montrer que $\{x \in S \mid \dim_R(S,x) = d\}$ est un semi-algébrique fermé dans S. La clôture de Zariski de S est une union finite de composantes irréductibles $V_1(K) \cup \ldots \cup V_m(K)$ où les V_i sont des variétés irréductibles réelles de dimension (réelle) inférieure ou égale à d.

On a : $\{x \in S \mid \dim_R(S,x) = d\} = \bigcup_{i=1}^{m} \{x \in S \cap V_i(K) \mid \dim_R(S \cap V_i(K),x) = d\}$.

En effet, si on a $\dim_R(S,x) = d$ c'est que x est le centre d'un ordre α sur un $K(V_i)$ où V_i est de dimension d et $\alpha \in S$; mais alors $\dim_R(S \cap V_i(K),x) = d$. Ceci montre une inclusion, et l'autre est claire.

On est ainsi amené à montrer la chose suivante:

LEMME 8.13. Si S est un semi-algébrique de $V(K)$ où V est irréductible réelle de dimension d, $\{x \in S \mid \dim_R(S,x) = d\}$ est un semi-algébrique fermé dans S.

PREUVE. Soit $M = \{y \in V(K) \mid \dim_R(V,y) = d\}$. D'après 8.9 M est un semi-algébrique fermé dans $V(K)$. Si $\operatorname{Sing} V$ est le lieu singulier de V, on a $\tilde{V} - \tilde{M} \subset \widetilde{\operatorname{Sing} V}$ et comme $\widetilde{\operatorname{Sing} V}$ ne peut contenir aucun ordre sur $K(V)$, $\operatorname{Spec}_R K(V)$ est contenu dans $\tilde{M}$.

Soit maintenant x un point de S tel que $\dim_R(S,x) = d$. On a un ordre α sur $K(V)$ dans S centré en x. Si U est un voisinage* quelconque de x dans $V(K)$, on a $\alpha \in \widetilde{M \cap U \cap S}$ et donc $M \cap U \cap S$ est d'intérieur non vide par 8.4. Réciproquement si $M \cap U \cap S$ est d'intérieur non vide pour tout voisinage semi-algébrique U de x dans $V(K)$, c'est que $\tilde{S} \cap \operatorname{Spec}_R K(V)$ continent une générisation de x, et $\dim_R(S,x) = d$.

Finalement $\dim_R(S,x) = d$ ssi

$$\forall\, \varepsilon \in K,\ \varepsilon > o \longrightarrow \exists\, z \in V(K)\ \exists\, \eta \in K,\ \eta > o$$
$$\forall\, y \in V(K)\ (|y-z| < \eta \longrightarrow |y-x| < \varepsilon \wedge y \in M \cap S).$$

On voit sans peine que ceci définit un fermé semi-algébrique de $V(K)$.

---

*semi-algébrique

PROPOSITION 8.14. Soient S un ensemble semi-algébrique, x un point de S. $\dim_R(S,x)$ est la dimension de la clôture de Zariski d'un voisinage suffisamment petit de x dans S.

PREUVE. C'est un corollaire de 8.11 et 8.12.

9°) L'INVARIANCE BIRATIONNELLE DES COMPOSANTES CONNEXES

Dans cette partie V est une variété affine irréductible réelle, réellement complète et non-singulière sur K réel clos. Nous allons voir que les composantes connexes de $\tilde{V}$ peuvent se lire sur le corps de fractions $K(V)$. Ceci montrera leur caractère d'invariant birationnel.

LEMME 9.1. $\mathrm{Spec}_R K(V)$ est dense dans $\tilde{V}$.

PREUVE. Tout point x de $V(K)$ est centre d'un ordre sur $K(V)$ puisque $\dim_R(V,x) = \dim_R V$.

Nous rappelons qu'a tout ordre α sur $K(V)$ est associé à une place réelle ϕ_α sur $K(V)$ (voir le début du 7°).*

PROPOSITION 9.2. La décomposition de V en composantes connexes induit une partition de $\mathrm{Spec}_R K(V)$ en ouverts-fermés qui est la seule partition en ouverts-fermés compatible avec la relation d'equivalence "avoir même place associée".

PREUVE. i) Soit D un ouvert-ferme de $\tilde{V}$. Soient α et β deux ordres sur $K(V)$ tels que $\phi_\alpha = \phi_\beta$. Si $\alpha \in D$, son centre appartient aussi à D. Comme c'est aussi le centre de β, on a $\beta \in D$. Ceci montre que la partition de $\mathrm{Spec}_R K(V)$ obtenue a bien la propriété annoncée.

ii) Soit W_1 un ouvert fermé de $\mathrm{Spec}_R K(V)$ et W_2 son complémentaire. On a des ouverts constructibles $\tilde{U}_1$ et $\tilde{U}_2$ de $\tilde{V}$ tels que $\tilde{U}_i \cap \mathrm{Spec}_R K(V) = W_i$. $\tilde{U}_1$ et $\tilde{U}_2$ sont disjoints (lemme 9.1). On peut les supposer réguliers $(\mathrm{int}(\mathrm{adh}(\tilde{U}_i)) = \tilde{U}_i)$ car $\mathrm{int}(\mathrm{adh}(\tilde{U}_1)) \cap \tilde{U}_2 = \emptyset$. Soit x un point de $F = V(K) - (U_1 \cup U_2)$, s'il en existe.

LEMME 9.3. Si d est la dimension de V, on a $\dim_R(F,x) = d-1$.

Admettons le lemme. On a une chaîne $\alpha = \alpha_1 \subset \dots \subset \alpha_d = x$ avec $\alpha \in F$.

Soit $\mathfrak{p} = \alpha \cap -\alpha$. $\mathfrak{p}$ est de codimension 1 dans $K[V]$ donc principal, et on a ainsi une place réelle $K(V) \longrightarrow k(\mathfrak{p}),\infty$

---

*Nous conviendrons ici que la donnée de ϕ_α contient celle de l'ordre induit par α sur le corps résiduel.

discrète de rang 1. Il y a deux ordres sur $K(V)$ compatibles avec cette place et induisant l'ordre α sur $k(\mathfrak{p})$. Soient α' et α'' ces deux ordres. On a bien sûr $\phi_{\alpha'} = \phi_{\alpha''}$. On a nécessairement $\alpha' \in \tilde{U}_1$ et $\alpha'' \in \tilde{U}_2$ (ou vice versa) puisque $\alpha \in adh(\tilde{U}_1)$, $\alpha \in adh(\tilde{U}_2)$ et que, en dehors de α, α' et α'' sont les seules générisations de α. Finalement on a montré que si la partition (W_1, W_2) est compatible avec la relation d'équivalence "avoir même place associée", W_1 et W_2 peuvent se relever en deux ouverts-fermés complémentaires de V. Ceci achève la démonstration.

PREUVE DU LEMME. Supposons que l'on ait $\dim_R(F,x) \leqslant d-2$. On peut alors, en prenant un voisinage suffisamment petit de x et en utilisant 6.4 et 8.12, se ramener à la situation suivante: U est une boule ouverte de $A^d(K)$, U_1 et U_2 deux ouverts semi-algébriques disjoints dans U et $F = U - (U_1 \cup U_2)$ est tel que $\sup_{y \in F} \dim_R(F,y) \leqslant d-2$. Soit p une projection de $A^d(K)$ sur $A^{d-1}(K)$ telle qu'au-dessus de tout point de $A^{d-1}(K)$ il n'y ait qu'un nombre fini de points de F. Soit H la clôture de Zariski de $p(F)$. La dimension de H est inférieure ou égale à $d-2$. Soient y et z deux points de $U-F$, t un point de U tel que $p(t) \notin H$. Le segment $]p(y), p(t)[$ coupe $p(F)$ en un nombre fini de points $p(y_1), \ldots, p(y_m)$ avec $y_i \in U-F$. De même $]p(t), p(z)[$ coupe $p(F)$ en $p(z_1), \ldots, p(z_n)$ avec $z_j \in U-F$. Soit L la ligne brisée

$$p(y), p(y_1), \ldots, p(y_m), p(t), p(z_1), \ldots, p(z_n), p(z).$$

L ne coupe pas F et $\tilde{L}$ est connexe. On a montré ainsi que $\widetilde{U-F}$ est connexe, ce qui est absurde.

REFERENCES

1. M. Artin, B. Mazur: On periodic points, Annals of Math. 81, 82-99, 1965.

2. J. Bochnak, G. Efroymson: Real algebraic geometry and the 17th Hilbert problem, Math. Annalen 251, 213-242, 1980.

3. R. Brown: Real places and ordered fields, Rocky Mountain J. of Math 1, 633-636, 1971.

4. G. W. Brumfiel: Partially ordered rings and semi-algebraic geometry, L.M.S. Lecture Note Series 37, Cambridge University Press, 1979.

5. M. Coste, M. F. Coste-Roy: Topologies for real algebraic geometry, dans A. Kock ed: Topos theoretic methods in geometry, Various publications series 30, Aarhus Universitet, 1979.

6. M. Coste, M. F. Coste-Roy: Le spectre étale réel d'un anneau est spatial, C. R. Acad. Sc. Paris Serie A 290, 91-94, 1980.

7. M. F. Coste-Roy: Le spectre réel d'un anneau, Thèse, Université Paris-Nord, 1980.

8. G. Efroymson: Local reality on algebraic varieties, J. of Algebra 29, 137-145, 1976.

9. G. Efroymson: Substitution in Nash functions, Pacific J. of Math. 63, 137-145, 1976.

10. M. Knebusch: Symmetric bilinear forms over algebraic varieties, dans G. Orzech ed: Conference on quadratic forms 1976, Queen's Papers in Pure and Applied Math. 46, 1977.

11. W. Krull: Allgemeine Bewertungstheorie, J. reine angew. Math. 167, 160-196, 1931.

12. T. Y. Lam: Ten Lectures on quadratic forms over fields, dans Conference on Quadratic Forms 1976.

13. S. Lang: The theory of real places, Annals of Math. 57, 378-391, 1953.

14. J. J. Risler: Une caracterisation des ideaux des varietes algebriques reelles, C. R. Acad. Sci. Paris Serie A 271, 1171-1173, 1970.

15. H. Delfs, M. Knebusch: Semialgebraic topology over a real closed field I, Preprint, Regensburg, 1980.

16. C. Delzell: A constructive, continuous solution to Hilbert's 17th problem, and other results in semi-algebraic geometry, Thesis, Stanford, 1980.

DEPARTEMENT DE MATHEMATIQUES
UNIVERSITE PARIS-NORD
AVENUE J. B. CLEMENT
93430 VILLETANEUSE

Contemporary Mathematics
Volume 8, 1982

SEMIALGEBRAIC TOPOLOGY OVER A REAL CLOSED FIELD

Hans Delfs and Manfred Knebusch

We fix a real closed field R. By a variety over R we always mean a scheme of finite type over R. This paper gives a short survey about our theory of semialgebraic spaces over R. Semialgebraic spaces seem to be the adequate generalization of the classical notion of semialgebraic sets over R. Copying the classical definition we can consider also semialgebraic subsets of arbitrary varieties over R. Introducing the category of semialgebraic spaces we get rid of the inconvenience that every semialgebraic set is embedded in a variety. The basic definitions are given in §1. They can be found, as well as a lot of foundational material, in the paper [DK II]. The application to the theory of Witt rings outlined in §3 is contained in [DK I]. The results on triangulation and cohomology of affine semialgebraic spaces are contained in the thesis of the first author ([D]). We omit here nearly all proofs and refer the reader to these papers. But we emphasize that in all these proofs Tarski's principle is never used to transfer statements from the field $\mathbb{R}$ to other real closed base fields.

Contents

1980 Mathematics Subject Classification 14G30, 12D15, 58A07

0271-4132/81/0000-0674/$05.50

§ 1 Basic definitions

For any variety V over R we denote by V(R) the set of R-rational points of V.

Definition 1: Let V = Spec(A) be an affine variety over R. A subset M of V(R) is called a *semialgebraic subset of V*, if M is a finite union of sets

$$\{x \in V(R) \mid f(x) = 0,\ g_j(x) > 0,\ j = 1,\dots,r\}$$

with elements $f, g_j \in A$.

The ordering of R induces a topology on the set V(R) of real points of every affine R-variety V, hence on every semialgebraic subset M of V. We call this topology the *strong topology*.

Definition 2: Let V, W be affine varieties over R and M, N be semialgebraic subsets of V resp. W.

A map $f : M \to N$ is called *semialgebraic with respect to* V *and* W, if f is continuous in strong topology and if the graph G(f) of f is a semialgebraic subset of $V \times_R W$. The semialgebraic maps from M to R with respect to V and $A_R^1 = \operatorname{Spec} R[X]$ are called the *semialgebraic functions on* M with respect to V.

Definition 3: A *restricted topological space* M is a set M together with a family $\mathring{\mathfrak{S}}(M)$ of subsets of M, called the open subsets of M, such that the following conditions are satisfied:

i) $\emptyset \in \mathring{\mathfrak{S}}(M)$, $M \in \mathring{\mathfrak{S}}(M)$

ii) $U_1 \in \mathring{\mathfrak{S}}(M)$, $U_2 \in \mathring{\mathfrak{S}}(M) \Rightarrow U_1 \cup U_2 \in \mathring{\mathfrak{S}}(M)$.

Notice the essential difference to the usual topological spaces: Infinite unions of open subsets are in general not open.

We consider every restricted topological space M as a site in the following sense:
The category of the site has as objects the open subsets $U \in \mathring{\mathfrak{S}}(M)$ of M and as morphisms the inclusion maps between such subsets.

The coverings $(U_i \mid i \in I)$ of an open subset U of M are the <u>finite</u> systems of open subsets of M with $U = \bigcup_{i \in I} U_i$.

<u>Example</u>: Let V be an affine R-variety and M be a semialgebraic subset of V. Let $\mathring{\mathfrak{S}}(M)$ be the family of all subsets of M which are open in M in the strong topology and which are in addition semialgebraic in V. $(M, \mathring{\mathfrak{S}}(M))$ is a restricted topological space. We call this topology of M the <u>semialgebraic topology</u> (with respect to V) and denote this site by M_{sa}.

<u>Definition 4</u>: A <u>ringed space over</u> R is a pair $(M, \mathcal{O}_M)$ consisting of a restricted topological space M and a sheaf $\mathcal{O}_M$ of R-algebras on M. A morphism $(f,\vartheta) : (M,\mathcal{O}_M) \to (N,\mathcal{O}_N)$ between ringed spaces consists of a continuous map $f : M \to N$ (i.e. the preimages of all open subsets of N are open) and a family $(\vartheta_V)_{V \in \mathring{\mathfrak{S}}(N)}$ of R-algebra-homomorphisms which are compatible with restriction.

<u>Example</u>: Let M be a semialgebraic subset of an affine R-variety V equipped with its semialgebraic topology. For every open semialgebraic subset U of M let $\mathcal{O}_M(U)$ be the R-algebra of semialgebraic functions on U with respect to V. Then $(M,\mathcal{O}_M)$ is a ringed space over R. It is called a <u>semialgebraic subspace</u> of V.

<u>Definition 5</u>:

i) An affine semialgebraic space over R is a ringed space $(M,\mathcal{O}_M)$ which is isomorphic to a semialgebraic subspace of an affine R-variety V.

ii) A semialgebraic space over R is a ringed space $(M,\mathcal{O}_M)$ which has a (finite) covering $(M_i \mid i \in I)$ by open subsets M_i, such that $(M_i, \mathcal{O}_M|M_i)$ is for all $i \in I$ an affine semialgebraic space.

iii) A morphism between semialgebraic spaces is a morphism in the category of ringed spaces.

As is shown in [DK II, §7], a morphism $(f,\vartheta) : (M,\mathcal{O}_M) \to (N,\mathcal{O}_N)$ between semialgebraic spaces is completely determined by f : For $V \in \mathring{\mathfrak{S}}(N)$ and $g \in \mathcal{O}_N(V)$ we have $\vartheta_V(g) = g \circ f$. Hence we write simply f instead of (f,ϑ). These morphisms are also called semialgebraic maps. The morphisms $f : M \to N$ between semialgebraic subspaces M,N of affine varieties V,W are just the semialgebraic maps from M to N with respect to V and W ([DK II, §7]).

Let M be a semialgebraic space. We denote by $\mathfrak{S}(M)$ the smallest family of subsets of M which fulfills the following conditions:

i) $\mathring{\mathfrak{S}}(M) \subset \mathfrak{S}(M)$

ii) $A \in \mathfrak{S}(M) \Rightarrow M - A \in \mathfrak{S}(M)$

ii) $A, B \in \mathfrak{S}(M) \Rightarrow A \cup B \in \mathfrak{S}(M)$.

The elements of $\mathfrak{S}(M)$ are called the <u>semialgebraic subsets</u> of M. In the special case $M = V(R)$ this definition coincides with Definition 1.

The <u>strong topology</u> on M is the topology in the usual sense which has $\mathring{\mathfrak{S}}(M)$ as a <u>basis</u> of open sets. Thus the open sets in the strong topology are the unions of arbitrary families in $\mathring{\mathfrak{S}}(M)$. In the special case that M is a semialgebraic subspace of variety V this topology is of course the same as the strong topology in the previous sense.

Since now notions like "open", "closed", "dense", ... always refer to the strong topology. In this terminology the sets $U \in \mathring{\mathfrak{S}}(M)$ are the open semialgebraic subsets of M and their complements $M - U$ are the closed semialgebraic subsets of M.

It is easy to show that for any two semialgebraic maps $f : M \to N$ and $g : L \to N$ the fibre product $M \times_N L$ exists in the category of semialgebraic spaces. If N is the one point space we write simply $M \times L$ for this product.

Every semialgebraic subset $A \in \mathfrak{S}(M)$ of a semialgebraic space M is in a natural way equipped with the structure of a semialgebraic space:

The elements $U \in \mathring{\mathfrak{S}}(A)$ are those subsets of A which are open in A (in strong topology) and which are semialgebraic in M. A semialgebraic function f on $U \in \mathring{\mathfrak{S}}(A)$ is a function $f : U \to R$, which is continuous in strong topology and whose graph is a semialgebraic subset of $M \times R$.

A is then a subobject of M in the category of semialgebraic spaces.

§ 2 Some results in the theory of semialgebraic spaces

Definition 1: A semialgebraic space M over R is separated, if it fulfills the usual Hausdorff condition: Any two different points x and y of M can be separated by open disjoint semialgebraic neighbourhoods.

Definition 2: A semialgebraic space M over R is called complete, if M is separated and if for all semialgebraic spaces N over R the projection

$$M \times N \to N$$

is closed, i.e., every closed semialgebraic subset A of $M \times N$ is mapped onto a closed semialgebraic subset of N.

Since the closed semialgebraic subsets of the projective line $P^1_R(R) = R \cup \{\infty\}$ which do not contain the point ∞ are finite unions of closed bounded intervals of R, we obtain immediately

Proposition 2.1. Let M be a complete semialgebraic space over R. Then every semialgebraic function $f : M \to R$ attains a minimum and a maximum on M.

Complete semialgebraic spaces are the substitute for compact spaces in usual topology.

Theorem 2.2 ([DK II, 9.4]).
Let M be a semialgebraic space over the field $\mathbb{R}$ of real numbers. Then M is complete if and only if M is a compact topological space.

Theorem 2.3 ([DK II, 9.6]).
Let V be a complete R-variety. Then the semialgebraic space $V(R)$ is also complete.

We consider every semialgebraic subset $M \subset R^n$ of $A^n_R = \operatorname{Spec} R[X_1,\dots,X_n]$ as a semialgebraic subspace of A^n_R.

Theorem 2.4 ([DK II, 9.4])
A closed and bounded semialgebraic subset of R^n is a complete semialgebraic space.

If R is not the field of real numbers, the spaces R^n and hence all spaces V(R), where V is an affine R-variety, are totally disconnected. To get reasonable connected components, one must find another notion of connectedness.

Definition 3: Let M be a semialgebraic space. A (semialgebraic) path in M is a semialgebraic map $\alpha : [0,1] \to M$ from the unit interval in R to M. Two points P,Q of M are connectable, if there is a path α in M with $\alpha(0) = P$ and $\alpha(1) = Q$.

Every semialgebraic space M splits under the equivalence relation "connectable" into path components.

Theorem 2.5 ([DK II, 11.2]):
Let M be a semialgebraic space. Then M has a finite number of path components. Each of these components is a semialgebraic subset of M.

It follows from a well known theorem of Tarski that the closure $\bar{N}$ of a semialgebraic subset N of a semialgebraic space M in strong topology is again a semialgebraic subset of M.

Theorem 2.6 (Curve Selection Lemma; [DK II, 12.1]):
Let N be a semialgebraic subset of a semialgebraic space M and let P be a point in the closure $\bar{N}$ of N in M (in the strong topology). Then there exists a path $\alpha : [0,1] \to M$ with $\alpha(0) = P$ and $\alpha(]0,1]) \subset N$.

An immediate consequence of the last two theorems is

Corollary 2.7 The finitely many path components of a semialgebraic space M are closed and hence open semialgebraic subsets of M.

Theorems 2.5 and 2.6 can be easily derived from the triangulation theorem (4.1), but it is not necessary to use such a strong result. The proofs in [DK II] are rather easy and elementary.

Definition 4 ([B], p.249): A semialgebraic space M is called connected if it is not the union of two non empty open semialgebraic subsets.

The images of (path-)connected semialgebraic spaces are obviously again (path-)connected. Since the unit interval in R is connected, we see that any path connected space is connected. Corollary 2.7 implies that also the converse is true.

Corollary 2.8: A semialgebraic space is connected if and only if it is path connected.

From now on we say simply "component" instead of "path component". The number of components is in the following sense a birational invariant.

Theorem 2.9 ([DK II, 13.3]):
Let V and W be birationally equivalent smooth complete varieties over R. Then V(R) and W(R) have the same number of components.

The proof of Theorem 2.9 in [DK II] is a straightforward adaption of classical arguments and illustrates that it is possible by our theory to transfer quite a lot of geometric ideas, familiar in the case $R = \mathbb{R}$, to arbitrary real closed base fields. A quite different proof of Theorem 2.9 has been given by M.F. Coste-Roy ([C]).

§ 3 An application to Witt rings

Let X be a divisorial variety over an arbitrary (not necessarily real closed) field k. (Notice that all regular and all quasiprojective varieties over k are divisorial). We consider the signatures of X, i.e. the ring homomorphisms from the Witt ring W(X) of bilinear spaces over X to the ring of integers $\mathbb{Z}$ (cf. [K] for the general theory and meaning of signatures).

It is shown in [K, $\overline{V}$.1] that any signature factors through some point x of X, i.e., there exists a commutative diagram

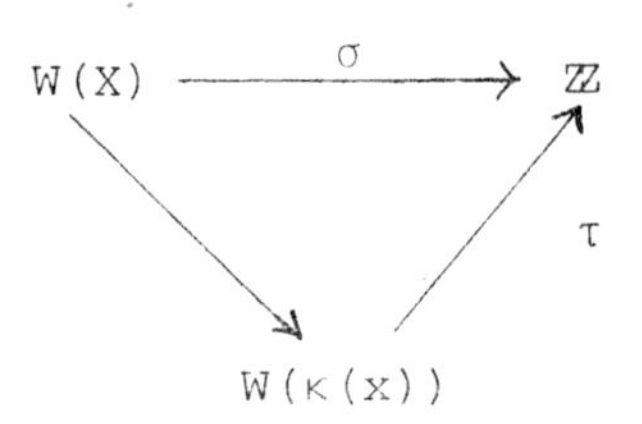

with $W(X) \to W(\kappa(x))$ the natural map from W(X) to the Witt ring of the residue class field $\kappa(x)$. Using our theory we are able to prove

Theorem 3.1 ([DK I ,5.1]):
Every signature σ of X factors through a closed point x of X.

Theorem 3.1 is first proved in the case that k = R is a real closed field. The proof runs essentially along the same lines as the proof in the special case that $R = \mathbb{R}$ [K, Chap $\overline{V}$]. It is based on Theorem 2.5 which states that X(R) has only a finite number of components. Using the theory of real closures of schemes ($[K_1]$), it is possible to extend Theorem 3.1 to arbitrary base fields.

§ 4 Triangulation of affine semialgebraic spaces

To avoid confusion what is meant by a triangulation we first give two definitions.

Definition 1: An open non degenerate n-simplex S over R is the interior (in strong topology) of the convex closure of n + 1 affine independent points $e_0,\dots,e_n$ in some space R^m, called the vertices of S, i.e.

$$S = \left\{ \sum_{i=0}^{n} t_i e_i \mid t_i \in R,\ t_i > 0,\ \sum_{i=0}^{n} t_i = 1 \right\}.$$

Definition 2. A simplicial complex over R is a pair $(X, X = \bigcup_{k=1}^{r} S_k)$ consisting of a semialgebraic subset X of some affine space

R^m and a decomposition $X = \bigcup_{k=1} S_k$ of X into <u>disjoint open nondegenerate</u> simplices S_k, such that the following condition is fulfilled:
The intersection $\overline{S_k} \cap \overline{S_l}$ of the closures of any two simplices S_k, S_l is either empty or is a face (defined as usual) of $\overline{S_k}$ as well as of $\overline{S_l}$.

We can now state the triangulation theorem which says that finitely many semialgebraic subsets of an affine semialgebraic space can be triangulated simultaneously.

<u>Theorem 4.1</u> ([D, 2.2]):
Let $\{M_j\}_{j \in J}$ be a finite family of semialgebraic subsets of an affine semialgebraic space M. Then there exists a semialgebraic isomorphism

$$\phi : X = \bigcup_{k=1}^{r} S_k \xrightarrow{\sim} \bigcup_{j \in J} M_j ,$$

such that each set M_j is a union of certain images $\phi(S_k)$ of simplices S_k of X.

In the proof of Theorem 3 one easily retreats to the case that the sets M_j are bounded semialgebraic subsets of some R^n. Then induction on n is used, the case n = 1 being trivial. The main problem is to find a substitute for the analytic tools used in the classical proofs for $R = \mathbb{R}$ (cf. [H]).

§ 5 Cohomological dimension

The following "inequality of Lojasciewicz" can be proved with help of Theorem 2.4.

<u>Lemma 5.1</u> ([D, 3.2]): Let M be a closed and bounded semialgebraic subset of R^n and f,g be semialgebraic functions on M. Assume that for all $x \in M$ $f(x) = 0$ implies $g(x) = 0$. Then there is a constant $C > 0$, $C \in R$, and a natural number m, such that for all $x \in M$

$$|f(x)| \geq C \, |g(x)|^m .$$

From Lemma 4.1 one can derive in a similar way as it is done in a special case for $R = \mathbb{R}$ in [BE]

Theorem 5.2 ([D, 3.3]):
Let M be a semialgebraic subset of an affine R-variety V = Spec A and let U be an open semialgebraic subset of M. Then U is a finite union of sets

$$\{x \in M \mid f_i(x) > 0, \ i = 1,\dots,r\}$$

with $f_i \in A$, $i = 1,\dots,r$.
Other proofs of this fact have been given by Delzell[De, Chap.II] and M.F. Coste-Roy ([C]).

A sheaf F on a semialgebraic space M assigns to every open semialgebraic subset U of M an abelian group F(U) such that the usual compatibilities with respect to restriction and the sheaf condition are fulfilled. Since we admit only finite coverings in the semialgebraic topology M_{sa} of M (cf. §1), the sheaf condition must hold only for finite coverings.

It is clear from Grothendieck's definition what the cohomology groups $H^q(M,F)$ of M with coefficients in an abelian sheaf F on M are ([G],[A]):
Choose a resolution

$$0 \to F \to J^0 \to J^1 \to J^2 \to \ \ldots\ldots$$

of F by injective sheaves J^k, apply the global section functor and take the cohomology groups of the arising complex:

$$H^q(M,F) = \mathrm{Ker}(J^q(M) \to J^{q+1}(M)) \ / \ \mathrm{Im}(J^{q-1}(M) \to J^q(M)).$$

It follows from Theorem 5.2 that an affine semialgebraic space has similar separating properties as a usual paracompact topological space. Using this fact we can prove

Theorem 5.3 ([D, 5.2]):
Let M be an affine semialgebraic space and F be an abelian sheaf on M. Then the canonical homomorphism

$$\check{H}^p(M,F) \to H^p(M,F)$$

from Čech- to Grothendieck cohomology is for all $p \geq 0$ an isomorphism.

Every semialgebraic space M has a certain well defined dimension ([DK II, §8]). If M is a semialgebraic subset of an R-variety V, dim M is simply the dimension of the Zariski-closure of M in V. Since every affine semialgebraic space can be triangulated (Theorem 4.1) and Grothendieck-cohomology coincides with Čech-cohomology, similar arguments as in [Go, II.5.12] show that the cohomological dimension does not exceed the topological dimension.

Theorem 5.4 ([D, 5.9]):
Let M be an affine semialgebraic space of dimension n. Then for all abelian sheaves F on M

$$H^q(M,F) = 0$$

for all $q > n$.

§ 6 The homotopy axiom in semialgebraic cohomology

Let M be a semialgebraic space and G be an abelian group. G yields the constant sheaf G_M on M: For an open semialgebraic subset U of M

$$G_M(U) = \prod_{\pi_0(U)} G ,$$

where $\pi_0(U)$ is the finite set of components of U.

We denote by [0,1] the unit interval in R.

In the classical theory homotopic maps induce the same homomorphisms in cohomology. This is also true in semialgebraic topology over an arbitrary real closed field, at least in the affine case.

Theorem 6.1 ([D, 7.1]): Let $f_0, f_1 : M \rightrightarrows N$ be homotopic semialgebraic maps between affine semialgebraic spaces, i.e. there exists a semialgebraic map $H : M \times [0,1] \to N$, such that $H(-,0) = f_0$,

$H(-,1) = f_1$. Then f_0 and f_1 induce the same homomorphisms in cohomology:

$$f_0^* = f_1^* : H^q(N,G_N) \to H^q(M,G_M).$$

We proof Theorem 6.1 by use of Alexander-Spanier-cohomology which is defined in a similar manner as in the classical case (cf. [S]). The sheaves of Alexander-Spanier-cochains yield a resolution of G_M by flask sheaves (defined as in the topological case). Flask resolutions can be used to determine cohomology. Thus Alexander-Spanier-cohomology is the same as Grothendieck-cohomology. The existence of infinitely small elements in a non archimedian field rises many difficulties in the proof of Theorem 6.1, compared with the classical case. For example, we cannot make intervals and triangles "arbitrarily small" by barycentric or even "linear" subdivision. Our proof is based on a careful investigation of the roots of a system of polynomials.

We use Theorem 6.1 to identify the semialgebraic cohomology groups $H^q(M,G_M)$ with certain simplicial cohomology groups. For the rest of this section let M be an affine semialgebraic space and G be a fixed abelian group. Consider a triangulation

$$\phi : X = \bigcup_{i=1}^{r} S_i \xrightarrow{\sim} M$$

of M (§4). For technical reasons we have to assume that ϕ is a barycentric subdivision of another triangulation of M.

We then associate to ϕ the following abstract simplicial complex K: The set V(K) of vertices of K consists of all vertices of X lying in X. (Notice that X is not necessarily closed).
A subset $e_0,\dots,e_q$ of V(K) is a simplex of K, if $e_0,\dots,e_q$ are the vertices of a simplex S_i, $i \in \{1,\dots,r\}$.

In the usual way we form the simplicial cohomology groups $H^q(K,G)$.

<u>Proposition 6.2</u> ([D,8.4]):

$$H^q(M,G_M) = H^q(K,G) \quad \text{for all } q \geq 0.$$

In particular, the simplicial cohomology groups $H^q(K,G)$ do not depend on the chosen triangulation of M.

The proof of Proposition 6.2 uses the description of $H^q(M,G_M)$ as Čech-cohomology and Theorem 6.1 which implies that the covering $\{St(e)\}_{e\in V(k)}$ of M, consisting of the star neighbourhoods of the vertices of M with respect to ϕ, is a Leray-covering for the sheaf G_M.

It is now possible to define also homology groups.

<u>Definition 1</u>: $H_q(M,G) := H_q(K,G)$ is called the q-th <u>homology group</u> of M with coefficients in G.

This definition does not depend on the chosen triangulation of M as follows from the corresponding fact for cohomology (Proposition 6.2). The homology groups $H_q(M,G)$ are functorial in M since, according to Theorem 4.1, every semialgebraic map between affine semialgebraic spaces can be "approximated" by a simplicial map (cf. [D, §8]).

Proposition 6.2 implies also that for $R = \mathbb{R}$ the semialgebraic cohomology groups coincide with the usual (singular) cohomology groups determined with respect to strong topology. For homology this is true by definition.

Another immediate consequence of the simplicial interpretation of homology and cohomology is that the groups are invariant under change of the base field. We will illustrate this in a special case.

Let L be a real closed field containing R. Assume, M is a semialgebraic subset of R^n. We choose a description of M by finitely many polynomial inequalities and equalities. Let M_L denote the semialgebraic subset of L^n defined by the same inequalities and equalities. By use of Tarski's principle - here clearly legitimate and unavoidable - we see that the set M_L is independent of the choice of the description of M.
It also follows from Tarski's principle that the triangulation

$$\phi : X = \bigcup_{i=1}^{r} S_i \xrightarrow{\sim} M$$

yields a triangulation

$$\phi_L : X_L = \bigcup_{i=1}^{r} S_{iL} \xrightarrow{\sim} M_L$$

of M_L. The associated abstract complex of ϕ_L is also K and we get with Prop. 6.2 and Def. 1:

$$H^q(M,G) = H^q(M_L,G),$$

$$H_q(M,G) = H_q(M_L,G).$$

Example: The n-sphere $S_R^n = \{(x_0,\ldots,x_n) \in R^{n+1} \mid \sum_{i=0}^{n} x_i^2 = 1\}$ has the same homology and cohomology as the n-sphere $S_{\mathbb{R}}^n$ over $\mathbb{R}$, since S_R^n and $S_{\mathbb{R}}^n$ both can be obtained from S_{R_0} by base extension, with R_0 the real closure of $\mathbb{Q}$. Thus

$$H_0(S_R^n,G) \cong H_n(S_R^n,G) \cong H^0(S_R^n,G) \cong H^n(S_R^n,G) \cong G,$$

$$H_q(S_R^n,G) = 0,\ H^q(S_R^n,G) = 0 \quad \text{for } q \neq 0,n.$$

§ 7 The duality theorems

As an example of our theory we want to explain that in a certain sense the classical duality theorems for manifolds remain true over an arbitrary real closed field. We consider a semialgebraic space M over R.

Definition 1. M is an *n-dimensional semialgebraic manifold* if every point $x \in M$ has an open semialgebraic neighbourhood which is isomorphic to an open semialgebraic subset of R^n.

Example 1: It follows from the implicit function theorem for polynomials (cf. [DK II, 6.9]) that every open semialgebraic subset U of the set V(R) of real points of an n-dimensional *smooth* R-variety V is an n-dimensional semialgebraic manifold ([DK II, §13]).

From now on we assume that M is *affine* and *complete*.
We choose a triangulation

$$\phi : X = \bigcup_{i=1}^{r} S_i \xrightarrow{\sim} M$$

and associate to ϕ an abstract simplicial complex K as in §6. Since M is complete, X is closed and bounded in its embedding space R^m.

We follow in our notation the book of Maunder on algebraic topology ([M]). For any point $x \in M$ we denote by $N_\phi(x)$ the union $\bigcup_{\phi^{-1}(x)\in\overline{S}_i} \phi(\overline{S}_i)$ of all closed "simplices" of M with respect to ϕ containing x. The union $Lk_\phi(x) := N_\phi(x) \setminus (\bigcup_{\phi^{-1}(x)\in\overline{S}_i} \phi(S_i))$ of those "simplices" of $N_\phi(x)$ which do not contain x is called the <u>link</u> of x.

<u>Definition 2</u>: M is called a homology-n-manifold if for all $x \in M$

$$H_q(Lk_\phi(x), \mathbb{Z}) = \begin{cases} \mathbb{Z} & q = 0,n \\ 0 & q \neq 0,n \end{cases}$$

Definition 2 does not depend on the chosen triangulation ϕ (cf. [M]).

<u>Example 2</u>: Every affine complete n-dimensional semialgebraic manifold is a homology-n-manifold ([D, §10]).

<u>Subexample 2 a</u>. The Zariski-open subset U of the m-dimensional projective space $\mathbb{P}^m_R$ over R, obtained by removing the hypersurface $X_0^2 + X_1^2 + \ldots + X_m^2 = 0$ from $\mathbb{P}^m_R$, is affine and has the same real points as $\mathbb{P}^m_R$. Hence the space V(R) of real points of a projective R-variety V is affine. It is also complete by Theorem 2.3. Thus if V is projective, smooth and has dimension n the space V(R) is a homology-n-manifold.

We return to our affine and complete semialgebraic space M and assume in addition that M is a homology-n-manifold.

<u>Definition 3</u>: If M is connected, we call M <u>orientable</u> if $H_n(M, \mathbb{Z}) = \mathbb{Z}$. In general M is called orientable, if each component of M is orientable.

<u>Example</u>: The n-sphere S^n_R over R is an orientable homology-n-manifold (cf. §6).

<u>Theorem 7.1</u> (Poincaré-duality):
Assume M is orientable. Then there are canonical isomorphisms

$$H^q(M,\mathbb{Z}) \xrightarrow{\sim} H_{n-q}(M,\mathbb{Z}).$$

If M is not orientable, there are still isomorphisms

$$H^q(M,\mathbb{Z}/2) \xrightarrow{\sim} H_{n-q}(M,\mathbb{Z}/2) .$$

The proof is very easy: Consider a realization $|K|_{\mathbb{R}}$ of the abstract complex K over $\mathbb{R}$, i.e., a closed simplicial complex over $\mathbb{R}$ with associated abstract complex K. Then $|K|_{\mathbb{R}}$ is an (orientable) homology-n-manifold over $\mathbb{R}$ and the classical Poincaré-Duality applies to $|K|_{\mathbb{R}}$. But semialgebraic (co-)homology of M and singular (co-) homology of $|K|_{\mathbb{R}}$ coincide both with the (co-)homology of the abstract complex K.

In a similar way other duality theorems can also be transferred to an arbitrary real closed field.

Using relative homology- and cohomology groups, one derives from Poincaré-duality

Theorem 6. (Alexander-duality)
Let A be a semialgebraic subset of the n-sphere S^n_R. Then there are isomorphisms

$$\tilde{H}^q(A) \xrightarrow{\sim} \tilde{H}_{n-q-1}(S^n_R - A),$$

where $\tilde{H}^q$(resp. $\tilde{H}_q$) denotes the reduced cohomology (resp.homology) group.

As an application we get the generalized Jordan curve theorem over any real closed field.

Corollary 7. Let M be a semialgebraic subset of S^{n+1}_R $(n \geq 1)$. Assume M is a homology-n-manifold with k components. Then $S^{n+1}_R - M$ has k+1 connected components. In particular: If M is semialgebraically isomorphic to S^n_R then $S^{n+1}_R - M$ has 2 components. In this case M is the common boundary of these two components.

Proof: From Alexander-duality we obtain

$$\tilde{H}_o(S^{n+1}_R - M,\mathbb{Z}) \cong \tilde{H}^n(M,\mathbb{Z}) \cong H^n(M,\mathbb{Z})$$

and by Poincaré-duality

$$H^n(M,\mathbb{Z}/2) \cong H_o(M,\mathbb{Z}/2) \cong \prod_{\pi_o(M)} \mathbb{Z}/2.$$

The statement concerning the boundary is proved by similar arguments.

References

[A] M. Artin, "Grothendieck topologies", Harvard University, 1962.

[B] G.W. Brumfiel, "Partially ordered rings and semialgebraic geometry", Cambridge University Press (1979).

[BE] J. Bochnak, G. Efroymson, Real algebraic geometry and the 17th Hilbert Problem, Math. Ann. 251, 213-241 (1980).

[C] M.F. Coste-Roy, Spectre réel d'un anneau et topos étale réel, Thèse, Université Paris Nord (1980).

[D] H. Delfs, Kohomologie affiner semialgebraischer Räume, Thesis, Regensburg (1980).

[DK I] H. Delfs, M. Knebusch, Semialgebraic topology over a real closed field I: Paths and components in the set of rational points of an algebraic variety, to appear in Math. Z.

[DK II] H. Delfs, M. Knebusch, Semialgebraic topology over a real closed field II: Basic theory of semialgebraic spaces, preprint, Regensburg (1980).

[De] C.N. Delzell, A constructive, continuous solution to Hilbert's 17th Problem, and other results in semialgebraic geometry, Thesis, Stanford University, 1980.

[G_o] R. Godement, "Théorie des faisceaux", Hermann, Paris (1958).

[G] A. Grothendieck, Sur quelques points d'algèbre homologique, Tohoku Math. Journ. A. IX (1957), 119-221.

[H] H. Hironaka, Triangulation of algebraic sets, Proc. Amer. Math. Soc., Symp. in Pure Math. 29 (1975), 165-185.

[K] M. Knebusch, "Symmetric bilinear forms over algebraic varieties". In: Conference on quadratic forms (Kingston 1976), 103-283. Queen's papers in Pure Appl. Math. 46 (1977).

[K_1] M. Knebusch, Real closures of algebraic varieties, ibid., 548-568.

[M] C.R.F. Maunder, "Algebraic topology", Van Nostrand Reinhold Company, London (1970).

[S] E.H. Spanier, "Algebraic topology", McGraw Hill Book Company, New York (1966).

FACHBEREICH MATHEMATIK DER UNIVERSITÄT
REGENSBURG
WEST GERMANY

Contemporary Mathematics
Volume 8, 1982

A FINITENESS THEOREM FOR OPEN SEMI-ALGEBRAIC SETS, WITH APPLICATIONS TO HILBERT'S 17TH PROBLEM

Charles N. Delzell

ABSTRACT. We prove a conjecture of G. Brumfiel: an open semi-algebraic set may be written as a finite union of finite intersections of sets of the form $f^{-1}((0,\infty))$ ($f \in K[X_1,\ldots,X_n]$, K an ordered field). Refinements, and applications to various continuous solutions to Hilbert's 17th problem, are given.

INTRODUCTION

Let K be an ordered field and R a real closed order-extension field, endowed with its usual order topology, generated by the open intervals (a,b). Write K^+ and R^+ for the non-negative elements of K and R. Let $X = (X_1,\ldots,X_n)$ be indeterminates, and $x = (x_1,\ldots,x_n) \in R^n$. Throughout, all indices range over finite sets whose sizes are rarely specified, but which will be computable if K is computable.[1] For $\{f_i\} \subseteq K[X]$ let[2]

$$U\{f_i\} = \{x \in R^n \mid \bigwedge_i f_i(X) > 0\},$$

$$W\{f_i\} = \{x \in R^n \mid \bigwedge_i f_i(X) \geq 0\}, \quad \text{and}$$

$$Z\{f_i\} = \{x \in R^n \mid \bigwedge_i f_i(X) = 0\}.$$

1980 Mathematics Subject Classification. 14G30, 12D15, 12J15, 03D15, 03F55.

[1] A computable field is one whose field operations are computable, and whose order relation is decidable. Examples of computable fields include $\mathbb{Q}$, $\overline{\mathbb{Q}}$, but not $\mathbb{R}$.

[2] $\bigwedge_i$ [resp. $\bigvee_i$] means iterated conjunction [resp. disjunction], indexed by i.

A set is called a basic open semialgebraic (s.a.) set (more precisely, a basic open K-s.a. set), or simply a U, if it is of the form $U\{f_i\}$; and similarly with U and "open" replaced by W and "closed." A set $S \subseteq R^n$ is called (K-)s.a. if it is a finite union of finite intersections of basic open and closed (K-)s.a. sets.

The main result in §1 is a proof of a "finiteness theorem for open s.a. sets," conjectured by G. Brumfiel ("Unproved Proposition" 8.1.2 [1979][3]):

THEOREM 1.1 (the finiteness theorem): If $S \subseteq R^n$ is s.a., then

(a) S is open if and only if $S = \bigcup_i U\{g_{ij}\}$, some $\{g_{ij}\} \subseteq K[X]$; equivalently,

(b) S is closed if and only if $S = \bigcup_i W\{g_{ij}\}$, some $\{g_{ij}\} \subseteq K[X]$.

If K is computable, the g_{ij} are computable from the presentation of S as a s.a. set.

The equivalence of (a) and (b) follows by taking complements and distributing. Theorem 1.1 would be trivially true if the index set of i were allowed to be infinite; hence the name "finiteness theorem." An extensive theory of s.a. sets is developed in Chapter 8 of [Brumfiel 1979] without this theorem. "It would be nice to have a simple proof of 8.1.2 right at the beginning. On the other hand, all the results we will prove in order to circumvent 8.1.2 are results we would want anyway."

The statement of the theorem is deceptively simple for s.a. sets in R^1, and deceptively difficult for sets in R^n, $n \geq 2$. Our proof proceeds by (1) giving a more delicate analysis of the case $n = 1$ (with parameters), and (2) using the "Good Direction Lemma," for a straightforward algebraic proof by induction.

---

[3]Brackets refer to entries in the bibliography, e.g., "[Brumfiel 1979];" if the author is clear from the context, we bracket only the year, e.g., "[1979]." A year in brackets is the year of publication, not the year of discovery.

In §1 we also explain how 1.1 is an improvement of the Tarski-Seidenberg theorem for quantifier-elimination.

In §2 we apply 1.1 to a particularly important closed s.a. set, namely P_{nd}, the convex cone of coefficients of positive semidefinite forms in $(X_0,\ldots,X_n)$ of degree d. This leads to a quick proof and refinement (2.1) of Daykin's [1960] "piecewise-rational" solution to Hilbert's 17th problem; Daykin's original construction, which was long and difficult, was obtained by working out Kreisel's [1960] sketch of a constructivization (using proof theory) of Artin's original solution to the 17th problem.

Then we make a finer analysis of P_{nd} than that given by 1.1: we show (2.2) that P_{nd} is a single W if and only if $d \le 2$. Using 2.2, we easily settle a question raised in the early sixties by G. Kreisel: "Is there a polynomially varying solution to Hilbert's 17th problem?" (answer (2.3): yes if and only if $d \le 2$). We then discuss how a still finer analysis of P_{nd} (Conjecture 2.5) would, if successful, answer affirmatively a broader, more important question of Kreisel: "Is there a continuously varying solution to Hilbert's 17th problem?" (Conjecture 2.4).

1. THE FINITENESS THEOREM

A function from one (K-)s.a. set to another is called (K-)s.a. if its graph is a (K-)s.a. set (in the product space). All our sets and functions will be understood to be s.a. and, if K is computable, defined by certain real roots of polynomials with computable coefficients; this will always follow from (if nothing else) the Tarski-Seidenberg theorem, which we can formulate in two ways, (1) logically and (2) geometrically.

(1) Elementary formulas in the first order language of ordered fields are those which are expressible using the usual symbols $0, 1, +, \cdot$, $=, <$, logical connectives $\wedge, \vee, \neg, \longrightarrow, \exists, \forall$, and variables $x_1, x_2, \ldots$, where quantification is over R (as opposed to allowing quantification, say, over the power set of R, or over certain subsets such as $\mathbb{N}$). The Tarski-Seidenberg theorem gives an algorithm which, when applied to an

elementary formula, eliminates one quantifier at a time, producing finally a logically equivalent quantifier-free formula. A consequence is that if we prove an elementary statement for one real closed extension R of K (perhaps by using "transcendental" properties of R, if R is $\mathbb{R}$), then it is true for any real closed extension.

(2) In geometric terms, an s.a. set $S \subseteq R^n$ is just one which is definable by a quantifier-free formula ϕ with n free variables. The Tarski-Seidenberg theorem demonstrates that the projection of S to R^{n-1} is also s.a., by eliminating the quantifer "$\exists x_n$" from $\exists x_n \phi(x_1,\ldots,x_n)$. For a proof of the Tarski-Seidenberg theorem, see, for example, the appendix of [Brumfiel 1979], which explains in geometric terms a proof due to Paul Cohen [1969].

The finiteness theorem can be viewed as an improvement of the Tarski-Seidenberg theorem: Suppose a closed s.a. set F is defined by an elementary formula; the Tarski-Seidenberg algorithm eliminates its quantifiers (showing that F is s.a.) but leaves a mixture of both relations, $<$ and $\leq$, <u>obscuring</u> the fact that F is closed; the finiteness theorem (b) eliminates the quantifiers but leaves only the $\leq$ relation (and no negations), <u>revealing</u> the fact that F is closed. This qualitative improvement of the Tarski-Seidenberg theorem is more satisfying than the (basically unsuccessful attempts at) quantitative improvements given in recent years (e.g., by Collins [1974] and Monk [1975]) expressed in terms of the (still large) amount of time and space needed to carry out the elimination. Indeed, Fischer and Rabin [1974] have shown that <u>every</u> decision method, deterministic or non-deterministic, for the elementary theory of real closed fields has a maximum computing time which dominates 2^{cN}, where N = length of the input formula and c is some positive constant, and is therefore unfeasible for all but the simplest problems. On the other hand, an efficient <u>description</u> of the elimination procedure is possible, and was given by Cohen [1969].

There is a trade-off in complexity between polynomials and s.a. functions: s.a. functions are complicated while polynomials are simple. But while representing S as a union of U's of polynomials as in the theorem is complicated, representing S as a "union" of U's of s.a. functions is simple: S is in fact a single U of a single s.a. function: $U\{\mathrm{dist}(x,R^n - S)\}$, where $U\{f\}$, for f s.a., has the obvious definition.

COROLLARY 1.2: A s.a. set S is relatively open in an s.a. set $T \subseteq R^n$ if and only if $S = T \cap \bigcup_i U\{g_{ij}\}$, and similarly with "closed" and "W" in place of "open" and "U."

PROOF OF 1.2 FROM 1.1: We may assume $T \neq S$. $S = T \cap U\{\mathrm{dist}(x,T - S)\}$, and $U\{\mathrm{dist}(x,T - S)\}$ is open (and s.a.; see also 8.13.12 of [Brumfiel 1979]) in R^n,[4] hence it is a union of U's by 1.1. Q.E.D.

The projective analogue of the theorem may be formulated in terms of <u>cones</u>, i.e., sets S in R^n such that $x \in S \longrightarrow cx \in S, \forall c > 0$.

COROLLARY 1.3: If the S in the theorem is also a cone, then we may take the $\{g_{ij}\}$ to be homogeneous.

PROOF OF 1.3 FROM 1.1: We may assume that S does not contain the origin, for otherwise by openness, $S = R^n = U\{1\}$. Thus

$$S = R^+ \cdot \bigcup_{\substack{1 \le k \le n \\ \ell = 0,1}} (S \cap Z\{X_k - (-1)^\ell\}) .$$

Identifying $Z\{X_k - (-1)^\ell\}$ with $Z\{X_k\}$, apply 1.1 to each $S \cap Z\{X_k - (-1)^\ell\}$ to write it as $\bigcup_i U\{g_{k\ell ij}\}$, with $\{g_{k\ell ij}\} \subseteq K[X_0,\ldots,X_{k-1},X_{k+1},\ldots,X_n]$. Homogenize the $g_{k\ell ij}$ by multiplying their monomial terms by suitable powers of X_k. Then $R^+ \cdot (S \cap Z\{X_k - (-1)^\ell\}) = U\{(-1)^\ell X_k\} \cap \bigcup_i U\{g_{k\ell ij}\}$, so that S is the union over k and ℓ of such unions of U's of homogeneous polynomials. Q.E.D.

---

[4]Proof: For any non-empty s.a. set S, $\mathrm{dist}(x,S)$ is a (uniformly) continuous function of x, since it is the infimum of the equicontinuous family of functions $\{\mathrm{dist}(x,y) \mid y \in S\}$.

PROOF OF 1.1: For $R^n \supseteq S$ s.a., define $\dim S = \max\{m \in N \mid S$ contains an (s.a.) homeomorphic image of a non-empty open s.a. subset of $R^n\}$ if $S \neq \emptyset$; define $\dim \emptyset = -1$. (Cf. §§8.9-10 of [Brumfiel 1979] for invariant definitions of dimension.) Let $X' = (X_1,\ldots,X_{n-1})$ and $x' = (x_1,\ldots,x_{n-1}) \in R^{n-1}$. We shall prove 1.1 with the help of a stratification lemma (1.7) whose proof requires

LEMMA 1.4 (the good direction lemma): If T is s.a. and nowhere dense in R^n (equivalently, $\dim T < n$), then there exists $v \in S^{n-1}$ (the unit sphere in R^n) such that, writing $\Pi_v: T \longrightarrow R^{n-1}$ for projection in the v direction into any subspace complementary to $R \cdot v$, we have $\forall x' \in R^{n-1}$, $\Pi^{-1}(x')$ is a discrete set. In fact, the other v (the "bad" v) forms a set of $\dim < n-1$ in S^{n-1}.

REMARK: It makes no difference which complementary subspace we use.

COROLLARY 1.5: $\dim \Pi_v T = \dim T$ for good v.

PROOF OF 1.4: For a "transcendental" proof, we could observe that the lemma is an elementary statement which is true for the case $R = \mathbb{R}$,[5] and thus is true for all real closed R, by Tarski-Seidenberg.

However, as Brumfiel discusses in the Introduction and 8.1 of [1979], there are philosophical and practical reasons for seeking proofs which make only "elementary" as opposed to transcendental applications of the Tarski-Seidenberg theorem. Accordingly, we now give an elementary algebraic proof.

Let $v_n = (0,\ldots,0,1) \in S^{n-1}$, and suppose the set of bad directions contains a non-empty (relatively) open set $U \subseteq S^{n-1}$. We shall show that T must then contain a non-empty open set, contrary to hypothesis. We may assume $U \cap U\{X_n\} \neq \emptyset$. If we replace U by $U \cap U\{X_n\}$, then (1) we may identify U with $\Pi_{v_n} U$ (used at the end of the proof), and

---

[5](See, e.g., [Hironaka 1975a] for a proof; in fact, when $R = \mathbb{R}$, the lemma is true even for T semi-analytic.)

(2) simultaneously $\forall v \in U$, set $S_v = \{x' \in R^{n-1} \mid \Pi_v^{-1}(x')$ is not discrete$\}$. Then the hypothesis of the following lemma is satisfied:

LEMMA 1.6 ("s.a. choice function"): Let $U \subseteq R^n$ be s.a. and $\forall v \in U$ let $\emptyset = S_v \subseteq R^m$ (some m) be s.a., described by a bounded number of polynomials of bounded degrees, in some fixed sequence. Furthermore, assume that the coefficients (in some order) of these polynomials are s.a. functions of v. Then we may construct a s.a. "choice function" $c: U \longrightarrow R^m$ such that $\forall v \in U$, $c(v) \in S_v$.

PROOF of 1.6: Induction on m. For $m = 1$, write $S_v = \cup_i I_i$ where $\{I_i\}$ = the connected components (which depend on v) of S_v, i.e., disjoint intervals in R^1, possibly infinite, open, or closed, some just points. Let I be the left-most interval. Then we may define c by

$$c(v) = \left\{ \begin{array}{rl} 0 & \text{if } I = R^1 \\ a - 1 & \text{if } \bar{I} = (-\infty, a] \\ a + 1 & \text{if } \bar{I} = [a, \infty) \\ \text{midpoint of } I & \text{if } I \text{ bounded or a point} \end{array} \right\}$$

(c is s.a. by Tarski-Seidenberg).

For $m > 1$, let $\Pi_v': S_v \longrightarrow R^1$ be projection onto the first coordinate; By the inductive hypothesis, we define $c_1: U \longrightarrow R^1$ such that $c(v) \in \operatorname{im} \Pi_v'$, and $c_2: U \longrightarrow R^{m-1}$ such that the map $c: U \longrightarrow R^m$ given by $c(v) = (c_1(v), c_2(v))$ has the required properties. This proves 1.6. Q.E.D..

Returning to the proof of 1.4, apply 1.6. For all $v \in U$ write $\Pi_v^{-1}(c(v)) = \{c(v) + tv \mid t \in I(v)\}$, thereby defining $I(v) \subseteq R$. Let $\{I_i(v)\}$ be the connected components of $I(v)$, i.e., points or intervals. Let $I_1(v)$ be the left-most of those intervals which are not points ($I_1(v)$ exists by hypothesis). Let $e_1(v) < e_2(v)$ be the endpoints (possibly $\pm\infty$) of $I_1(v)$.

Shrinking U if necessary, we may assume, since c is s.a., that c is a certain real root of a fixed Y-irreducible polynomial $f \in K[X'][Y]$, i.e., $f(x',c(x')) = 0\ \forall x' \in U$ (identifying U with $\Pi_{v_n} U$). Shrinking U again, e_1 and e_2 are either constantly $\pm\infty$, or are also roots of such polynomials. By the implicit function theorem (proved algebraically in 8.7.2 of [Brumfiel 1979]), c is C^1 off the discriminant locus $\Pi_{v_n} Z\{f,\partial f/\partial Y\} = Z\{\text{resultant of } f, \partial f/\partial Y\} \subseteq R^{n-1}$ of f. Similarly e_1 and e_2 are C^1 on a dense open subset. Therefore shrinking U one more time, we construct an open interval $I \neq \emptyset$ such that $\forall v \in U$, $I \subseteq I_1(v)$. We therefore have a C^1 map $p: U \times I \longrightarrow T$ defined by $p(x',t) = c(x') + t\cdot(x',1) = (c_1(x') + tx_1,\ldots,c_{n-1}(x') + tx_{n-1},t)$. The derivative of p is

$$dp = \begin{bmatrix} \frac{\partial c_1}{\partial x_1} + t & \frac{\partial c_1}{\partial x_2} & \cdots & \frac{\partial c_1}{\partial x_{n-1}} & x_1 \\ \frac{\partial c_2}{\partial x_1} & \frac{\partial c_2}{\partial x_2} + t & \cdots & \frac{\partial c_2}{\partial x_{n-1}} & x_2 \\ \vdots & \vdots & & \vdots & \vdots \\ \frac{\partial c_{n-1}}{\partial x_1} & \frac{\partial c_{n-1}}{\partial x_2} & \cdots & \frac{\partial c_{n-1}}{\partial x_{n-1}} + t & x_{n-1} \\ 0 & 0 & \cdots & 0 & 1 \end{bmatrix}.$$

For any fixed $x' \in U$, $|dp_{(x',t)}| = |dc_{x'} + tI_{n-1}|$ (here I_{n-1} = the $(n-1)\times(n-1)$ identity matrix), and for all but a discrete set of t, this determinant is $\neq 0$. Therefore, except at those t, p is locally onto (i.e., im p, hence T, contains a non-empty open set), by the inverse function theorem (proved algebraically in §8.13 III of [Brumfiel 1979]). This completes the algebraic proof of 1.4. Q.E.D.

We now continue the proof of 1.1 by introducing a "topographic stratification" (1.7) of s.a. sets; to state it, we need the following notation. For $0 \leq m \leq n$, write $X' = (X_1,\ldots,X_m)$, $x' = (x_1,\ldots,x_m) \in R^m$ and let $\Pi_m: T \longrightarrow R^m$ be projection onto the first m coordinates, i.e., $x \longmapsto x'$. $U = \bigsqcup_i T_i$ will mean that $U = \cup_i T_i$ and that the T_i are disjoint s.a. sets.

LEMMA 1.7 (topographic stratification): Let $0 \le \dim T \le m \le n$ and let $\Pi_m: T \longrightarrow R^m$ be projection onto the first m coordinates. Then we may choose the last $n - m$ coordinates so that

(a) $\dim \Pi_m T = \dim T$,

(b) $\Pi_m T = \bigsqcup_i T_i$, some T_i such that

(c) $\left[\begin{array}{l} \bigwedge_i \text{ there exists } J_i \subseteq \mathbb{N}^{n-m} \text{ such that for } j \in J_i \text{ and } m < k \le n, \\ \text{there is a s.a. function } c_{ijk}: T_i \longrightarrow R \text{ with } \forall x' = (x_1,\dots,x_m) \in T_i, \\ \qquad \Pi_m^{-1}(x') = \{(x', c_{i,j,m+1}(x'),\dots,c_{ijn}(x')) \mid j \in J_i\} \end{array}\right].$

REMARK: It will be evident from the proof of 1.7 that we may further arrange that if $j = (j_{m+1},\dots,j_n)$ and $j' = (j_{m+1},\dots,j_\ell,j'_{\ell+1},\dots,j'_n) \in J_i$, and $j_{\ell+1} < j'_{\ell+1}$, then throughout T_i, $c_{i,j,\ell+1} < c_{i,j',\ell+1}$ and for $k \le \ell$, $c_{ijk} = c_{ij'k}$. Thus the graphs of the s.a. functions $(c_{i,j,m+1},\dots,c_{ijn})$: $T_i \longrightarrow R^{n-m}$ (for all i, and $\forall j \in J_i$) form a partition of T. Since all s.a. functions are piecewise real algebraic analytic, we could therefore also arrange for each statum to be a real algebraic analytic manifold of some dimension $\le \dim T$. The term "topographic" was proposed by Andreotti (cf. [Lojasiewicz 1964], footnote 11) to describe a similar situation. Henkin [1960] stated a slightly weaker form of this result; he gave no proof, but indicated how a model-theoretic proof could be given. This stratification is also similar to Collins' [1974] cylindrical algebraic decomposition.

PROOF OF 1.7: We use "reverse induction" on m. The statement is vacuous for $m = n$, so assume it has been established for some $m > \dim T$; we establish it for $m - 1$. Using 1.4 with m in place of n, pick a good direction $v \in R^m$ (with non-zero m^{th} coordinate) for $\Pi_m T$ (possible, since $m > \dim T = \dim \Pi_m T$), and adjust the X_m-axis so as to be parallel to v. Let $\Pi'_i: T \longrightarrow R^{m-1}$ be projection onto the first $m - 1$ coordinates (where the T_i are given by the inductive hypothesis). Using Tarski-Seidenberg, write $\Pi_i^! T_i = \sqcup_\ell T_{i\ell}$ such that there exist $p_{i\ell} \in \mathbb{N}$ and s.a. functions $c'_{i\ell k}: T_{i\ell} \longrightarrow R$ $(1 \le k \le p_{i\ell})$ such that $\forall x'' = (x_1,\ldots,x_{m-1}) \in T_{i\ell}$, $c'_{i\ell 1}(x'') < \cdots < c'_{i\ell p_{i\ell}}(x'')$, and $\Pi_i^{!-1}(x'') = \{(x'',c'_{i\ell k}(x'')) \mid 1 \le k \le p_{i\ell}\}$. Write $\Pi_{m-1}T = \sqcup_a T'_a$ in such a way that each $T_{i\ell}$ is a union of some of the T'_a; set $I_a = \{(i,\ell) \in \mathbb{N}^2 \mid T'_a \subseteq T_{i\ell}\}$. By the inductive hypothesis, $\forall x'' \in T'_a$,

$$\Pi_{m-1}^{-1}(x'') = \{(x'',c'_{i\ell k}(x''),c_{i,j,m+1}(x'',c'_{i\ell k}(x'')),\ldots,c_{ijn}(x'',c'_{i\ell k}(x''))) \mid (i.\ell) \in I_a, 1 \le k \le p_{i\ell}, j \in J_i\} .$$

But this is just the statement of 1.7 for $m - 1$, after a change of notation.

Q.E.D.

With this stratification procedure we can now prove the main inductive step (1.8) in the proof of 1.1; in fact, taking $T = S$ in 1.8, 1.1(a) follows immediately.

LEMMA 1.8: If $T \subseteq S \subseteq R^n$ are s.a. and S is open, then $T \subseteq \bigcup_i U\{g_{ij}\} \subseteq S$, for some $\{g_{ij}\} \subseteq K[X]$, computable from the presentations of T and S (if K is computable).

PROOF OF 1.8: Induction on $m = \dim T$. For $m = -1$, just take $\{g_{ij}\} = \{0\}$. For $m \ge 0$, assume 1.8 has been proved for subsets of S of $\dim < m$; we prove it for T. Apply 1.7 to this T and m. Set $\delta(x) = \max\{\delta \mid \bigwedge_{i=1}^n |y_i - x_i| < \delta \longrightarrow y \in S\}$ $(\delta(x) < \infty$ unless $S = R^n)$. We have $T \subseteq \bigcup_i V_i \subseteq S$, where V_i is

$$\{x \mid x' \in T_i \wedge \bigvee_{j \in J_i} \bigwedge_{m<k\le n} |x_k - c_{ijk}(x')| < \delta(x',c_{i,j,m+1}(x'),\ldots,c_{ijn}(x'))\}.$$

(This uses the fact that $\delta > 0$ throughout S, which is the only consequence of the openness of S that we use.) It would suffice to enclose each V_i in a union of U's contained in S. However, we do not quite achieve this; instead we shall enclose in a union of U's in S, all but a subset of $\dim < m$ of $\Pi_m^{-1}(T_i)$; by the inductive hypothesis, this will prove 1.8.

We need a definition and one more lemma, in which we analyze a parametrized version of 1.1, for the case $n = 1$. The formulation is simplified if we agree that $1 \cdot \infty = \infty$, $(-1) \cdot \infty = -\infty$, $a + \infty = \infty$, and $a - \infty = -\infty$.

DEFINITION: For $A \subseteq R^n$ s.a., we shall call a function $c: A \longrightarrow R \cup \{\pm\infty\}$ s.a. of degree $\leq d$ if $A = \sqcup_i A_i$, for some A_i such that $\bigwedge_i$ either

(1) $\forall x \in A_i$, $c(x) = \infty$, or $\forall x \in A_i$, $c(x) = -\infty$, or

(2) $\left[\begin{array}{c} \exists k \in \mathbb{N} \text{ and } \exists P_i \in K[X,Y] \text{ of } Y\text{-degree} \leq d, \text{ such that } \forall x \in A_i, \\ P_i(x,Y) = 0 \text{ has a finite and constant number } j_i \\ \text{of real roots } c_{i1}(x) < \cdots < c_{ij_i}(x), \text{ and } c(x) = c_{ik}(x). \end{array}\right]$

LEMMA 1.9: If $c: A \longrightarrow R \cup \{\pm\infty\}$ is s.a. of degree $\leq d$, and $C = \{(x,y) \mid x \in A \wedge y > c(x)\}$, then we may construct $\{g_{ijk}\} \subseteq K[X,Y]$ and A_i such that $A = \sqcup_i A_i$ and $C = \cup_i [(A_i \times R) \cap \cup_j U\{g_{ijk}\}]$.

PROOF OF 1.9: Induction on d. For $d = 0$, take, for $i = 0,1$, $A_i = c^{-1}((-1)^i \cdot \infty)$ and $g_{i11}(X,Y) = (-1)^{i+1}$. For $d > 0$, assume that 1.9 has been proved for s.a. functions of $\deg \leq d - 1$, and that c is s.a. of $\deg \leq d$, presented as in the Definition. We shall prove it for c. $\bigwedge_i$, for $0 \leq e \leq d$, and for $j = 0,1$, set

$$A_{iej} = \left\{ x \in A_i \;\middle|\; \begin{array}{c} \text{either } e = 0 \;\; c(x) = (-1)^j \cdot \infty, \\ \text{or } e = \text{smallest integer s.t. } \dfrac{\partial^{e+1} P_i}{\partial Y^{e+1}}(x,c(x)) \neq 0, \\ \text{and } (-1)^j \cdot \dfrac{\partial^{e+1} P_i}{\partial Y^{e+1}}(x,c(x)) > 0 \end{array} \right\}.$$

Define $b: A \longrightarrow R \cup \{\pm\infty\}$ by cases: for i such that $c(A_i) = \pm\infty$, let $b(x) = c(x)$ $\forall x \in A_i$; for the other i, define $b(x)$ throughout each A_{iej} to be the largest root $< c(x)$ of $\partial^{e+1} P_i / \partial Y^{e+1}(x,Y)$ if this exists, and $-\infty$ otherwise. Set $B = B_1 \cup B_2 \cup B_3$, where

$$B_1 = \bigcup_{cA_i = -\infty} A_{i01} \times R, \quad B_2 = \{(x,y) | x \in A \wedge y > b(x)\},$$

$$B_3 = \bigcup_{cA_i \subseteq R} [(A_{iej} \times R) \cap U\{(-1)^j \frac{\partial^e p_i}{\partial Y^e}\}].$$

Define $a: A \longrightarrow R \cup \{\pm\infty\}$ by cases: for i such that $c(A_i) = \pm\infty$, let $a(x) = c(x)$ $\forall x \in A_i$; for the other i, define $a(x)$ throughout each A_{iej} to be the smallest root $> c(x)$ of $\partial^{e+1}p_i/\partial Y^{e+1}(x,Y)$ if this exists, and ∞ otherwise. By calculus, we have $C = B \cup \{(x,y) \mid x \in A \wedge y > a(x)\}$. By the inductive hypothesis, C can be written in the required form. This proves 1.9. Q.E.D.

Returning to the proof of 1.8, write V_i as a finite union of finite intersections of sets of the form

$$\{x \mid x' \in T_i \wedge (-1)^\ell x_k > (-1)^\ell c_{ijk}(x') - \delta(x', c_{i,j,m+1}(x'), \ldots, c_{ijn}(x'))\}$$

(for $m < k \leq n$, $j \in J_i$, and $\ell = 0,1$). 1.9 applies to each of these sets (taking $A = T_i$, etc.) to give $\{g_{iruv}\} \subseteq K[X]$ and $T_i = \bigsqcup_r T_{ir}$ such that $V_i = \bigcup_r [(T_{ir} \times R^{n-m}) \cap \bigcup_u U\{g_{iruv}\}]$ (in fact, $\bigwedge_i$ no more than one of the variables $X_{m+1}, \ldots, X_n$ will occur in any of the g_{iruv}). Each T_{ir}, being s.a., can be written as $\bigcup_s (Z\{f_{irst}\} \cap U\{f'_{irst}\})$, some $\{f_{irst}, f'_{irst}\} \subseteq K[X']$. Therefore $V_i = \bigcup_{r,s} V_{irs}$, where $V_{irs} = [(Z\{f_{irst}\} \cap U\{f'_{irst}\}) \times R^{n-m}] \cap \bigcup_j U\{g_{iruv}\}$. For those r,s such that $\bigwedge_t f_{irst} = 0$, V_{irs} is evidently a union of U's. For the other r,s, $\dim \Pi_m^{-1}(Z\{f_{irst}\} \cap U\{f'_{irst}\}) = \dim (Z\{f_{irst}\} \quad U\{f'_{irst}\}) < m$. This, as remarked before the Definition, completes the proof of 1.8 and hence, as remarked before 1.8, completes the proof of 1.1. Q.E.D.

2. APPLICATIONS TO HILBERT'S 17TH PROBLEM

Now let $X = (X_0, \ldots, X_n)$ be indeterminates, and let $x = (x_0, \ldots, x_n) \in R^{n+1}$. $f \in K[X]$ is called <u>positive semidefinite</u>, or psd, (over R) if $\forall x \in R^{n+1}$, $f(x) \geq 0$. Let $\alpha = (\alpha_0, \ldots, \alpha_n) \in \mathbf{N}^{n+1}$ be a multi-index, let $|\alpha| = \Sigma\alpha_i$, fix an even $d \in \mathbf{N}$, let

$C = \langle C_\alpha \rangle_{|\alpha|=d}$ be $\binom{n+d}{n}$ indeterminates (in some fixed order), let $c = \langle c_\alpha \rangle_{|\alpha|=d}$ be an element of $R^{\binom{n+d}{n}}$, let $f \in \mathbb{Z}[C;X]$ be the general form of degree d in X with coefficients C (i.e., $f(C;X) = \Sigma_{|\alpha|=d} C_\alpha X^\alpha$, where $X^\alpha = X_0^{\alpha_0} \cdots X_n^{\alpha_n}$), and let

$$P_{nd} = \{c \in R^{\binom{n+d}{n}} \mid f(c;X) \text{ is psd (over } R) \text{ in } X\}.$$

Daykin [1960] showed how to compute effectively, from n and d alone, $p_{ij} \in \mathbb{Z}[C]$ and $r_{ij} \in \mathbb{Q}(C;X)$ (homogeneous in X) such that

(1) $\bigwedge_i f(C;X) = \sum_j p_{ij}(C) r_{ij}(C;X)^2$ and

(2) $\forall c \in P_{nd} \ \bigvee_i \ \bigwedge_j \left[\begin{array}{c} p_{ij}(c) \geq 0, \text{ and the denominator of} \\ r_{ij}(c;X) \text{ does not vanish identically in } X \end{array} \right]$.

This result includes Artin's original solution to Hilbert's 17th problem, since it says that a psd form with coefficients in K is a positively-weighted sum of squares (SOS) of homogeneous rational functions, with coefficients and weights also in K; in fact, these coefficients and weights are "piecewise-polynomial" functions of c. This representation is also a slight improvement of piecewise-rational (p.r.) SOS-representations found in the fifties by A; Robinson and L; Henkin: (1) in Daykin's representation, there is a primitive recursive bound on the number and degrees of the p_{ij} and the r_{ij} (namely,

$$2^{2^{\cdot^{\cdot^{\cdot^{2^{cd}}}}}},$$

where there are n 2's, and where c is a positive constant); Henkin's and Robinson's bounds were only general recursive. (2) In Daykin's representation the pieces on which the coefficients and weights are defined as rational functions are basic closed s.a. (namely, $W_i = W\{p_{ij}\}$), whereas the earlier pieces were only s.a.

We now prove and refine Daykin's result.

PROPOSITION 2.1: There exist $\{g_{ij}\} \subseteq \mathbb{Z}[C]$, and $\{h_{kiJ}\} \subseteq \mathbb{Q}[C;X]$, and $s_i \in \mathbb{N}$ $(k = 1,2)$ such that

$$(1) \qquad \bigwedge_i \; f = \frac{\Sigma_J \, g_{iJ}h^2_{1iJ}}{f^{2s_i} + \Sigma_J \, g_{iJ}h^2_{2iJ}} \quad \text{and}$$

$$(2) \qquad \forall c \in P_{nd} \; \bigvee_i \; \bigwedge_J \, g_{iJ}(c) \geq 0.$$

REMARK: To transform (1) into a (positively weighted) SOS of rational functions, just multiply the numerator and denominator of (1) by the denominator. Our proof of 2.1 uses G. Stengle's "Positivstellensatz" [1974 and 1979]: for $\{F,g_i\} \subseteq K[X]$, if $\forall x \in W\{g_i\}$, $F(x) \geq 0$, then

$$F = \frac{\Sigma_I c_{1I} g_I h^2_{1I}}{F^{2s} + \Sigma_I c_{2I} g_I h^2_{2I}},$$

some $s \in N$, $c_{jI} \in K^+$, and $h_{jI} \in K[X]$ $(j = 1,2)$, where the g_I are products of the g_i. Further, the h_{jI} can be chosen to be homogeneous if F and g_i are, though Stengle proved this only for the case $\{g_i\} = \{0\}$ [1979].

PROOF OF 2.1: By the finiteness theorem, we can write $P_{nd} = \bigcup_i W_i$, where $W_i = W\{g_{ij}\} \subseteq R^{\binom{n+d}{n}}$, for some $\{g_{ij}\} \subseteq \mathbb{Z}[C]$. For each i we apply the Positivstellensatz to f, which is nonnegative on $W_i \subseteq R^{\binom{n+d}{n}+n+1}$ (we are now viewing $\{g_{ij}\}$ as being in the larger ring $\mathbb{Z}[C;X]$). 2.1 follows immediately, taking the g_{iJ} to be products of the g_{ij}. Q.E.D.

Curiously we saw on p. 91 that Daykin's result already gives a union-of-W's representation of P_{nd}, without the finiteness theorem.

Let us now make a finer analysis of P_{nd} than that given by the finiteness theorem.

THEOREM 2.2: P_{nd} is a single W if and only if $d \leq 2$.

PROOF: For the "if" direction, we assume $d = 2$ and use induction on n. For $n = 1$, $P_{12} = W\{A, C, 4AC - B^2\}$ (writing $f(A,B,C;X,Y) = AX^2 + BXY + CY^2$). To prove P_{n2} is a single W for $n \geq 1$, we may suppose, inductively, that the condition for a quadratic form in $X_0, \ldots, X_{n-1}$ to be psd is a conjunction of non-strict inequalities in C. Write

$$f(X_0, \ldots, X_n) = f_2(X_0, \ldots, X_{n-1}) + f_1(X_0, \ldots, X_{n-1})X_n + f_0X^2,$$

where $\deg f_i = i$ $(i = 0,1,2)$. Then f is psd if and only if f_2, f_0, and $4f_0f_2 - f_1^2$ are all psd in $X_0, \ldots, X_{n-1}$; this is just a conjunction of three conjunctions, since these three forms are quadratic (except the constant form f_0, for which the psd property is an "improper" conjunction, namely, with only one conjunct).

For the "only if" part we use <u>reductio ad absurdum</u>: If $d \geq 4$ and $P_{nd} = W\{g_i(C)\}$, then $P_{14} = W\{g_i(C',0)\}$ (i.e., we set some of the C equal 0), which gives

$$\begin{aligned} W\{g_i(B,C,0)\} &= \{(b,c) \in R^2 \mid X^4 + bX^2Y^2 + cY^4 \text{ is psd}\} \\ &= \{(b,c) \mid \forall x \geq 0,\ \forall y \geq 0,\ x^2 + bxy + cy^2 \geq 0\} \\ &= \{(b,c) \mid b^2 - 4c \leq 0 \vee -b + \sqrt{b^2 - 4c} \leq 0\} \\ &= W\{4C - B^2\} \cup W\{B,C\}. \end{aligned}$$

But now we see that this set (striped in the figure below) cannot be written as a single $W\{g_i\}$, since one of the g_i would have to be divisible by (precisely) an odd power of $B^2 - 4C$, which would make it change sign across even the dotted part of the parabola. Q.E.D.

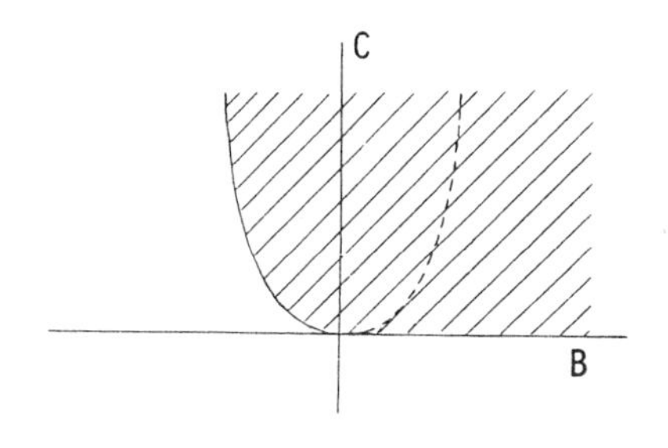

We apply 2.2 to settle a question raised in the early sixties by G. Kreisel: "Do there exist $\{g_i\} \subseteq \mathbb{Q}\,C$ and $\{h_i\} \subseteq \mathbb{Q}(C;X)$ such that $f = \Sigma_i\, g_i h_i^2$ and $\forall c \in P_{nd}$, all $g_i(c) \geq 0$?" Corollary 2.3 below shows that the answer to this question is yes iff $d \leq 2$.

COROLLARY 2.3: For $d \leq 2$, there are $s \in \mathbb{N}$, $\{g_J\} \subseteq \mathbb{Z}[C]$, and $\{h_{kJ}\} \subseteq \mathbb{Q}[C;X]$, such that

(1) $$f = \frac{\Sigma_J\, g_J h_{1J}^2}{f^{2s} + \Sigma_J\, g_J h_{2J}^2}$$ and (2) $\forall c \in P_{nd}$, all $g_J(c) \geq 0$;

for $d > 2$, there do not exist $e_i \in \mathbb{N}$, $p_j \in K[C]$, and functions (even discontinuous and non-s.a.) $a_{ij}\colon P_{nd} \longrightarrow R^{m_i}$ such that $\forall c \in P_{nd}$

(3) $$f(c;X) = \sum_j p_j(c) \left[\frac{f_1(a_{1j}(c);X)}{f_2(a_{2j}(c);X)}\right]^2, \text{ and}$$

(4) $$\bigwedge_j \left[p_j(c) \geq 0 \text{ and, (if } c \neq 0) \ a_{2j}(c) \neq 0 \in K^{m_2}\right],$$

even if we also allow the rational functions to be discontinuous in X. Here $m_i = \binom{n+e_i}{n}$ (where $i = 1,2$ and $e_1 = d/2 + e_2$) and f_i is the general form of degree e_i in X.

PROOF: For $d = 2$ we just combine 2.2 and the proof of 2.1, with a single W_i, so that we may drop the i.

For $d > 2$, we note that if 2.3 were false, then we could conclude that $P_{nd} = W\{p_j\}$ ($\subseteq$ by (4) and $\supseteq$ by (3)), contradicting 2.2. Q.E.D.

Kreisel's question as stated above was part of a broader question: "Is there a continuous solution to Hilbert's 17th problem?"[6] In view of 2.3, the best we can hope for is the following

CONJECTURE 2.4: There exist $s \in \mathbb{N}$ and continuous $\mathbb{Q}$-piecewise-rational functions $a_{ij}\colon P_{nd} \longrightarrow R^{m_i}$ and $p_{ij}\colon P_{nd} \longrightarrow R^+$, such that $\forall c \in P_{nd}$,

[6] The question appeared in print, e.g., on pp. 115-6 of [Kreisel 1977a] and in footnote 1 of [1977b].

$$(*) \qquad f(c;X) = \frac{\Sigma_j \, p_{1j}(c) f_1(a_{1j}(c);X)^2}{f(c;X)^{2s} + \Sigma_j \, p_{2_j}(c) f_2(a_{2_j}(c);X)^2}$$

Here $m_i = \binom{n+e_i}{n}$ (where $i = 1,2$, $e_1 = ds + d/2$, and $e_2 = ds$) and f_i is the general form of degree e_i in X. By "$\mathbb{Q}$-piecewise-rational" we mean that P_{nd} has been written as a finite union of $\mathbb{Q}$-W's upon each of which a_{ij} and p_{ij} are described by rational functions $\varepsilon \, \mathbb{Q}(C)$. The $\mathbb{Q}$-s.a. descriptions of the a_{ij} and the p_{ij} are computable from n and d.

(*) leads to a SOS of functions which are homogeneous and rational in X, and continuous simultaneously in c and x for $(c,x) \varepsilon P_{nd} \times R^{n+1}$, for since the denominator of (*) can vanish only where f does, the "squeeze" or "sandwich" theorem implies that the rational functions extend (namely, by 0) continuously.

In [Delzell 1980 and "to appear"], we construct continuous $\mathbb{Q}$-s.a. functions a_{ij} and p_{ij} such that (*) holds. Thus if K is real closed, the coefficients and weights in the SOS-representation will be in K; taking $K = \mathbb{R}$, we obtain the first constructive, in particular, intuitionistic, solution to Hilbert's 17th problem over $\mathbb{R}$, since (1) the $\mathbb{Q}$-s.a. descriptions of the a_{ij} and p_{ij} are computable from n and d, and (2) while elements of $\mathbb{R}$ can be given only as approximations, we can approximate $a_{ij}(c)$ and $p_{ij}(c)$ by approximating c, by continuity. (More generally, the last sentence holds with $\mathbb{R}$ replaced by any real closed field with a dense computable subfield.)

A proof of 2.4 awaits further refinements of the finiteness theorem:

CONJECTURE 2.5: $P_{nd} = \cup_i W_i$, where the W_i are $\mathbb{Q}$-W's which are $\mathbb{Q}$-p.r. neighborhood-retracts; this means there are open s.a. sets $U_i \supseteq W_i$ and $\mathbb{Q}$-p.r. retractions $r_i \colon U_i \longrightarrow W_i$.

S.a. retractions were used in our s.a. continuous solution of (*) to help us glue together the piecewise-continuous solutions of 2.1;

s.a. r_i are easy to obtain for <u>any</u> closed s.a. set W_i. As explained in [1980], 2.5 implies 2.4.

We know of two necessary conditions on the W_i in 2.5: (1) they must contain plenty of rational points (i.e., points with coordinates in $\mathbb{Q}$) —it would suffice[7] for their interiors to cover $\mathrm{Int}(P_{nd})$, and (2) they must not have certain cusps with irrational "slopes." B. Casler has shown me an encouraging result that overlaps 2.5: If K is a countable subfield of $\mathbb{R}$ (e.g., $\mathbb{Q}$) and $R = \mathbb{R}$, then from a retraction $r: U \longrightarrow W$ (not necessarily s.a.) onto a closed s.a. set W with dense interior, we can construct a new retraction $r': U \longrightarrow W$ which takes points with coordinates in K to points with coordinates in K. Again, as explained in [Delzell 1980], this leads to continuous but not s.a. a_{ij} and p_{ij} in (*) which take points with coordinates in K to points with coordinates in K, solving the 17th problem continuously for those special K.

BIBLIOGRAPHY

Bochnak, J., and Efroymson, G., Real Algebraic Geometry and the 17th Hilbert Problem, Math. Ann. (1980).

Brumfiel, G., Partially Ordered Rings and Semi-Algebraic Geometry. Lecture Note Series of the London Math. Soc. (Cambridge Univ. Press, Cambridge, 1979).

Cohen, P. J., Decision Procedures for Real and P-adic Fields, Commun. Pure & Applied Math. 22, (1969), 131-51.

Collins, G. E., Quantifier Elimination for Real Closed Fields by Cylindrical Algebraic Decomposition, Second GI Conf. on Automata Theory & Formal Languages, Lecture Note Series in Computer Science 33 (Springer-Verlag), Berlin, 1975, pp. 134-83; preliminary report appeared in SIGSAM Bull. 8 (3), issue #31, 1974, pp. 80-90; synopsis in SIGSAM Bull. 10 (1), 1976, pp. 10-2.

Coste, M. and M. F., Topologies for Real Algebraic Geometry, Topos theoretic Methods in Geometry, in Various Publications Series 30, A. Kock, ed. (Matematisk Institut, Aarhus Univ., 1979).

Daykin, Thesis, Univ. of Reading, 1960 (unpublished); cited by Kreisel, G., A Survey of Proof Theory, J. Symb. Logic 33, (1968), 321-88.

Delzell, C. N., A Constructive, Continuous Solution to Hilbert's 17th Problem, and Other Results in Semi-Algebraic Geometry, Ph.D. Dissertation, Stanford University, 1980.

[7](For the purpose of proving 2.4.)

A Constructive, Continuous Solution to Hilbert's 17th Problem, in preparation. See also a preliminary abstract in AMS Abstracts, Jan. 1981.

Efroymson, G., Substitution in Nash Functions, Pacific J. Math. 63, (1976), 137-45.

Fischer, M. J., and Rabin, M. O., Super-Exponential Complexity of Presburger Arithmetic, M.I.T., MAC Tech. Memo. 43, (Feb. 1974). Also in Complexity of Computation. (Proc. Sympos., New York, 1973), pp. 27-41. SIAM-AMS Proc., Vol. VII, Amer. Math. Soc., Providence, R.I., 1974.

Henkin, L., Sums of Squares, Summaries of Talks Presented at the Summer Institute of Symbolic Logic in 1957 at Cornell University, (Institute Defense Analyses, Princeton, 1960), 284-91.

Hironaka, H., Triangulations of Semi-Algebraic Sets, Proc. Symp. in Pure Math. 29, (Amer. Math. Soc., Providence, 1975), 165-85.

Subanalytic Sets, Number Theory, Algebraic Geometry, and Commutative Algebra—in Honor of Yasuo Akizuki, (Kusunoki, Y., et al., eds.) (Kinokuniya Book-Store Co. Ltd., Tokyo, 1975), 453-93.

Kreisel, G., Sums of Squares, Summaries of Talks Presented at the Summer Institute of Symbolic Logic in 1957 at Cornell University, (Institute Defense Analyses, Princeton, 1960), 313-20.

A Survey of Proof Theory, J. Symb. Logic 33, (1968), 321-88.

On the Kind of Data Needed for a Theory of Proofs, Logic Colloquium 1976 (Gandy, R. O., and Hyland, J. M. E., eds.), (North-Holland Publishing Co., Amsterdam, 1977), 111-28. MR58#21397.

Review of L. E. J. Brouwer, Collected Works, Vol. I., Philosophy and Foundations of Mathematics (A. Heyting, ed.), Bull. Amer. Math. Soc. 83 (1977), 86-93.

Lojasiewicz, S., Triangulations of Semi-Analytic Sets, Ann. Scuola Norm. Sup. Pisa 18, (1964), 449-74.

Ensembles Semi-analytiques, Lecture Note (1965) at I.H.E.S., Bures-sur-Yvettes, Reproduit No. A66.765, Ecole Polytechnique, Paris. (An English version of p. 65ff. appears in [Hironaka 1975b].)

Monk, L. G., Elementary-Recursive Decision Procedures, Ph.D. Thesis (U. C. Berkeley, 1975).

Stengle, G., A Nullstellensatz and a Positivstellensatz for Semi-Algebraic Geometry, Math. Ann. 207, (1974), 87-97.

Integral Solution of Hilbert's 17th Problem, Math. Ann. 246, (1979), 33-39.

DEPARTMENT OF MATHEMATICS
LOUISIANA STATE UNIVERSITY
BATON ROUGE, LOUISIANA 70803

Current Address:
Department of Mathematics
Maharishi International University
Fairfield, Iowa 52556

Contemporary Mathematics
Volume 8, 1982

AN (ALMOST TRIVIAL) LOCAL-GLOBAL PRINCIPLE FOR THE REPRESENTATION OF -1 AS A SUM OF SQUARES IN AN ARBITRARY COMMUTATIVE RING

Ludwig Bröcker, Andreas Dress and Rudolf Scharlau

Throughout this paper R denotes a commutative ring with a unit element $1 \in R$. R has "finite level" if there exist $n \in \mathbb{N}$ and $x_1,\ldots,x_n \in R$ with $-1 = x_1^2 + \cdots + x_n^2$, in which case we define the level $s(R)$ of R to be the minimal n for which such $x_1,\ldots,x_n \in R$ exist. Otherwise we define $s(R) = \infty$.

For a field K it is a classical result, due to Artin and Schreier, that $s(K) = \infty$ if and only if K is formally real, i.e., can be ordered. In this note we give a rather elementary proof that $s(R) = \infty$ if and only if there exists a ring homomorphism $R \longrightarrow K$ into an ordered field K.

We will state a number of simple applications and in particular discuss for which domains R it can happen that $s(R) = \infty$, but $s(K) < \infty$ for the quotient field $K = \mathrm{Quot}(R)$ of R (cf. [1] and [5]).

Only after the first draft of this paper had been written by the second and the third named author, they learned about the existence of the preprint [4] written by the first named author, which approaches closely related problems in about the same way. After some hesitation, we decided to go on and to publish this note as a joined paper, too, basically to popularise the amazing simplicity of the arguments involved.

At no extra cost, we can formulate our main result for more general quadratic forms than sums of squares as follows.

THEOREM 1. Let $A \subseteq R$ be a subset such that $1 \in A$ and $A \cdot A \subseteq A$. The following statements are equivalent.

(i) There exists a representation $-1 = \sum_{i=1}^{n} a_i x_i^2$ for some $n \in \mathbb{N}$, $a_i \in A$, $x_i \in R$.

1980 Mathematics Subject Classification 10C04, 13K05.

0271-4132/81/0000-0676/$03.00

(ii) For every prime ideal $p < R$, there exists a representation $-1 = \sum_{i=1}^{n} (a_i + p)\, x_i^2$, $n \in \mathbf{N}$, $a_i \in A$, $x_i \in R(p)$. Here $R(p)$ denotes the quotient field of R/p.

(iii) For every prime ideal p and any ordering of $R(p)$, not all $a + p$ $(a \in A)$ are positive or zero.

(iv) For any ring homomorphism $\phi : R \longrightarrow K$ of R into an ordered field K there exists some $a \in A$ with $\phi(a) < 0$.

Before proving theorem 1, we make a definition and introduce some notation. We call an ideal $a \leq R$ <u>A-real</u>, if $\sum a_i x_i^2 \in a$, $a_i \in A$, $x_i \in R$ implies $a_i\, x_i^2 \in a$ for all i. In particular, a prime ideal $p < R$ is A-real, if and only if -1 cannot be represented in the form

$$\sum_{i=1}^{n} (a_i + p)\, x_i^2 \quad \text{with} \quad n \in \mathbf{N}, \quad a_i \in A \quad \text{and} \quad x_i \in R(p) = \mathrm{Quot}(R/p).$$

Thus, to prove (i) $\Longrightarrow$ (ii), we will have to show that the impossibility of representing -1 in the form $\sum_{i=i}^{n} a_i x_i^2$ with $n \in \mathbf{N}$, $a_i \in A$, $x_i \in R$ implies the existence of an A-real prime ideal $p < R$.

For this purpose, we introduce the sets

$$\boxed{A} = \left\{ \sum_{i=1}^{n} a_i\, x_i^2 \mid n \in \mathbf{N}, \quad a_i \in A, \quad x_i \in R \right\}$$

$$\boxed{A}_0 = \boxed{A} \cap (- \boxed{A})$$

$$q_A : \text{the ideal generated by } \boxed{A}_0 .$$

Note that $\boxed{A}_0$ is an additive subgroup of R, because $\boxed{A}$ is closed under addition.

The proof of theorem 1 is essentially contained in the following series of lemmata.

LEMMA 1. $2\, q_A \subseteq \boxed{A}_0$.

PROOF. We only need to show that $2xy \in \boxed{A}_0$ for $x \in R$, $y \in \boxed{A}_0$. But

$$2xy = (x+1)^2 y - x^2 y - y,$$

and $\pm z^2 y \in \boxed{A}_0$ for all $z \in R$.

LEMMA 2. $-1 \in \boxed{A}$ if and only if $q_A = R$.

PROOF. $-1 \in \boxed{A}$ implies $-1 \in \boxed{A}_0 \subseteq q_A$, so $q_A = R$. On the other hand, $1 \in q$ implies $2 \in \boxed{A}_0$ by lemma 1, so $-2 \in \boxed{A}$, and finally $-1 = -2+1^2 \in \boxed{A}$.

LEMMA 3. q_A is an A-real ideal.

PROOF. Let $q = \sum a_i x_i^2 \in q_A$. It is enough to show that $-a_j x_j^2 \in \boxed{A}$ for all j, because this implies $a_j x_j^2 \in \boxed{A}_0 \subseteq q_A$. Now, $2q \in \boxed{A}_0$ by lemma 1, so $-2q \in \boxed{A}$. It follows that

$$a_j x_j^2 = -(-2q + \sum_i a_i x_i^2 + \sum_{i \neq j} a_i x_i^2) \in -\boxed{A}.$$

LEMMA 4. If $a \leq R$ is A-real, then also

$$\mathrm{rad}(a) = \{x \in R \mid \exists\, k \in \mathbb{N} \text{ s.th. } x^k \in a\}$$

is A-real.

PROOF. Let $\sum_1^n a_i x_i^2 \in \mathrm{rad}(a)$, i.e., $(\sum a_i x_i^2)^k \in a$ some $k \in \mathbb{N}$. We have

$$\left(\sum_i^n a_i x_i^2\right)^k = \sum_i^m b_j y_j^2$$

for appropriate m, $b_j \in A$, $y_j \in R$, among which all terms $a_i^k \left(x_i^k\right)^2$ appear. a is A-real, so $a_i^k x_i^{2k} \in a$ for all i, which means that $a_i x_i^2 \in \mathrm{rad}(a)$. The following lemma is crucial.

LEMMA 5. Suppose $a < R$ is A-real, $a \neq R$. If p is minimal among the prime ideals containing a, then p is A-real.

PROOF. A prime ideal p contains a if and only if it contains $\mathrm{rad}(a)$, so we may assume $a = \mathrm{rad}(a)$ by lemma 4. Let $\sum a_i x_i^2 \in p$. By a well-known lemma ([3], Chap. II, §2, Proposition 12), there exists a $y \in R \backslash p$ such that $y \cdot \sum a_i x_i^2 \in \mathrm{rad}(a) = a$. A fortiori we have $\sum a_i (x_i y)^2 \in a$, so $a_i \cdot (x_i y)^2 \in a \subseteq p$ for all i. It follows that $a_i x_i^2 \in p$, because $y^2 \notin p$. Finally we need the following general lemma from the theory of orderings of fields.

LEMMA 6. Let $P \subseteq F \backslash \{0\}$ be closed under addition and multiplication in the field F, and assume P contains all squares in $F \backslash \{0\}$. Then for some ordering of F all members of P are positive.

PROOF OF THEOREM 1. The implications "(i) $\Longrightarrow$ (ii) $\Longrightarrow$ (iii) $\Longleftrightarrow$ (iv)" are obvious and "(iii) $\Longrightarrow$ (ii)" follows from lemma 6.

Now suppose that (ii) holds, i.e., that no prime ideal $p < R$ is A-real. By lemma 5, no proper ideal can be A-real, so, by lemma 3, $q_A = R$ and thus, by lemma 2, $1 \in \boxed{A}$, which is assertion (i).

In most of the following remarks and corollaries we restrict ourselves to the case $A = \{1\}$ and leave it to the reader to formulate more general results.

REMARK 1. The generalization of theorem 1 ((ii) $\Longrightarrow$ (i)), concerning the representation of an arbitrary unit instead of -1 as a sum of squares does not hold: let R be the ring of integers of a totally real number field. It is not even true that a unit y (or any element in R) is a sum of squares in R provided it is totally positive and a sum of squares in all completions $\hat{R}_p$, $p \leq R$ a prime ideal.

A counterexample is $5 + 2\sqrt{6} \in \mathbb{Z}[\sqrt{6}]$. (For $p \nmid 2$, all elements in $\hat{R}_p$ are sums of squares. The conditions for the p dividing 2 can be collected by requiring that $y \equiv x^2 \bmod 2R$ for some $x \in R$.)

The generalization of (ii) $\Longrightarrow$ (i) is generally false also for orders in non-totally real fields K. This has essentially local reasons, depending on powers of the primes dividing 2. For instance, for $R = \mathbb{Z}[i]$ any element is a sum of squares in K and modulo any prime ideal. But i cannot be a sum of squares in R, because it is not congruent to a square modulo $2R = p^2$, $p = R(1+i)$.

But we can state the following corollary which contains a local-global principle for the representation of arbitrary units by sums of squares under the assumption that all fields R/m, m maximal, be formally real.

COROLLARY 1. Let A be as in theorem 1, suppose A consists of units and for all maximal ideals $p < R$ there exists an ordering of $R(p)$ such that $A+p \subseteq R(p)$ consists of positive elements. If $b \in R$ is a unit such that

$$b+p = \sum_1^n (a_i+p)\, x_i^2, \quad n \in \mathbb{N}, \quad x_i \in R(p)$$

has a solution for all prime ideals $p < R$, then

$$b = \sum_1^n a_i\, x_i^2, \quad n \in \mathbb{N}, \quad x_i \in R$$

has a solution.

PROOF. For the subset $B := \bigcup_{i \in \mathbb{N}_0} (-b)^i A \subseteq R$, the condition (ii) of theorem 1 is satisfied. By theorem 1, there exists an equation $-1 = \sum b_j t_j^2$, $b_j \in B$, $t_j \in R$, i.e.,

$$-1 = \sum a_i x_i^2 - b \sum b_j y_j^2, \quad a_i, \; b_j \in A, \quad x_i, \; y_j \in R,$$

i.e.,
$$b \sum b_j y_j^2 = \sum a_i x_i^2 + 1 = : z .$$

We have $z \notin m$ for every maximal ideal $m < R$, for otherwise $-1 \equiv \sum a_i x_i^2$ had a solution mod m. So $c = \sum b_j y_j^2$ is a unit, and

$$b = \left(\sum a_i \left(\frac{x_i}{c}\right)^2 + \left(\frac{1}{c}\right)^2\right) \left(\sum b_j y_j^2\right)$$

is of the form $\sum c_k z_k^2$ for appropriate $c_k \in A$, $z_k \in R$.

REMARK 2. If we merely require the quotient field Quot(R) and all R/m for maximal m to be of finite level, R itself does not have to be of finite level. See the proof of corollary 3 below.

The next corollary is known (see [6]) and a direct, equally elementary proof can be given (see [8], Chap. III, §2, Prop. 4, p. 190).

COROLLARY 2. If all localizations R_m of R at maximal ideals m have finite level, then R has finite level.

PROOF. If p is any prime ideal and m any maximal ideal containing p, then $R_m \subseteq R_p$, so R_p has finite level and in particular $R(p) = R_p/pR_p$ has finite level.

Baeza has remarked in [1], that corollary 2 implies the following: Let R be a Prüfer domain such that its quotient field $Q(R)$ has finite level. Then R has finite level.

The next corollary has also been mentioned in [1]. The proof outlined in [1], reducing to the local case by corollary 2 and then applying Witt ring techniques (cf. [5], 3.1), appears to be more complicated than ours.

COROLLARY 3. Let R be any regular ring. If the quotient field of R has finite level, then R has finite level.

For the convenience of the reader, we collect the necessary facts about regular rings, for details see [9], Chapitre IV. A local ring R is regular iff R is noetherian and the minimal number of generators of the maximal ideal equals the dimension $\dim R$ of R, i.e., the maximal length of strictly increasing chains of prime ideals. Any regular R is necessarily necessarily an integral domain. An arbitrary R is regular iff it is noetherian and all localizations R_m at maximal ideals m are regular. For regular R in fact all R_p, p prime, are regular. If R is regular local with maximal ideal m and p is any element in $m \setminus m^2$, then $R/(p)$ is regular of dimension $\dim R - 1$ and $R_{(p)}$ is a discrete valuation ring.

PROOF OF COROLLARY 3. For any $\mathfrak{p} \lhd R$ we have to show that $R(\mathfrak{p})$ is of finite level. W.l.o.g. we may assume that $R = R_{\mathfrak{p}}$. We proceed by induction on Dim R. Choose $p \in \mathfrak{p} \setminus \mathfrak{p}^2$.

Then $R_{(p)}$ is a discrete valuation ring, so its residue class field $\overline{R_{(p)}}$ is of finite level, since $\mathrm{Quot}(R_{(p)}) = \mathrm{Quot}(R)$ is assumed to be of finite level. Let $\overline{R}$ and $\overline{\mathfrak{p}}$ be the image of R and $\mathfrak{p}$ in $\overline{R_{(p)}}$, respectively. Then $\overline{R} \cong R/(p)$ is regular with $\mathrm{Quot}(\overline{R}) = \overline{R}_{(p)}$ of finite level and $\dim \overline{R} = \dim R - 1$, thus—by induction— $\overline{R}(\overline{\mathfrak{p}}) \cong R(\mathfrak{p})$ is of finite level.
Q.E.D.

The next corollary shows that it happens quite often that a domain is of infinite level, but its quotient field is of finite level. All extensions of the rational function field $\mathbb{Q}(x)$ contain such rings.

COROLLARY 4. Let K be a field of finite level. All subrings R of K with quotient field $\mathrm{Quot}(R) = K$ have finite level if and only if K is of positive characteristic or K is algebraic over $\mathbb{Q}$.

PROOF. If $\mathrm{char}(K) = p > 0$, every subring $R \subseteq K$ contains the prime field $\mathbb{F}_p$, and -1 is a sum of squares in $\mathbb{F}_p$. If $\mathbb{Q} \subseteq K$ and K is algebraic over $\mathbb{Q}$, it is well-known that $\mathrm{char}(R/\mathfrak{p}) \neq 0$ for every subring $R \subseteq K$ and any prime ideal $\mathfrak{p} \lhd R$ different from $\{0\}$. So theorem 1 shows that R is of finite level iff $K = \mathrm{Quot}(R)$ is of finite level.

Finally, suppose that some $x \in K$ is transcendental over $\mathbb{Q}$. Given the subring $\mathbb{Q}[x] \subseteq K$ and the maximal ideal $x\,\mathbb{Q}[x]$, there exists a valuation ring B in K with maximal ideal m such that $K \supset B \supset \mathbb{Q}(x)$ and $\mathbb{Q}[x] \cap m = x\,\mathbb{Q}[x]$. The ring $R := \mathbb{Z}+m$ has quotient field K and infinite level, because $R/m \simeq \mathbb{Z}$.

Note that the maximal ideals in R are given by $m + p\mathbb{Z}$, where p runs through the rational primes, so they all have quotient fields of finite level. This proves remark 2 above.

For the following we need the concept of ultrafilters. An <u>ultrafilter</u> on a set I is a set A of subsets of I such that

$$U \subseteq V,\quad U \in A \Longrightarrow V \in A$$

$$U, V \in A \Longrightarrow U \cap V \in A$$

$$\emptyset \notin A$$

$$U \in A \quad \text{or} \quad I \setminus U \in A \quad \text{for all} \quad U \subseteq I.$$

Trivial examples are given by the subsets that contain a fixed element, but only for finite I do all ultrafilters occur in this way. If $(K_i)_{i \in I}$ is an infinite family of fields and A an ultrafilter on I, then

$$p_A = \{ x = (x_i) \mid U(x) \in A \},$$

where $$U(x) = \{ i \mid x_i = 0 \}$$

is a prime ideal in $R \prod_{i \in I} K_i$ and in fact a maximal ideal. All $p < R$ are of this form. The fields R/p_A with nontrivial A are called the <u>ultra products</u> of $(K_i)_{i \in I}$.

Part a) of the next corollary immediately follows from theorem 1; part b) is clear from a) and the preceding remarks.

COROLLARY 5. a) Let R be a ring such that $s(R(p)) < \infty$ for all prime ideals $p < R$. Then there exists a constant s such that $s(R(p)) \leq s$ for all p.

b) Let $(K_i)_{i \in I}$ be a family of fields, such that $s(K_i)$ is not bounded. Then $(K_i)_{i \in I}$ possesses a formally real ultraproduct.

Up to now, all our results were formulated in completely elementary terms. We now turn to the theory of quadratic forms and prove an essentially known theorem which makes the implication (ii) $\Longrightarrow$ (i) of theorem 1 more precise (for $A = \{1\}$). Let $W(R)$ denote the Witt ring of nondegenerate symmetric bilinear forms on finitely generated projective R-modules. Remember that a <u>signature</u> of R is a homomorphism of $W(R)$ into $\mathbb{Z}$ (sending 1 into 1). It is well-known that R possesses a signature if and only if R is not of finite level (cf. [8]). So the following theorem is in fact stronger than theorem 1.

THEOREM 2. Let σ be any signature of R. Then there exists a prime ideal $p < R$ and a signature τ of $Q(R/p)$ such that σ equals the canonical map $W(R) \longrightarrow W(R(p))$ followed by τ. In other words, all signatures are of the form $W(R) \longrightarrow W(K) \simeq \mathbb{Z}$ for some ring homomorphism $R \longrightarrow K$ of R into some real closed field K.

PROOF. By theorem 1 in [6], there exists a maximal ideal $m < R$ and a signature σ' of the localization R_m such that the canonical map followed by σ' equals σ. So we may assume that R is local. In this case the theorem is proved in appendix B of [7], where a possible p is given explicitly. However, we want to deduce the assertion as a corollary of theorem 1.

To this end, introduce $A := \{a \in R^* \mid \sigma \langle a \rangle = 1\}$, where $\langle a \rangle$ denotes the one-dimensional form given by a. An equation

$$-1 = \sum_{i=1}^{n} a_i x_i^2 \text{ would imply that } \langle a_1,\dots,a_n\rangle = \langle -1\rangle + \phi \text{ in } W(R)$$

for some (n-1)-dimensional ϕ, and applying σ to both sides would give a contradiction. So by theorem 1, there exist a $p < R$ and an ordering of $R(p)$ such that $A + p \subseteq R(p)$ consists of positive elements only. Let τ be the signature of $R(p)$, corresponding to this ordering (i.e., $\tau\langle y\rangle = 1 \iff y > 0$ for $y \in R(p) \setminus \{0\}$.) If $\tilde{\tau}$ denotes the corresponding signature of R, both $\tilde{\tau}$ and σ take the value 1 on the forms $\langle a\rangle$, $a \in A$. But these forms generate $W(R)$, because $R^* = A \cup -A$, and the one-dimensional forms generate $w(R)$. So $\tilde{\tau} = \sigma$.

REFERENCES

1. R. Baeza: Über die Stufe von Dedekind-Ringen, Arch. Math. 33 (1979), 226-231.

2. N. Bourbaki: Algèbre, Hermann, Paris (1964).

3. N. Bourbaki: Algèbre commutative, Hermann, Paris (1961).

4. L. Bröcker: Positivitätsbereiche in kommutativen Ringen, to appear in Abh. math. Sem. Hamburg.

5. J.-L. Colliot-Thélène: Formes quadratiques sur les anneaux semi-locaux réguliers, Bull. Soc. Math. France 59 (1979), 13-31.

6. A. Dress: The weak local global principle in algebraic K-theory, Communications in algebra 3(7) (1975), 615-661.

7. M. Knebusch: Real closures of commutative rings I, J. reine angew. Math. 274/75 (1975), 61-89.

8. M. Knebusch: Symmetric bilinear forms over algebraic varieties, in, Conference on quadratic forms 1976, Queen's paper in pure and applied mathematics No. 46 (1977).

9. J.-P. Serre: Algèbre locale. Multiplicités, Lecture Notes in Math. 11 (1964).

1ST AUTHOR:

MATHEMATISCHES INSTITUT
WESTFALISCHE WILHELMS-UNIVERSITAT
44 MUNSTER

2ND AND 3RD AUTHORS:

FAKULTAT DER MATHEMATIK
BOX 8648
UNIVERSITAT BIELEFELD
(D48) BIELEFELD

Contemporary Mathematics
Volume 8, 1982

THE NASH RING OF A REAL SURFACE

Gustave A. Efroymson

INTRODUCTION.

The object of this paper is to show how to extend results obtained previously about planar Nash rings to reasonable non-singular real surfaces in R^n. The main tool used is linear projection from R^n to R^2 so that the planar results can be applied to give the results wanted. We describe these results in general terms now; for a more thorough treatment, see below.

Let N_S be the ring of Nash functions on S and A_S be the ring of real analytic functions on S. Theorem 1 essentially says that if p is a height one prime in N_S, then PA_S is also a prime ideal. This should mean that the ring of Nash functions is just as good as the ring of analytic functions on S, at least algebraically. In the planar case, this result was used to show that if $f \in N_S$ is ≥ 0 on S, then f can be represented as sum of two squares of functions $f_1, f_2 \in N_S$, i.e., $f = f_1^2 + f_2^2$. In the case where S is as above, one can show Theorem 2: If $f \in N(S)$ is ≥ 0 on S, then f can be represented as a sum of four squares of functions f_i, $i = 1,\ldots,4$ in N_S.

Now for reasonable surfaces. In Section 1 we consider S to be a non-singular real analytic component of a real algebraic surface V (where V need not be non-singular). Then in Section 2, we will consider more general surfaces which are semi-algebraic real non-singular open surfaces in R^n with some additional properties which should be verified in most cases.

SECTION 1.

Let S be a real non-singular component of a real algebraic surface $V \subseteq R^n$. Then S is also a semi-algebraic set. The theory of [1] applies and we can define Nash functions on S. We use [1] as a reference for what follows. We need some definitions and results. So recall that $f: S \longrightarrow R$ is a Nash function on S if f is analytic and algebraic, i.e., f is a real analytic function on S with semi-algebraic graph. Let

---

1980 Mathematics Subject Classification. 14G30, 58A07, 12J15.

N_S = the ring of Nash functions on S. Note that $f \in N_S$ iff $f \neq 0$ anywhere on S. Moreover, the ring N_S is Noetherian.

Now since S is non-singular, locally S is Nash isomorphic to an open set in R^2 and so if $f \in N_S$, the ideal (f) will locally (and thus globally) be $\prod_i P_{C_i} P_{C'_j}$ where each C_i is a real analytic component of a real algebraic curve. And P_{C_i} = the ideal of Nash functions vanishing on C_i, and C'_j is a complex analytic component of an algebraic curve on the complexification S_C of S, and C'_j has one real support point. Here $P_{C'_j}$ = ideal of real Nash functions which vanish on the complex set C'_j. The reference for this is [3]. Now we know that for any such C_i or C'_j, we get a prime ideal P_{C_i} or $P_{C'_j}$. What is not so obvious is that if $C_i \neq C_j$, then $P_{C_i} \neq P_{C_j}$. To see this we must show $\exists$ g with $g(C_i) = 0$, $g(C_j) \neq 0$.

THEOREM 1. Let C_1 and C_2 be real analytic components of real algebraic curves D_1 and D_2 respectively. We assume $C_1 \neq C_2$ (but we could have $D_1 = D_2$). Then $P_{C_1} \neq P_{C_2}$. Moreover, if C_1 or C_2 is complex with isolated point real support, the same holds.

PROOF. We use linear projection from R^n to R^2 which we can take to be $(x_1,\ldots,x_n) \xrightarrow{\pi} (\ell_1,\ell_2)$ where ℓ_1 and ℓ_2 are linearly independent homogeneous linear polynomials in $x_1,\ldots,x_n$. If C_1 is an open curve, it will have two ends which are at ∞ in R^n and we want to make sure that π takes the ends to ∞ also. By a suitable transformation $x_1 \longrightarrow X_1 + \lambda(x_3,\ldots,x_n)$, we can achieve $x_1(\text{end}) = \infty$, if it isn't so at the start. Next, we note that if $(v_1 \wedge v_2)_p$ gives the tangent space to S at P, then $(\ell_1,\ell_2)(v_1 \wedge v_2) \neq 0$ implies that π will be a local isomorphism near P from S to R^2. So for given ℓ_1,ℓ_2 we will usually have a local isomorphism except at a set of codimension 1 on S and thus there will be only a finite number of such points on C_1. At these points, we can, by a Nash isomorphism of $\mathbb{R}^n$, arrange that these points are non-singular on C_1 and that the tangent line to C_1 is mapped 1-1 by π near these points. Finally there may be points $P_1 \in C_1$, $P_2 \in C_2$ where $\pi(P_1) = \pi(P_2)$ but $P_1 \neq P_2$. Then we can also arrange that these are non-singular points on C_1 and C_2 and that the tangent lines are mapped to distinct lines in R^2. From all this it will follow that $\pi(C_1)$ is an analytic component of an algebraic curve distinct from $\pi(C_2)$ and so we can find $g \in N_R^2$ by Theorem 1 of [3], so that $g(\pi(C_1)) = 0$ and $g(\pi(C_2)) \neq 0$. Then $g(\ell_1,\ell_2)$ will be in P_{C_1} and not in P_{C_2}. So we have $P_{C_1} \neq P_{C_2}$.

In the case of the C_j' where we have one real support point P, we just choose π so that it is a local isomorphism near P and then apply the results in [2].

PROPOSITION 1. Let $C = C_1$ as above. Then P_C can be generated by ≤ 3 elements.

PROOF. From the above argument, we take $g_1 = g$ and then note that since at all but a finite number of points on C, π_1 is a local isomorphism $S \longrightarrow R^2$, since g_1 generates $P_{\pi_1(C)}$, $g(\ell_1,\ell_2)$ will generate P_C except at possibly this finite number of points. So we now choose another π_2 so that π_2 is a local isomorphism at these points and so that $\pi_1^{-1}(\pi_1(C)) \cap \pi_2^{-1}(\pi_2(C)) = C \cup$ a finite number of points. This will happen generically, so can be arranged. Then choose π_3 so that $\pi_3^{-1}(\pi_3(C))$ misses these points and we will have $(g_1,g_2,g_3) = P_C$ since (g_1,g_2) generates P_C everywhere on C and there are no points in the support of (g_1,g_2,g_3) of C.

PROPOSITION 2. If $C = C'$ as above is a complex curve on $S_{\mathbb{C}}$, the complexification of S, and C has real support one point, then $P_C = (g_1^2 + g_2^2 + g_3^2 + g_4^2)$ for suitable g_i in N_S.

PROOF. Choosing π_1 a local isomorphism at P on $S_{\mathbb{C}}$ we find that there are elements h_1, h_2 in N_R^2 so that $P_{\mathbb{C}\pi_1(C)} = (h_1 + ih_2)$. Let $g_i = h_i(\ell_1,\ell_2)$. Then choose π_2 also a local isomorphism at P and so that $\pi_2^{-1}(\pi_2(P)) \cap \pi_1^{-1}(\pi_1(P)) = P$. Then choosing g_3 and g_4 like g_1 and g_2, we can show that $g_1^2 + g_2^2 + g_3^2 + g_4^2$ generates P_C. For the only common zero of $g_1,\ldots,g_4$ on S is P, and since both $g_1 + ig_2$ and $g_3 + ig_4$ are local generators of $P_{\mathbb{C},C}$ at P, we see that $g_3 + ig_4 = (g_1 + ig_2)(u_1 + iu_2)$ so that $\sum_{i=1}^{4} g_i^2 = (g_1^2 + g_2^2)(1 + u_1^2 + u_2^2)$ and so also generates the ideal P_C at P.

THEOREM 2. Let S be as above. Then if $f \in N_S$ has $f \geq 0$ on N_S, then f can be represented as the sum of 4 squares in N_S.

PROOF. Let $(f) = \prod_{i,j} P_{C_i} P_{C'_j}$ with the C_i real, and the C'_j complex with one point real support. Then as in [4], it is easy to see that each P_{C_i} must occur to an even power in the product. For let t be a local parameter for C_1 at a point P where P is in the support of C_1 and no other C_j, and then $f = ut^m$ near P where u is a unit at P. Then unless m is even f will change sign at P contradicting $f \geq 0$. But m is the power of P_{C_1} that appears in the product representation of (f). Then, as Bochnak has noted, it is easy to see that if $P_C = (g_1,\ldots,g_r)$ then $P_C^2 = (g_1^2 + \ldots + g_r^2)$. For one of the g_i, say g_1, generates P_C at any given point P of C. So $g_i = u_i g_1$ for u_i a unit at P, $i > 1$. And as above, $\sum_{i=1}^{r} g_i^2 = g_1^2(1 + \sum_{i=1}^{r} u_i^2)$ implies what we want, since this can be done for some g_i at each $P \in C$. Now use Proposition 2, together with the fact that in any commutative ring if f_1 and f_2 can be represented as a sum of 4 squares, so can $f_1 f_2$, and we are done.

SECTION 2: GENERAL S.

We will be more informal in this section since the result itself is rather informal. The intent is to show that for most cases one would think of, the restrictions on S in Section 1 are unnecessary.

So consider the case where S is a non-singular real semi-algebraic surface in R^n. As can be seen, the problem with extending the results of Section 1 is in making sure that if C is a real component of an algebraic curve on S with ends P_1, P_2 on $\overline{S}\backslash S$, then we can send P_1 and P_2 to infinity by a map of R^n to itself which is a Nash isomorphism on S. Then the method of Section 1 will apply. It is easy to send one of P_1, P_2 to infinity. But to proceed we will assume that for every such P_1, P_2 on $\overline{S}\backslash S$, there exists an arc D from P_1 to P_2 such that $D \subseteq R^n\backslash S$, and $D = \cup C_i$ where each C_i is a connected piece of a curve D_i such that there exist Nash isomorphisms $\phi_i: R^n \longrightarrow R^n$ taking D_i to a line. To see that this is sufficient, we first send P_1 to infinity. Then let C_1 be the part of D going to infinity. We map D_1 to the x_n-axis so that C_1 is mapped to the negative x_n-axis by a Nash isomorphism of R^n with itself (which we assumed exists). Then use the map:

$$x_i \longrightarrow x_i/\lambda\sqrt{2}, \quad i < n, \quad x_n \longrightarrow \lambda/\sqrt{2},$$

where

$$\lambda = x_n^2 + \sqrt{\sum_{i=1}^{r} x_i^2}.$$

This map is a Nash isomorphism except on the negative x_n-axis and maps the rest of R^n to the set $x_n > 0$. Then we take the sequence of maps

$$x_n \longrightarrow x_n/\sqrt{1 + x_n^2} = y \quad \to 2y - 1 = z \longrightarrow z/\sqrt{1 - z^2} = x_n'.$$

This sends the positive x_n-axis to the whole x_n'-axis sending $0 \longrightarrow \infty$ and so induces an isomorphism of $x_n > 0$ with R^n. Moreover, the image of C_2, the next piece of D, will now have one end at infinity and we can proceed by induction until P_2 is at infinity.

SECTION 3: COMMENTS AND AN EXAMPLE

J. Bochnak and J. J. Risler [2] have proved results like those considered here for the case of real analytic functions on a real analytic surface S. Their proofs are different because they can use local data to construct global functions and we can't do that until we prove Theorem 1. Nevertheless, there are several ways that their methods can be used and we would like to take that up now. First a definition.

DEFINITION. For S a real surface as previously, let $q(S)$ = least integer k for which every f in $N(S)$ with $f \geq 0$ on S, can be written as a sum of k squares of elements in $N(S)$. Let $p(S)$ be the number similarily defined for analytic functions.

Theorem 2 states that $q(S) \leq 4$ for non-singular surfaces considered there. Bochnak and Risler, loc. cit. prove $p(S) \leq 4$ for certain analytic manifolds S (although they only state $p(S) \leq 7$). For surfaces S with trivial class group, which condition is equivalent to $H^1(S,Z_2) = 0$, they show $p(S) \leq 3$ and for S^2, the two sphere, they show $p(S^2) = 3$. It is easy to see that their proofs can be adapted to the Nash case and we sketch this now.

First an example to show that $p(S^2) \geq 3$.

EXAMPLE. Let $S^2 = \{(x,y,z) \mid x^2 + y^2 + z^2 = 1\}$. Let $f = x^2 + y^2 + (z-1)^2$ be a function on S^2. Then we claim that f can't be written as a sum of two squares f_1, f_2 in $N(S^2)$. For if this representation exists, then at $(0,0,1) = P$, we find that x and y are local coordinates and that $f = 2(1-/\sqrt{1-x^2-y^2})$ in these coordinates. So in local power series, $f = x^2 + y^2 +$ higher degree terms. Now, if $f_1 = 0$ and $f_2 = 0$ intersect at P, then they must intersect at some other point which would be a zero of f. But if $f_1 = 0$ and $f_2 = 0$ are tangent at P, and the tangent line were $t(x,y) = 0$ in local coordinates, then $t^2 = x^2 + y^2$ which is impossible. Note that similar examples should work more generally.

To see that $g(S) \leq 3$, for surfaces S with trivial class group, we refer back to our proof of Theorem 2. There we represented $(f) = \prod_{i,j} P_{C_i} P_{C'_j}$ where each P_{C_i} must occur to an even power. Then since we have trivial class group $\prod_i P_{C_i} = (h^2)$ for some $h \in N_S$. Since $h^2 \mid f$, there exists ϕ in N_S with $\phi h^2 = f$, and ϕ has only a finite number of points $P_1,\ldots,P_r$ as real support. Now choose our linear projection π so that it is a local isomorphism at each P_i and so no two P_i are mapped to the same point. Then, as in our proof, we find for each i, g_{i1} and g_{i2} so that $g_{i1}^2 + g_{i2}^2$ generates $P_{C'_i}$ locally at P_i. But then $\prod_i (g_{i1}^2 + g_{i2}^2) = h_1^2 + h_2^2$ will locally generate $P_{C'_i}$ at each P_i. Next, take $h_1^2 + h_2^2 + \phi^2$ and it will vanish only at $P_1,\ldots,P_r$ and still locally generate each $P_{C'_i}$. Finally, we see that $f = h^2(h_1^2 + h_2^2 + \phi^2)$ will be our representation of f as a sum of three squares. Again it should be noted that this proof is just an adaptation of that in [2].

BIBLIOGRAPHY

1. Bochnak, J., and Efroymson, G., Real Algebraic Geometry and the Hilbert 17th Problem, Math. Ann. 251, 213-241 (1980).

2. Bochnak, J., and Risler, J. J., Le théorème des zéros pour les variétés analytique réeles de dimension 2. Ann. Sci. École Norm Sup. 8, 353-364 (1975).

3. Efroymson, G., Nash Rings on Planar Domains, Trans. A.M.S. 249 (2), 435-445 (1979).

4. Efroymson, G., Sums of Squares on Planar Nash Rings, Pac. J. Math., to appear.

DEPARTMENT OF MATHEMATICS
UNIVERSITY OF NEW MEXICO
ALBUQUERQUE, NEW MEXICO 87131

Contemporary Mathematics
Volume 8, 1982

EXTENSION OF AN ORDER TO A SIMPLE TRANSCENDENTAL EXTENSION

Robert Gilmer[1]

ABSTRACT. If $A \cup B$ is a partition of the ordered field K such that $A < B$, then we show that the number of extensions of the order on K to a simple transcendental extension $K(t)$, with $A < t < B$, is $|S| + 1$, where S is the set of elements θ of the real closure of K such that $A < \theta < B$.

Let $(K, +, \cdot, \leq)$ be an ordered field. Let t denote an indeterminate over K, and let $(K^*, +, \cdot, \leq)$ be the real closure of K. If A and B are subsets of K, we write $A < B$ if $a < b$ for all $a \in A$, $b \in B$; if $c \in K$, then we write $A < c$ or $c < B$ instead of $A < \{c\}$ or $\{c\} < B$, respectively. A pair (A,B) of subsets of K is called a _gap_ in K if $K = A \cup B$ and $A < B$. A gap (A,B) in K is said to be _filled_ in an ordered extension field $(L, +, \cdot, \leq)$ of K if there exists $x \in L$ such that $A < x < B$. A combination of Lemma 13.12 of [2] and Exercises 15 and 16 of [1] yields the following result.

THEOREM 1. _If K is real closed, then for each gap (A,B) in K, there exists a unique extension of the order on K to $K(t)$ so that $A < t < B$._

Lemma 13.12 of [2] motivated us to raise the following questions, labelled as (Q1) and (Q2).

(Q1) _Given a gap (A,B) in K, can the order on K be extended to $K(t)$ in such a way that $A < t < B$?_

(Q2) _If the answer to (Q1) is affirmative, then in how many ways can the order be extended?_

Question (Q1) has an affirmative answer. Szczerba establishes this result in [7], and in [6], Scott states the following more general result.

THEOREM 2. _If $\{(A_\lambda, B_\lambda)\}$ is a family of gaps in K, then there exists an ordered extension field L of K such that each of the gaps (A_λ, B_λ) is filled in L._

1980 Mathematics Subject Classification. 12J15.

[1]Supported by NSF Grant 7903123.

In view of Theorem 2, we turn to a consideration of (Q2). It is well known that there is a unique extension of the order on K to $K(t)$ in which t is infinitely large — that is, $t > K$. Under this ordering, a polynomial $f_0 + f_1 t + \ldots + f_n t^n \in K[t]$ with $f_n \neq 0$ is positive if an only if $f_n > 0$. A consideration of this order on $K(t)$ allows us to answer (Q2) in several cases.

THEOREM 3. <u>Fix</u> $a \in K$ <u>and denote by</u> $<$ <u>the order on</u> $K(t)$ <u>extending the order on</u> K <u>in which</u> t <u>is infinitely large</u>.

(1) $\{x \in K | x < (a + t^{-1})\} = \{x \in K | x \leq a\}$, <u>and</u> $\{x \in K | x < (a - t^{-1})\} = \{x \in K | x < a\}$.

(2) <u>If</u> $<_1$ <u>is an order on</u> $K(t)$ <u>extending the order on</u> K, <u>if</u> $\{x \in K | x <_1 t\} = \{x \in K | x \leq a\}$, <u>and if</u> $u \in K(t)$ <u>is determined by the equation</u> $a + u^{-1} = t$ (that is, $u = (t - a)^{-1}$), <u>then</u> u <u>is a field generator of</u> $K(t)$ <u>over</u> K <u>and</u> u <u>is infinitely large under</u> $<_1$.

(3) <u>If</u> $<_2$ <u>is an order on</u> $K(t)$ <u>extending the order on</u> K, <u>if</u> $\{x \in K | x <_2 t\} = \{x \in K | x < a\}$, <u>and if</u> $v \in K(t)$ <u>is determined by the equation</u> $a - v^{-1} = t$ (that is, $v = (a - t)^{-1}$), <u>then</u> v <u>is a field generator of</u> $K(t)$ <u>over</u> K <u>and</u> v <u>is infinitely large under</u> $<_2$.

We omit the routine verification of the details of the proof of Theorem 3. As an immediate corollary to Theorem 3, we obtain the following statement.

THEOREM 4. <u>Let</u> (A,B) <u>be a gap in</u> K. <u>If</u> A <u>or</u> B <u>is empty, if</u> A <u>has a largest element, or if</u> B <u>has a smallest element, then there exists a unique extension of the order on</u> K <u>to</u> $K(t)$ <u>in such a way that</u> $A < t < B$.

We remark that under the hypothesis of Theorem 4, there exists no element $\theta \in K^*$ so that $A < \theta < B$. For A or B empty, this statement is well known, and in the case where A has a largest element a or B has a smallest element b, it follows as in the proof of Theorem 3 that either $(\theta - a)^{-1}$ or $(b - \theta)^{-1}$ is infinitely large, and hence transcendental over K. We shall subsequently refer to this remark in stating Theorem 8.

Using Theorem 3, we proceed to outline a constructive proof of the result that (Q1) has an affirmative answer (neither [6] nor [7] provides a constructive proof). Thus, if B is empty, we take the ordering on $K(t)$ in which t is infinitely large; if A is empty, the ordering on $K(t)$ in which $-t$ is infinitely large satisfies the requirements of (Q1). If A has a largest element a, then we take the ordering on $K(t) = K((t - a)^{-1})$ in which $(t - a)^{-1}$ is infinitely large, and if B has a smallest element b, the ordering on $K(t)$ in which $(b - t)^{-1}$ is infinitely large works. The remaining case is where A and B are nonempty, A has no largest element and B has no smallest element. In this case, we define $P = \{f(t) \in K[t] | f(t)$ is nonnegative on an initial segment of $B\}$ and show that P induces an order on $K[t]$, hence on $K(t)$, with $A < t < B$. It is clear that $P + P \subseteq P$ and $PP \subseteq P$. Moreover, $P \cap (-P) = \{0\}$, for if $f \in P \cap (-P)$, then f is identically zero on an initial segment of B, and each such segment is infinite since B contains no smallest element. We observe that $P \cup (-P) = K[t]$, for if not, then there exists a nonzero element $f \in K[t]$ such that f is neither nonpositive nor nonnegative on an initial segment of B; consequently, we can find a decreasing sequence $b_0 > b_1 > b_2 > \ldots$ of elements of B such that $f(b_{2i}) > 0$ and $f(b_{2i+1}) < 0$ for each $i \geq 0$. It then follows from the intermediate value theorem for real closed fields [4, p. 278] that f has infinitely many roots in K^*, and hence $f = 0$. This contradiction shows that $P \cup (-P) = K[t]$. Thus, P determines an order on $K(t)$. If $u \in K$, it is clear that $u \in P$ if and only if u is nonnegative in the ordering on K, so the order on $K(t)$ induced by P is an extension of the order on K. Finally, since $t - a$ is positive on B for each a in A and $b - t$ is nonnegative on the initial segment of B determined by b for each b in B, it follows that $A < t < B$.

The remaining case of (Q2) is the same as the case considered above — that in which A and B are nonempty, A has no largest element, and B has no smallest element. Theorem 5 addresses this remaining case of (Q2). The proof of Theorem 5 uses the following lemma.

LEMMA 1. _Assume that_ $\theta \in K^*$, $\theta \notin K$. _There is an extension of the order on_ K _to_ $K(t)$ _such that_ $\{x \in K | x < t\} = \{x \in K | x < \theta\}$.

PROOF. We remark that in the statement of Lemma 1, the symbol $<$ is being used in two different senses: in $\{x \in K | x < t\}$, it represents the extended order relation on $K(t)$, and in $\{x \in K | x < \theta\}$, it refers to the order on K^*.

Let f be an irreducible polynomial over K such that $f(\theta) = 0$; note that f has degree at least 2. If $g \in K[t] - \{0\}$, then there is a unique integer k such that f^k divides g in $K[t]$, while f^{k+1} does not divide g in $K[t]$. We define g^* to be g/f^k; since f does not divide g^*, then $g^*(\theta)$ is a nonzero element of $K(\theta)$. Let $P = \{g \in K[t] | g = 0$, or $g \neq 0$ and $g^*(\theta) > 0\}$. We prove that P induces an order on $K[t]$. It is clear that $K[t] = P \cup (-P)$ and that $P \cap (-P) = \{0\}$. If $g,h \in P - \{0\}$, then let r and s be the integers such that $g^* = g/f^r$ and $h^* = h/f^s$, where $r \leq s$. If $r < s$, then $g + h = f^r(g^* + f^{s-r} h^*)$, where f does not divide $g^* + f^{s-r} h^*$; consequently, $(g + h)^* = g^* + f^{s-r} h^*$, $(g + h)^*(\theta) = g^*(\theta) > 0$, and $g + h \in P$. If $r = s$, then $g + h = f^r(g^* + h^*)$, where $(g^* + h^*)(\theta) = g^*(\theta) + h^*(\theta) > 0$. Therefore f does not divide $g^* + h^*$, $(g + h)^* = g^* + h^*$, and $g + h \in P$. This proves that $P + P \subseteq P$. Because f is irreducible, f does not divide g^*h^* so that the equality $gh = f^{r+s}g^*h^*$ implies that $(gh)^* = g^*h^*$. Thus $(gh)^*(\theta) = g^*(\theta)h^*(\theta) > 0$, $gh \in P$, and $PP \subseteq P$. Therefore P induces a total order on $K[t]$ and on $K(t)$. For an element $x \in K$, $(t - x)^* = t - x$ so that $t - x \in P$ if and only if $\theta - x > 0$ — that is, if and only if $\theta > x$. Consequently, $\{x \in K | x < t\} = \{x \in K | x < \theta\}$ as asserted. This completes the proof of Lemma 1.

THEOREM 5. <u>Assume that</u> (A,B) <u>is a gap in</u> K <u>such that</u> A <u>and</u> B <u>are nonempty</u>, A <u>has no largest element</u>, <u>and</u> B <u>has no smallest element</u>. <u>Let</u> $S = \{x \in K^* | A < x < B\}$.

(1) <u>If</u> S <u>is empty, then there exists a unique extension of the order on</u> K <u>to</u> $K(t)$ <u>so that</u> $A < t < B$.

(2) <u>If</u> S <u>is a singleton set</u>, <u>then there exist two distinct extensions of the</u> K<u>-order to</u> $K(t)$ <u>so that</u> $A < t < B$.

(3) <u>If</u> S <u>contains more than one element</u>, <u>then</u> S <u>is infinite and there are infinitely many extensions of the order on</u> K <u>to</u> $K(t)$ <u>so that</u> $A < t < B$.

<u>Proof</u>. To prove (1) and (2), consider an ordering $<_1$ on $K(t)$ that extends $<$. Let $(K(t)^*, +, \cdot, <_1)$ be a real closure of $(K(t), +, \cdot, <_1)$. The algebraic closure of K in $K(t)^*$ is a real closure of K; we denote it by K'. By Theorem 1, the induced order $<_1$ on $K'(t)$ is uniquely determined by the gap (A',B') in K', where $A' = \{x \in K' | x <_1 t\}$ and $B' = \{x \in K' | t <_1 x\}$. In case (1), $A' = \{x \in K' | x <_1 a$ for some $a \in A\}$ and $B' = \{x \in K' | b <_1 x$ for some $b \in B\}$. Moreover, since any two real closures of K are K-isomorphic, then the sets A' and B' are independent of the particular order $<_1$. It follows that A and B uniquely determine the ordering on $K'(t)$, and hence on $K(t)$, in case (1). In case (2) we retain the notation above and we let θ be the element of K' such that $A < \theta < B$. Either (2a) $A' = \{x \in K' | x \leq_1 \theta\}$ and $<_1$ is the ordering on $K'(t)$ in which $(t - \theta)^{-1}$ is infinitely large, or

(2b) $A' = \{x \in K' | x <_1 \theta\}$ and $<_1$ is the ordering on $K'(t)$ in which $(\theta - t)^{-1}$ is infinitely large. In case (2a), $A' = \{\theta\} \cup \{x \in K' | x <_1 a$ for some $a \in A\}$, and $B' = \{x \in K' | b <_1 x$ for some $b \in B\}$. Again these sets are independent of the particular order $<_1$ so that a unique induced order on $K(t)$ is determined in case (2a). Similar remarks apply in case (2b). Thus, in case (2), there are at most 2 distinct extensions of the order on K to $K(t)$. To see that there are two such extensions, we examine the proof of Lemma 1. If g is the minimal polynomial for θ over K, then taking $f = g$ in the proof of Lemma 1, we obtain an ordering $<_2$ on $K(t)$ extending the order on K such that $A <_2 t <_2 B$ and such that $0 <_2 g$. On the other hand, if $f = -g$ in the proof of Lemma 1, we obtain an order $<_3$ on $K(t)$ such that $A <_3 t <_3 B$ and such that $g <_3 0$.

In case (3), the set S is infinite and we can choose an infinite subset $\{\theta_i\}_{i=1}^{\infty}$ of S such that if f_i is the minimal polynomial for θ_i over K, then $f_i \neq f_j$ for $i \neq j$. We let $<_i$ be the ordering on $K(t)$ induced by considering θ_i and f_i in the proof of Lemma 1. For each i, we have $A <_i t <_i B$. To complete the proof, it suffices to show that $<_i$ and $<_j$ are distinct for $i \neq j$. Thus, choose $g,h \in K[t]$ so that $gf_i + hf_j = 1$ and let $k = gf_i - hf_j$. Neither f_i nor f_j divides k, so under $<_i$, the sign of k is determined by the sign of $g(\theta_i)f_i(\theta_i) - h(\theta_i)f_j(\theta_i)$ $= -h(\theta_i)f_j(\theta_i) = -1$, whereas the sign of k under $<_j$ is determined by the sign of $g(\theta_j)f_i(\theta_j) - h(\theta_j)f_j(\theta_j) = 1$; that is, $k <_i 0$ while $0 <_j k$. This completes the proof of Theorem 5.

It follows at once from Theorem 5 that a form of the converse of Theorem 1 is valid.

THEOREM 6. *If each gap (A,B) in K determines a unique extended order on $K(t)$ so that $A < t < B$, then K is real closed.*

Another consequence of Theorem 5 is the following result due to Hafner and Mazzola [3, Theorem 8].

THEOREM 7. *The following conditions are equivalent.*

(1) *K is dense in K^*.*

(2) *The number of different extensions to $K(t)$ of the ordering on K which induce the same ordering on $K \cup \{t\}$ is at most 2.*

Finally, a combination of Theorem 5 and the remark following Theorem 4 yields the following summary result stated in the abstract.

THEOREM 8. *Let (A,B) be a gap in K and let S be the set of elements θ of K^* such that $A < \theta < B$. The number of distinct orderings on $K(t)$ that extend the order on K and are such that $A < t < B$ is $|S| + 1$.*

BIBLIOGRAPHY

1. N. Bourbaki, Algebre, Chapitre VI, Groupes et corps ordonnes, Hermann, Paris, 1952.

2. L. Gillman and M. Jerison, Rings of continuous functions, Van Nostrand, Princeton, N. J., 1960.

3. P. Hafner and G. Mazzola, "The cofinal character of uniform spaces and ordered fields", Z. für Math. Logik und Grundlagend. Math. 17(1971), 377-384.

4. N. Jacobson, Lectures in abstract algebra, Vol. III, Van Nostrand, Princeton, N. J., 1964.

5. C. Massaza, "Sugli ordinamenti di un campo puramente transcendente di un campo ordinato", Rend. Mat. 1(1968), 202-218.

6. D. Scott, "On completing ordered fields", pp. 274-278 of Applications of model theory to algebra, analysis, and probability, Holt, Rinehart, and Winston, New York, 1969.

7. L. W. Szczerba, "Filling in gaps in ordered fields", Bull. Acad. Polon. Sci. Ser. Sci. Math. Astron. Phys. 18(1970), 349-352.

DEPARTMENT OF MATHEMATICS
FLORIDA STATE UNIVERSITY
TALLAHASSEE, FL 32306

Contemporary Mathematics
Volume 8, 1982

THEORIE DES MODELES ET FONCTIONS DEFINIES POSITIVES SUR LES VARIETES ALGEBRIQUES REELLES

Danielle Gondard

ABSTRACT. We give a short demonstration by a logic method of a Robinson's theorem on the decomposition of positive definite functions on an algebraic variety in sum of squares of rational functions on the variety.

Finally we give another example using the same method and the above results.

En 1900 à Paris, Hilbert invité à faire une conférence expose 23 problèms ouverts de mathématics [19].

Expliquons ce qu'était <u>le 17ème de ces problèmes</u>:

Soit $\mathbb{R}$ le corps des nombres réels, $\mathbb{R}[X_1, \ldots, X_n]$ l'anneau des polynômes à n variables et à coefficients dans $\mathbb{R}$, $\mathbb{R}(X_1, \ldots, X_n)$ son corps des fractions, c'est-à-dire le corps des fractions rationnelles à n variables et à coefficients dans $\mathbb{R}$.

Toute fraction rationnelle $f \in \mathbb{R}(X_1, \ldots, X_n)$ peut s'écrire sous la forme $f = \frac{f_1}{f_2}$ où f_1 et f_2 sont dans $\mathbb{R}[X_1, \ldots, X_n]$.

On dit <u>que</u> $f \in \mathbb{R}(X_1, \ldots, X_n)$ <u>est définie en</u> $(x_1, \ldots, x_n) \in \mathbb{R}^n$ si on peut écrire f sous la forme $f = \frac{f_1}{f_2}$ avec $f_2(x_1, \ldots, x_n) \neq 0$.

On dit <u>que</u> f <u>est définie positive</u> si en tout point $(x, \ldots, x_n) \in \mathbb{R}^n$ où f est définie $f(x_1, \ldots, x_n) \geq 0$.

Il est clair que toute somme de carrés $\sum_{i=1}^{p} f_i^2$ où $f_i \in \mathbb{R}(X_1, \ldots, X_n)$ est définie positive. Hilbert a cherché s'il existait d'autres éléments définis positifs dans $\mathbb{R}(X_1, \ldots, X_n)$.

1980 Mathematics Subject Classification 12D15, 12J15, 14G30.

Dans le cas $n = 0$, $\mathbb{R}(X_1, \ldots, X_n) = \mathbb{R}$ et f est définie positive si et seulement si f est positif dans $\mathbb{R}$, donc si et seulement si f est le carré d'un élément de $\mathbb{R}$.

Dans le cas $n = 1$, $f \in \mathbb{R}(X)$ est définie positive si et seulement si f est somme de deux carrés. On peut en faire une démonstration élémentaire que l'on trouvera dans [12] page 80 en utilisant la décomposition d'un polynôme en produit de facteurs irréductibles dans $\mathbb{R}(X)$ et l'égalité, vraie dans tout anneau commutatif:

$$(a^2 + b^2)(c^2 + d^2) = (ac - bd)^2 + (ad + bc)^2.$$

Dans le cas $n = 2$, l'article [20] de Hilbert et l'identité de Lagrange donnant le produit de deux sommes de quatre carrés comme somme de quatre carrés permettent de conclure que $f \in \mathbb{R}(X,Y)$ est définie positive si et seulement si f peut s'ecrire comme somme de quatre carrés.

Enonçons alors le 17ème problème de Hilbert dans le cas de n variables: soit $f \in \mathbb{R}(X_1, \ldots, X_n)$, définie positive; f est-elle somme de carrés de fractions rationnelles et si oui, de combien?

Le 17ème problème de Hilbert présente donc deux aspects, d'abord l'aspect qualitatif, et celui-ci étant résolu l'aspect quantitatif.

On peut alors se demander quels sont les corps K tels que l'on puisse se poser le même problème pour $K(X_1, \ldots, X_n)$.

Il est clair que pour pouvoir parler de positivité de $f(X_1, \ldots, X_n)$ il faut que le corps K soit ordonné.

Si le corps K est non commutatif, l'ensemble des éléments positifs de K contient l'ensemble des produits de carrés. Il existerait donc des éléments positifs de K non sommes de carrés et pour poser un problème analogue il faudrait considérer autre chose que les sommes de carrés.

Si K est un corps commutatif ordonné mais que K soit ordonnable de plusieurs façons distinctes, alors il existerait des éléments positifs pour l'ordre donné sur K mais négatifs dans un autre ordre. De tels éléments ne sont donc pas totalement positifs dans K et ne pourront donc pas s'exprimer comme somme de carrés d'éléments de K.

Cependant en modifiant de façon convenable la notion de fonction définie positive j'ai pu obtenir dans [13] le théorème suivant:

"Soit K un corps ordonnable tel que pour tout ordre de K, K soit dense (au sens de la topologie définie par les intervalles) dans sa clôture réelle, alors toute fonction $f \in K(X_1, \ldots, X_n)$ totalement définie positive (i.e., définie positive pour tout ordre de K) est somme de carrés d'éléments de $K(X_1, \ldots, X_n)$."

Finalement nous pouvons énoncer de manière générale le 17ème problème de Hilbert, sous la forme suivante:

Soit K un corps commutatif, ordonné mais ordonnable d'une seule manière, comme $\mathbb{Q}$, $\mathbb{R}$ et les corps ordonnés maximaux par exemple, soit $f \in K(X_1, \ldots, X_n)$ définie positive. Peut-on écrire f comme une somme finie de carrés d'éléments de $K(X_1, \ldots, X_n)$, et si oui, peut-on trouver $pf(K(X_1, \ldots, X_n))$ le plus petit entier (que nous appelerons constante de pfister) tel que toute fraction rationnelle de $K(X_1, \ldots, X_n)$ définie positive s'écrive sous la forme d'une somme de $pf(K(X_1, \ldots, X_n))$ carrés d'éléments de $K(X_1, \ldots, X_n)$?

Remarquons tout de suite que pour étudier ce problème on peut se réduire au cas $f \in K[X_1, \ldots, X_n]$ positive sur K^n, car si $f = \frac{f_1}{f_2}$ est définie positive alors $f_2^2 f = f_1 \cdot f_2 \in K[X_1, \ldots, X_n]$ est positive. Si on peut décomposer $f_1 \cdot f_2$ en some de carrés dans $K(X_1, \ldots, X_n)$ on pourra décomposer $\frac{f_1}{f_2}$ en divisant tous les carrés par f_2^2.

I - RAPPELS DE LOGIQUE ET EXEMPLE D'APPLICATION

Pour exprimer des énoncés de la théorie des corps commutatifs ordonnés maximaux on utilisera un langage L formé de:

0 et 1	comme symboles de constante
+ et $\cdot$	comme symboles fonctionnels à 2 variables
-	comme symbole fonctionnel à 1 variable
= et > 0	comme symboles relationnels.

Dans ce langage L on écrit le système A des axiomes de la théorie des corps commutatifs ordonnés maximaux

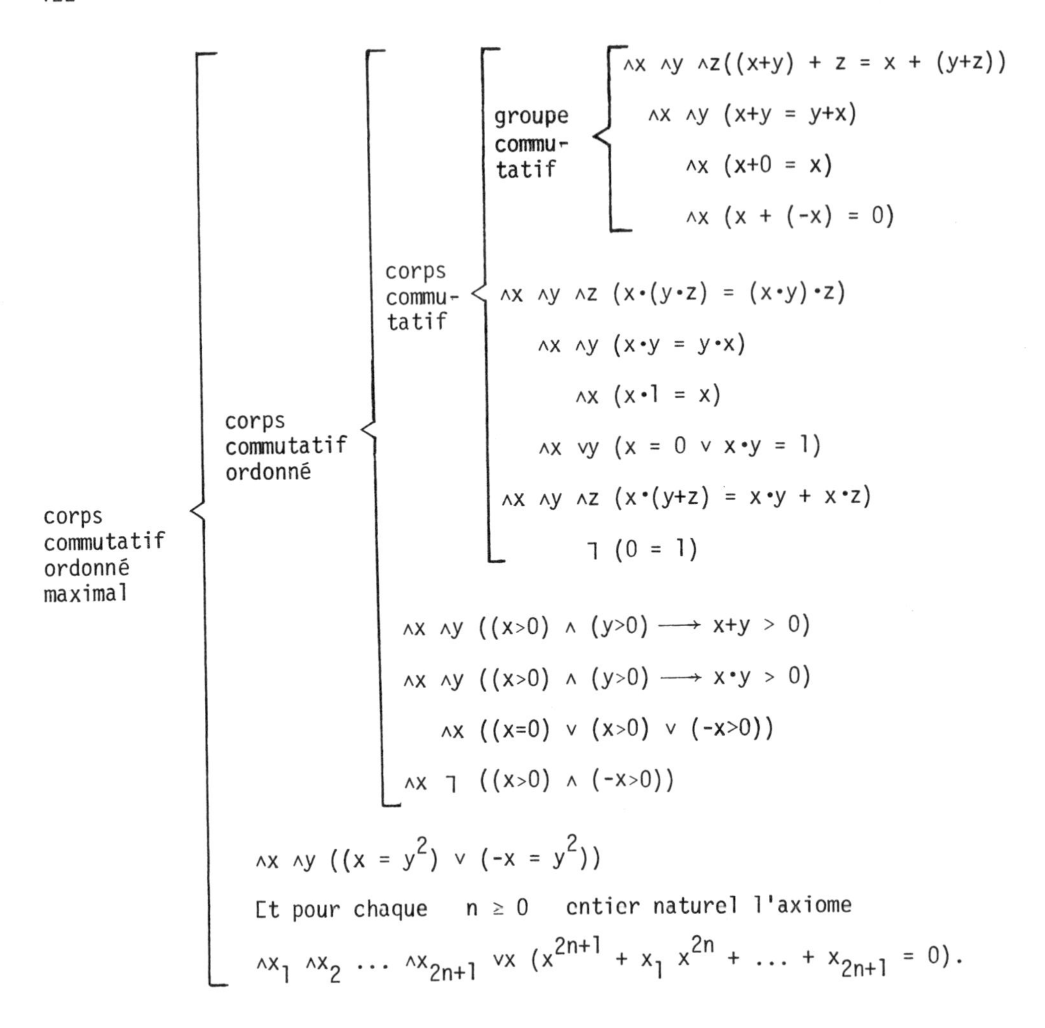

On sait que A permet l'élimination des quantificateurs. C'est-à-dire que pour toute formule F du langage L il existe une formule F', sans quantificateurs, du même langage L telle que $F \longleftrightarrow F'$ soit une conséquence de A.

Soit alors K un corps ordonné maximal. Soit L' le langage obtenu à partir de L en ajoutant aux symboles de constante des symboles correspondant aux éléments de K. Soit D_K le diagramme de K c'est-à-dire les énoncés qui lient les symboles de constante de L' (c'est-à-dire correspondant aux relations entre les éléments de K).

Soit A' le système d'axiomes du langage L' formé des formules de A et de celles de D_K. Alors les modèles de A' sont les corps ordonnés maximaux L qui sont extensions ordonnées de K.

L'élimination des quantificateurs pour le système A du langage L entraîne que le système d'axiomes A du langage L' est saturé, donc que pour toute formule close F de L', F ou $\neg F$ est une conséquence de A'.

Finalement si une formule close F est satisfaite dans un modèle de A elle le sera donc dans tous les modèles de A.

Nous utiliserons cette conclusion sous la forme suivante:

"Si une formule close F du langage L' est vraie dans le corps ordonné maximal K, elle est vraie dans tout corps ordonné maximal extension ordonnée de K".

THEOREME ET DEMONSTRATION: Nous allons démontrer le théorème suivant:

Soit K un corps ordonné maximal, soit $f \in K(X_1, \ldots, X_n)$ définie positive sur K. Alors f est somme de carrés d'éléments de $K(X_1, \ldots, X_n)$.

Par une remarque déjà faite, il suffit de démontrer le théorème pour $f \in K[X_1, \ldots, X_n]$.

L'hypothèse que $f \in K[X_1, \ldots, X_n]$ est (définie) positive sur K s'écrit.

$$\wedge x_1 \wedge x_2 \ldots \wedge x_n \, (f(x_1, \ldots, x_n) \geq 0).$$

Soit donc F cette formule close du langage L' qui est donc vraie sur K par hypothèse.

F sera donc vraie dans tout corps ordonné maximal extension ordonnée de K.

Considérons le corps $K(X_1, \ldots, X_n)$ muni d'un ordre arbitraire qui en fasse un corps ordonné. Il est bien clair que le corps $K(X_1, \ldots, X_n)$ est extension ordonnée de K. Soit alors $\overline{K(X_1, \ldots, X_n)}$ la clôture réelle de $K(X_1, \ldots, X_n)$. $\overline{K(X_1, \ldots, X_n)}$ est un corps ordonné maximal (bien sûr), extension ordonnée de K (il contient K et K n'a qu'un seul ordre), donc $\overline{K(X_1, \ldots, X_n)}$ est un modèle de A'.

La formule F vraie dans K est donc aussi vraie dans $\overline{K(X_1, \ldots, X_n)}$.

En choisissant $x_1 = X_1, \ldots, x_n = X_n$ nous obtenons donc que:

$$f(X_1, \ldots, X_n) \geq 0 \quad \text{dans} \quad \overline{K(X_1, \ldots, X_n)}.$$

Mais $f(X_1, \ldots, X_n) \in K(X_1, \ldots, X_n)$ et $\overline{K(X_1, \ldots, X_n)}$ est extension ordonnée de $K(X_1, \ldots, X_n)$ donc on a aussi $f(X_1, \ldots, X_n) \geq 0$ dans $K(X_1, \ldots, X_n)$.

Cette démonstration de $f(X_1, \ldots, X_n) \geq 0$ dans $K(X_1, \ldots, X_n)$ pouvant être faite pour tous les ordres possibles de $K(X_1, \ldots, X_n)$, alors $f(X_1, \ldots, X_n)$ est un élément totalement positif de $K(X_1, \ldots, X_n)$ et donc $f(X_1, \ldots, X_n)$ est une somme de carrés d'éléments de $K(X_1, \ldots, X_n)$.

II - FONCTIONS DEFINIES POSITIVES SUR LES VARIETES REELLES [16]

a) NOTATIONS ET DEFINITIONS: Soit K un corps ordonné maximal.

Soit $V \neq \emptyset$ une variété algébrique réelle irréductible de K^n, avec $n \geq 1$. V est l'ensemble des éléments de K^n qui annulent un nombre fini d'équations polynomiales. Ces polynômes engendrent un idéal $I(V)$ appelé idéal de la variété. $I(V)$ est un idéal premier réel [9].

Rappelons qu'un idéal I d'un anneau A est réel si est seulement si: $f_1^2 + \ldots + f_p^2 \in I$, où $f_i \in A$, entraîne $f_i \in I$.

Notons $K[V] = K[X_1, \ldots, X_n]/I(V)$. $K[V]$ est un anneau intègre réel et son corps des fractions, que nous noterons $K(V)$, est une extension ordonnable de K.

Si $f \in K[V]$, soit $F \in K[X_1, \ldots, X_n]$ telle que $f = \pi(F)$ où π désigne l'homomorphisme canonique de $K[X_1, \ldots, X_n]$ dans $K[V]$.

A f on peut associer la fonction polynomiale $\tilde{f}$ définie par: $\tilde{f} : V \longrightarrow K$, et est telle que $\tilde{f}(x_1, \ldots, x_n) = F(x_1, \ldots, x_n)$.

Autrement dit, $\tilde{f}$ est la restriction à $V \subset K^n$ de la fonction polynomiale associée à F et définie sur K^n. $\tilde{f}$ est indépendante du choix de F tel que $f = \pi(F)$, et l'application $f \longrightarrow \tilde{f}$ étant injective on peut identifier chaque $f \in K[V]$ à la fonction $\tilde{f}$ correspondante.

En conclusion $f \in K[V]$ peut être considérée comme la restriction à V de la fonction polynomiale associée à $F \in K[X_1, \ldots, X_n]$ où F est telle que $f = \pi(F)$.

On dit que $f \in K[V]$ est positive sur V lorsque $f(x_1, \ldots, x_n) \geq 0$ pour tout $(x_1, \ldots, x_n) \in V$.

On dit que $f \in K(V)$ <u>est définie en</u> $(x_1, \ldots, x_n) \in V$ si on peut écrire $f = \frac{f_1}{f_2}$ avec f_1 et f_2 dans $K[V]$ et $f_2(x_1, \ldots, x_n) \neq 0$.

Dans ce cas on pose $f(x_1, \ldots, x_n) = \frac{f_1(x_1, \ldots, x_n)}{f_2(x_1, \ldots, x_n)}$ ce qui est indépendant du choix de f_1 et f_2.

On dira alors que $f \in K(V)$ <u>est définie positive</u> lorsque $f(x_1, \ldots, x_n) \geq 0$ pour tout $(x_1, \ldots, x_n) \in V$ où f est définie.

b) THEOREME ET DEMONSTRATION: Dans [16] nous avons démontré le théorème suivant:

<u>Si</u> $f \in K(V)$ <u>est définie positive sur</u> V, <u>il existe</u> $h_1, \ldots, h_s$ <u>dans</u> $K(V)$ <u>telles que</u> $f = h_1^2 + \ldots + h_s^2$.

Remarquons tout d'abord qu'il suffit de faire la démonstration pour $f \in K[V]$, car si $f = \frac{f_1}{f_2}$ est définie positive sur V, alors $f_2^2\, f$ est positive sur V.

<u>Soient</u> $G_1, \ldots, G_r$ les générateurs de l'idéal $I(V)$ et <u>soit</u> $F \in K[X_1, \ldots, X_n]$ telle que $f = \pi(F)$.

L'hypothèse que f est définie positive sur V s'exprime de la manière suivante:

$$\wedge x_1 \wedge x_2 \ldots \wedge x_n \, (G_1(x_1, \ldots, x_n) = 0 \wedge \ldots \wedge G_r\,(x_1, \ldots, x_n) = 0 \rightarrow F(x_1, \ldots, x_n) \geq 0).$$

En reprenant ici les mêmes notations qu'au I nous aurons:

Soit H cette formule close du langage L'. La formule H est satisfaite dans le modèle K du système d'axiomes A', donc elle est vraie dans tout corps ordonné maximal qui soit extension ordonnée de K.

Choisissons un ordre quelconque sur $K(V)$, soit $\overline{K(V)}$ la clôture réelle de $K(V)$ ainsi ordonné. L'ordre de $\overline{K(V)}$ induit nécessairement celui de K qui est unique, donc $\overline{K(V)}$ est un corps ordonné maximal qui est extension ordonnée de K. $\overline{K(V)}$ est donc un modèle de A.

La formule H est donc vraie dans $\overline{K(V)}$.

Posons pour i de 1 à n, $\xi_i = \pi(X_i)$, $\xi_i \in K(V)$.

Calculons $G_j(\xi_1, \ldots, \xi_n) = G_j(\pi(X_1), \ldots, \pi(X_n)) = \pi G_j(X_1, \ldots, X_n)$ et $\pi G_j = 0$ car $G_j \in I(V)$.

Les ξ_i sont donc tels que pour tout j, $G_j(\xi_1, \ldots, \xi_n) = 0$.

La formule H vraie dans $\overline{K(V)}$ entraîne donc que $F(\xi_1, \ldots, \xi_n) \geq 0$ dans $\overline{K(V)}$, mais $F(\xi_1, \ldots, \xi_n)$ appartient à $K(V)$ donc $F(\xi_1, \ldots, \xi_n) \geq 0$ dans $K(V)$.

$F(\xi_1, \ldots, \xi_n) = F(\pi(X_1), \ldots, \pi(X_n)) = \pi F(X_1, \ldots, X_n) = f$. Donc $f \in K(V)$ est positive dans $K(V)$.

On peut démontrer que $f \in K(V)$ est positive dans $K(V)$ pour tout ordre de $K(V)$ donc f est totalement positive dans $K(V)$ et f est somme de carrés dans $K(V)$.

c) REMARQUE IMPORTANT: On peut se poser la <u>question</u> de savoir si <u>pour</u> $f \in K[V]$ <u>définie positive sur</u> V, il <u>existe</u> $F \in K[X_1, \ldots, X_n]$ <u>positive sur</u> K^n <u>et telle que</u> $f = \pi(F)$, ce qui entraînerait que le théorème précédent serait conséquence du 17ème problème de Hilbert.

La réponse à cette question est négative en général comme je l'ai montré dans [15].

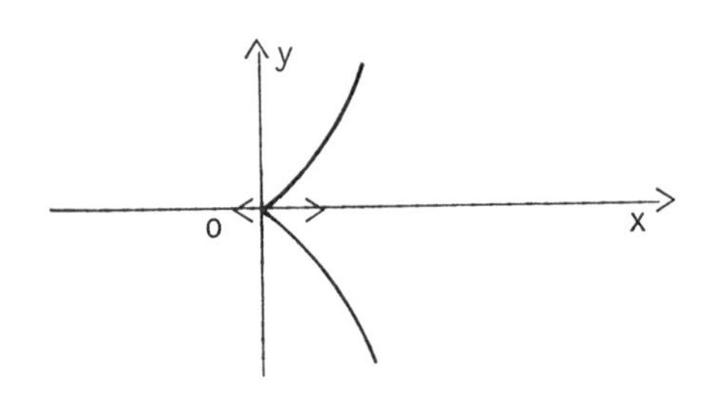

Soit en effet V la courbe de $\mathbb{R}^2$ définie par $X = t^2$, $Y = t^3$.

L'ensemble des polynômes s'annulant sur V, $I(V)$, est l'idéal engendré par $Y^2 - X^3$.

Soit $f \in \mathbb{R}[V]$, définie par $f(x,y) = x$ pour tout $(x,y) \in V$. f est donc bien positive sur V. Toutefois il n'existe aucun polynôme $F \in \mathbb{R}[X,Y]$ tel que $\pi(F) = f$ et F positif sur $\mathbb{R}^2$.

En effet on pourrait écrire F sous la forme:

$$F = X + (X^3 - Y^2)G \quad \text{avec} \quad G \in \mathbb{R}[X,Y].$$

Alors $F(x,0) = x + x^3 G(x,0)$ a le même signe que x quand x tend vers zéro. Donc F ne peut pas être positif sur $\mathbb{R}^2$ quelque soit le choix de G.

d) PROBLEME DE LA RECIPROQUE: On pouvait espérer que toute somme de carrés de $K(V)$ était une fonction définie positive sur V, mais ceci n'est pas vrai comme le montre le contre exemple suivant que m'a proposé Van Den Dries.

Soit I l'idéal de $\mathbb{R}[X,Y]$ engendré par $X^3 + X^2 + Y^2$. I est un idéal premier réel d'après un théorème de [32] ((h) page 30) car c'est un idéal principal engendré par un polynôme qui change de signe dans $\mathbb{R}^2$.

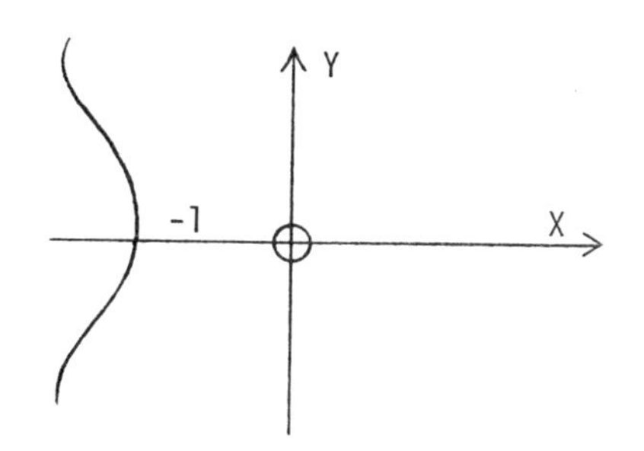

La variété V d'idéal I possède $(0,0)$ comme point isolé et peut être représenté comme ci-contre

Soit alors
$f(X,Y) = X^2 + Y^2 - \frac{1}{4}$.

Cette fonction est positive sur V sauf au point $(0,0)$ car
$X^2 + (- X^3 - X^2) - \frac{1}{4} = -X^3 - \frac{1}{4}$ et si $x \leq -1$ ceci est bien positif.

Soit maintenant $g(X,Y) = X^2(X^2 + Y^2 - \frac{1}{4})$, cette fonction est, elle, positive sur V tout entière, donc $\overline{g(X,Y)}$ est somme de carrés dans $K(V)$ et donc $\overline{f(X,Y)}$ l'est aussi en divisant tout par $\overline{X}^2$ qui n'est pas nul car $X^2 \notin I(V)$.

$\overline{f(X,Y)}$ <u>est donc une somme de carrés dans</u> $K(V)$ <u>mais n'est pas définie positive</u>.

e) ASPECT QUANTITATIF: Donnons ici quelques résultats que nous avons obtenus dans [16]:

Si la variété V est de dimension d, alors l'extension $K(V)$ de K est une extension de degré de transcendance d et $K(V) = L(\alpha)$, avec α algébrique sur L et $L = K(\xi_1, \ldots, \xi_d)$ où $\{\xi_1, \ldots, \xi_d\}$ désigne une base de transcendance.

THEOREME 1: <u>Si</u> V <u>est de dimension</u> d, <u>alors toute fonction</u> $f \in K(V)$ <u>définie positive sur</u> V <u>est la somme d'au plus</u> 2^d <u>carrés dans</u> $K(V)$.

Ceci résulte du lemme de Pfister suivant: [29].

LEMME: Soit K ordonné maximal, L/K une extension de degré de transcendance $d \geq 0$.

Soit $\theta = (1,a_1) \otimes (1,a_2) \otimes \ldots \otimes (1,a_d)$ une forme quadratique non singulière de dimension 2^d sur L.

Soit $b \in L$ un élément totalement positif, alors θ représente b sur L.

On sait que toute forme quadratique ϕ est équivalente à une <u>forme diagonale</u> ψ, définissable par

$$\psi(x) = \sum_{i=1}^{m} a_i x_i^2, \qquad \psi \text{ sera alors noté } \quad \psi = (a_1, \ldots, a_m).$$

ψ sera dite <u>non singulière</u> si $\psi = (a_1, \ldots, a_m)$ et que pour tout i, $a_i \neq 0$.

Si $\psi = (a_1, \ldots, a_m)$ et $\phi = (b_1, \ldots, b_p)$ alors

$$\psi \otimes \phi = (\ldots, a_i b_j, \ldots)_{\substack{i=1,\ldots,m \\ j=1,\ldots,p}}.$$

Enfin on dit que ϕ <u>de dimension</u> m <u>représente un élément</u> b <u>de</u> L <u>sur</u> L, s'il existe un vecteur u de L^m tel que $\phi(u) = b$.

La théorème résulte alors immédiatement du lemme;

prenons $L = K(V)$ et $a_1 = \ldots = a_d = 1$.

Soit f une fonction définie positive de $K(V)$, comme f est somme de carrés dans $K(V)$, alors f est totalement positive dans $K(V)$ et donc la forme quadratique $(1,1) \otimes \ldots \otimes (1,1)$ représente f dans $K(V)$. Donc f est somme de 2^d carrés dans $K(V)$.

Il résulte du théorème 1 qu'il existe un plus petit entier $pf(K(V))$ tel que chaque fonction définie positive sur V soit la somme de $pf(K(V))$ carrés de fonctions de $K(V)$.

Le théorème 1 s'énonce alors

<u>Si</u> V <u>est de dimension</u> d, $pf(K(V)) \leq 2^d$.

THEOREME 2: <u>Soit</u> V <u>de dimension</u> d; $K(V) = L(\alpha)$ <u>avec</u> α <u>algébrique sur</u> $L = K(\xi_1, \ldots, \xi_d)$ <u>et</u> $\{\xi_1, \ldots, \xi_d\}$ <u>une base de transcendance</u>.

<u>Supposons que</u> α <u>soit algébrique de degré impair</u> m. <u>Alors</u>

$\eta = 1 + \xi_1^2 + \ldots + \xi_d^2$ <u>n'est pas la somme de</u> d <u>carrés dans</u> $K(V)$ <u>et donc</u> $pf(K(V)) \geq d+1$.

<u>Si $m = 1$</u>. C'est le théorème de Cassels car $K(V) = L$. Il se démontre par récurrence en utilisant le lemme de Cassels [3] suivant:

LEMME: Pour $\lambda \in K$ et $car(K) \neq 2$, $X^2 + \lambda$ est somme de $n > 1$ carrés dans $K(X)$ si et seulement si

ou -1 est somme de $n-1$ carrés dans K
où λ est somme de $n-1$ carrés dans K.

Par récurrence on vérifie que c'est vrai pour $d = 0$ et on suppose le résultat vrai jusqu'à $d-1$. Si on avait $1 + \xi_1^2 + \ldots + \xi_d^2$ somme de d carrés dans $K(\xi_1, \ldots, \xi_d)$ alors d'après le lemme on aurait l'une des deux conséquences:

(i) $1 + \xi_1^2 + \ldots + \xi_{d-1}^2$ est somme de $d-1$ carrés dans $K(\xi_1, \ldots, \xi_{d-1})$, ce qui est faux d'après l'hypothèse de récurrence.

(ii) -1 est somme de $d-1$ carrés dans $K(\xi_1, \ldots, \xi_{d-1})$, ce qui est faux car $K(\xi_1, \ldots, \xi_{d-1})$ est ordonnable.

ENSUITE ON PROCEDE PAR RECURRENCE: C'est vrai pour $m = 1$.

On suppose l'hypothèse vraie jusqu'à $m-2$ (Impair inférieur à m). Donc $\eta = 1 + \xi_1^2 + \ldots + \xi_d^2$ n'est pas la somme de d carrés dans $K(W)$ où $K(W) = L(\beta)$ avec β algébrique sur L de degré P impair avec $P \leq m-2$, et $K(W)$ est le corps des fractions rationnelles d'une variété irréductible de K^n de dimension d.

Supposons alors que η soit la somme de d carrés dans $L(\alpha)$, alors:

$$\eta = \sum_{i=1}^{d} f_i^2(\alpha) \quad \text{où} \quad f_1, \ldots, f_d \in L[X]$$

et sont de degrés inférieurs strictement à m.

Le polynôme de $L[X]$, $\eta - \sum_{i=1}^{d} f_i^2$ s'annule sur l'élément α et c'est donc un multiple du polynôme minimal g de degré m de α sur L. Donc:

$$\eta - \sum_{i=1}^{d} f_i^2 = gh \quad \text{avec} \quad g \quad \text{et} \quad h \quad \text{dans} \quad L[X]$$

et g de degré égal à m.

Le premier membre de cette égalité est de degré égal à $2 \max\{\deg f_i\}$, car la somme des coefficients des termes de ce degré ne peut s'annuler dans le corps ordonnable L. (Dans un corps ordonnable $\Sigma x_i^2 = 0$ entraîne $x_i = 0$ pour tout i).

Donc nous avons:

$$m + \deg h = 2 \max\{\deg f_i\} \leq 2 (m-1).$$

Comme m est impair alors $\deg h$ est aussi impair et puisque $m + \deg h \leq 2 (m-1)$ alors

$$\deg h \leq m-2.$$

Il existe donc $h' \in L[X]$, irréductible dans $L[X]$ avec h' divise h, h' de degré impair et bien sûr

$$\deg h' \leq \deg h \leq m-2.$$

Soit alors γ une racine de h' et soit $L(\gamma)$ l'extension algébrique correspondante donc de degré impair inférieur ou égal à $m-2$.

Alors $L(\gamma)$ est un corps ordonnable, extension de type fini de K et c'est le corps des fonctions ra tionnelles d'une variété irréductible W de K^n, de dimension d.

(Il suffit pour cela d'utiliser le travail de Dubois et Efroymson dans [9].) Donc d'après l'hypothèse de récurrence η n'est pas somme de d carrés dans $L(\gamma) = K(W)$.

Pourtant

$$\eta - \sum_{i=1}^{d} f_i(\gamma) = g(\gamma)\ h(\gamma) = g(\gamma)\ h'(\gamma)\ q(\gamma) = 0,$$

car γ est racine de h'.

η serait donc somme de d carrés dans $L(\gamma)$ a qui est impossible.

On a donc bien obtenu $pf(K(V)) \geq d+1$ pour tous les cas où $K(V) = L(\alpha)$ avec α algébrique de degré impair sur $L = K(\xi_1, \ldots, \xi_d)$.

THEOREME 3: E. Witt a démontré dans [37] en 1934 que $pf(K(V)) = 2$ si $d = 1$.

On peut donc former la conjecture suivante:

CONJECTURE: Si V est de dimension d, $pf(K(V)) \geq d+1$.

Cette conjecture est donc démontrée dans les cas des théorèmes 2 et 3. Elle peut l'être également dans certains autres cas comme celui de V sphère de dimension d dans $\mathbb{R}^{d+1}$.

III - UN EXEMPLE D'UTILISATION DES METHODES ET RESULTATS PRECEDENTS

Nous allons maintenant parler d'un problème analogue que nous avons traité dans [17].

a) DEFINITIONS ET LEMME: Soit K un corps, $n \geq 1$ un entier, $\mu_n(K)$ l'anneau des matrices carrées d'ordre n à coefficients dans K. On identifie K au sous corps de $\mu_n(K)$ formé des matrices scalaires.

Soit $S_n(K)$ le sous-ensemble de $\mu_n(K)$ formé des matrices symétriques (groupe abélien additif).

Soit $A \in \mu_n(K)$, on note $K[A]$ la sous algèbre de $\mu_n(K)$ engendréé par A, $K[A] = \{f(A);\ f \in K[X]\}$. Remarquons que si $A \in S_n(K)$ alors $K[A] \subset S_n(K)$.

Soit maintenant K un corps ordonné, notons p l'ensemble des positifs de K.

Soit $\overline{K}$ la clôture réelle de K.

Définissons alors p_n l'ensemble des matrices $A = (A_{ij})$ de $S_n(K)$ telles que la forme quadratique sur $\overline{K}$ associée à A soit positive c'est-à-dire telles que

$$\forall (u_1, \ldots, u_n) \in \overline{K}^n, \quad \sum_{i,j=i}^{n} A_{ij}\, u_i\, u_j \geq 0$$

$A \in S_n(K)$ sera dite <u>positive</u> si $A \in p_n$.

p_n est alors l'ensemble des éléments positifs pour un ordre partiel sur $S_n(K)$ compatible avec la structure de groupe additif de $S_n(K)$ et prolongeant l'ordre de K car nous avons:

$$p_n \cap K = p,\ p_n + p_n \subset p_n,\ p_n \cap (-p_n) = \{0\}.$$

<u>Soit</u> K <u>un corps ordonnable</u>

Une matrice A de $S_n(K)$ sera dite <u>naturellement positive</u> si elle est positive (pour l'ordre p_n sur $S_n(K)$ associé à l'ordre p sur K) pour tout ordre de K.

Il nous faudra dans la suite utiliser un théorème de R. Ciampi [5] légèrement amélioré dans [17] sous la forme du lemme ci-dessous:

LEMME: Soit K un corps ordonnable, soit $A \in S_n(K)$.

Les conditions suivantes sont équivalentes:

(1) A est naturellement positive

(2) A est somme de carrés de matrices de $K[A]$

(3) A est somme de carrés de matrices de $S_n(K)$.

b) PROBLEME ET THEOREME: Soit K un corps ordonné, $m \geq 0$ et $n \geq 1$ des entiers. On dit que la matrice $F = (F_{ij}) \in S_n(K(X_1, \ldots, X_m))$ est définie en $(x_1, \ldots, x_m) \in K^m$ lorsque pour tout i et tout j de 1 à n on peut écrire $F_{ij} = \frac{G_{ij}}{H_{ij}}$, G_{ij} et H_{ij} étant dans $K[X_1, \ldots X_m]$ et $H_{ij}(x_1, \ldots, x_m) \neq 0$.

Soit $F \in S_n(K(X_1, \ldots, X_m))$ une matrice definie en $(x_1, \ldots, x_m)$.
On dit que F est positive en $(x_1, \ldots, x_m) \in K^m$ lorsque la matrice $F(x_1, \ldots, x_m) = (F_{ij}(x_i, \ldots, x_m))$ de $S_n(K)$ est une matrice positive.

On dira alors que $F \in S_n(K(X_1, \ldots, X_m))$ est définie positive lorsque F est positive en chaque point $(x_1, \ldots, x_m) \in K^m$ où F est définie.

Dans [17] nous avons d-montré le théorème suivant:

THEOREME: Soit R un corps ordonné maximal. Une matrice $F \in S_n(R(X_1, \ldots, X_m))$ est définie positive si et seulement si F est somme de carrés de matrices appartenant à $S_n(R(X_1, \ldots, X_m))$.

DEMONSTRATION: Supposons d'abord $F = \sum_{\ell=1}^{r} M_\ell^2$ où $F = (F_{ij})$ et $M_\ell = (M_{\ell ij})$ sont des matrices symétriques à coefficients dans $R(X_1, \ldots, X_m)$.

Soit alors $Q \in R(X_1, \ldots, X_m)$, $Q \neq 0$ tel que pour tout ℓ, tout i et tout j, QF_{ij}, $QM_{\ell ij} = N_{\ell ij}$ appartiennent à $R[X_1, \ldots, X_m]$.
Notant I la matrice identité de $\mu_n(R)$ alors

$$(Q\ I)^2\ F = \sum_{\ell=1}^{r} ((Q\ I)\ M_\ell)^2 = \sum_{\ell=1}^{r} N_\ell^2.$$

Donc $Q^2\ F_{ij} = \sum_{\ell=1}^{r} \sum_{h=1}^{n} N_{\ell ij}\ N_{\ell hj}$.

Calculons pour tout $(x_1, \ldots, x_m) \in R^m$ et pout tout $(u_1, \ldots, u_n) \in \overline{R}^n = R^n$:

$$\sum_{i,j=1}^{n} Q^2 F_{ij}(x_1, \ldots, x_m) u_i u_j =$$

$$= \sum_{i,j=1}^{n} \sum_{\ell=1}^{r} \sum_{h=1}^{n} N_{\ell ih}(x_1, \ldots, x_m) N_{\ell hj}(x_1, \ldots, x_m) u_i u_j$$

$$= \sum_{\ell=1}^{r} \sum_{h=1}^{n} (\sum_{i=1}^{n} N_{\ell ih}(x_1, \ldots, x_n) u_i)(\sum_{j=1}^{n} N_{\ell hj}(x_1, \ldots, x_m) u_j)$$

et comme N_ℓ est symétrique nous obtenons

$$\sum_{i,j=1}^{n} Q^2 F_{ij}(x_1, \ldots, x_m) u_i u_j = \sum_{\ell=1}^{r} \sum_{h=1}^{n} (\sum_{i=1}^{n} N_{\ell ih}(x_1, \ldots, x_m) u_i)^2.$$

Ceci pour tout $(x_1, \ldots, x_m) \in R^m$ et tout $(u_1, \ldots, u_n) \in \overline{R}^n = R^n$.

Donc $(Q^2 I)F$ est positive en tout $(x_1, \ldots, x_m) \in R^m$ et donc F est définie positive.

Réciproquement, supposons que $F \in S_n(R(X_1, \ldots, X_m))$ soit définie positive. Soit alors $Q \in R[X_1, \ldots, X_m]$, $Q \neq 0$ tel que $QF_{ij} \in R[X_1, \ldots, X_m]$ pour tous i et j.

Alors $(Q^2 I)F$ est positive en tout point de R^m.

Si on montre que $(Q^2 I)F$ est somme de carrés de matrices de $S_n(R(X_1, \ldots, X_m))$, il en sera de même pour F.

On peut donc supposer pour la démonstration que les coefficients F_{ij} de la matrice sont dans $R[X_1, \ldots, X_m]$ et que F est positive.

Ceci s'exprime par la formule suivante:

$$(*) \quad \wedge x_1, \ldots, \wedge x_m, \wedge u_1, \ldots, \wedge u_n \sum_{i,j=1}^{n} F_{ij}(x_1, \ldots, x_m) u_i u_j \geq 0.$$

Cette formule close (*) est vraie dans R qui est un modèle de A' (cf. notations du II), elle est donc vraie dans tout corps ordonné maximal, extension ordonnée de R et en particulier dans $\overline{R(X_1, \ldots, X_m)}$, clôture réelle de $R(X, \ldots, X_m)$ ordonné par un ordre quelconque (qui induit donc bien le seul ordre possible de R).

En choisissant $x_1 = X_1, \ldots, x_n = X_n$ on obtient que

$$\wedge u_1, \ldots, \wedge u_n, \sum_{i,j=1}^{n} F_{ij}\, u_i u_j \geq 0$$

est vraie dans $\overline{R(X_1, \ldots, X_m)}$.

Donc F est positive.

Ceci étant vrai pour tout ordre sur $R(X_1, \ldots, X_m)$, F est naturellement positive et d'après le lemme du paragraphe précédent F est somme de carrés de matrices de $S_n(R(X_1, \ldots, X_m))$.

c) ASPECT QUANTITATIF: Soit R ordonné maximal.

Posons $pf(S_n(R(X_1, \ldots, X_m)))$ le plus petit entier s'il existe, tel que toute matrice définie positive de $S_n(R(X_1, \ldots, X_m))$ soit somme de $pf(S_n(R(X_1, \ldots, X_m)))$ carrés de $S_n(R(X_1, \ldots, X_m))$.

Nous avons pu démontrer que:

THEOREME: <u>Pour tout</u> $n \geq 1$ <u>et tout</u> $m \geq 0$, <u>on a l'inégalité</u>

$$pf(R(X_1, \ldots, X_m)) \leq pf(S_n(R(X_1, \ldots, X_m))) \leq 2^m.$$

DEMONSTRATION DE LA MINORATION: Soit $F = (F_{ij})$ la matrice telle que F_{11} soit un polynôme de $R[X_1, \ldots, X_m]$ non somme de $pf(R(X_1, \ldots, X_m)) - 1$ carrés dans $R(X_1, \ldots, X_m)$, et $F_{ij} = 0$ pour tout $(i,j) \neq (1,1)$.

Alors F est symétrique et définie positive (clair).

Montrons que F n'est pas la somme de $pf(R(X_1, \ldots, X_m) - 1$ carrés de matrices $G_\ell = (G_{\ell ij})$ de $S_n(R(X_1, \ldots, X_m))$.

Si c'était vrai nous aurions: $F = \sum_{\ell=1}^{pf-1} G_\ell^2$ et donc

$$F_{ij} = \sum_{\ell=1}^{pf-1} \sum_{h=1}^{n} G_{\ell ih}\, G_{\ell hj} \quad \text{et en particulier}$$

$$F_{jj} = \sum_{\ell=1}^{pf-1} \sum_{h=1}^{n} G_{\ell jh}\, G_{\ell hj} = \sum_{\ell=1}^{pf-1} \sum_{h=1}^{n} (G_{\ell jh})^2 \quad \text{car } G_\ell \text{ est symétrique.}$$

Si $j \neq 1$ alors $F_{jj} = 0$ et donc $G_{\ell jh} = G_{\ell hj} = 0$.

Ceci pour tout ℓ tout h et tout $j \neq 1$. (On utilise le fait que le corps $R(X_1, \ldots, X_m)$ est ordonnable.)

Donc pour $j = 1$

$$F_{11} = \sum_{\ell=1}^{pf-1} \sum_{h=1}^{n} (G_{\ell 1h})^2 = \sum_{\ell=1}^{pf-1} (G_{\ell 11})^2$$

ce qui est impossible d'après l'hypothèse pour F_{11}.

DEMONSTRATION DE LA MAJORATION: Pour alléger les notations posons $K = R(X_1, \ldots, X_m)$.

K est ordonnable et le polynôme minimal p de F sur K se décompose en produit de polynômes irréductibles distincts $p = \prod_{i=1}^{t} p_i$.

$K[Y]$ étant principal et les p_i premiers entre eux on a les isomorphismes d'anneaux suivants:

$$K[F] \simeq K[Y]/(p) \simeq \prod_{i=1}^{t} K[Y]/(p_i)$$

qui appliquent $F \in K[F]$ sur $Y \bmod(p)$ de $K[Y]/(p)$ et sur $(Y \bmod(p_1), \ldots, Y \bmod(p_t))$ de $\prod_{i=1}^{t} K[Y]/(p_i)$.

Pour montrer que $Y \bmod(p_i)$ est somme de carrés dans $L_i = K[Y]/(p_i)$ il suffit de montrer que $Y \bmod(p_i)$ est élément totalement positif de L_i.

Or il y a bijection entre l'ensemble des ordres sur L_i et l'ensemble des K-isomorphismes Z de L_i dans les clôtures réelles $\overline{K}$ de K (par rapport à tous les ordres de K). Ainsi on doit donc montrer que $Z(Y \bmod(p_i)) \geq 0$ dans $\overline{K}$ pour chaque K-isomorphisme $Z : L_i \to \overline{K}$.

Comme $Y \bmod(p_i)$ est racine de $p_i \in K[Y]$ dans $L_i = K[Y]/(p_i)$, alors $Z(Y \bmod(p_i))$ est racine de p_i dans $\overline{K}$.

$Z(Y \bmod(p_i))$ est donc une des valeurs propres de F et F étant naturellement positive, $Z(Y \bmod(p_i))$ est positif dans $\overline{K}$, ceci pour tout ordre de K.

Comme $Z(Y \bmod(p_i)) \geq 0$ dans $\overline{K}$ pour chaque K-isomorphisme Z de L_i dans $\overline{K}$, alors $Y \bmod(p_i)$ est totalement positif dans L_i et donc est une somme de carrés dans L_i.

Comme L_i est extension algébrique ordonnable de K (puisque les valeurs propres sont dans $\overline{K}$), et donc, en utilisant les résultats du chapitre précédent, $Y \bmod(p_i)$, qui est totalement positif dans L_i, est somme de 2^m carrés dans L_i:

$$Y \bmod(p_i) = \sum_{j=1}^{2^m} \alpha_{ij}^2 \quad \text{où} \quad \alpha_{ij} \in L_i.$$

On en déduit:

$$(Y \bmod(p_1), \ldots, Y \bmod(p_t)) = (Y \bmod(p_i), 0, \ldots, 0) + \ldots$$

$$\ldots + (0, \ldots, 0, Y \bmod(p_t))$$

$$= \sum_{j=1}^{2^m} (\alpha_{1j}^2, 0, \ldots, 0) + \ldots + \sum_{j=i}^{2^m} (0, \ldots, 0, \alpha_{tj}^2)$$

$$= \sum_{j=1}^{2^m} (\alpha_{1j}, \ldots, \alpha_{tj})^2.$$

Donc par les isomorphismes indiqués ci-dessus F est somme de 2^m carrés de matrices de $K[F] \subset S_n(R(X_1, \ldots, X_m))$.

N.B.: Les résultats de ce paragraphe peuvent être aussi obtenus dans le cadre de paragraphe suivant [18].

d) CAS DES MATRICES HERMITIENNES: D. Ž. Djoković dans [7] en utilisant nos articles [17] et [13] a étendu nos théorèmes au cas des matrices hermitiennes sans utiliser l'élimination des quantificateurs. Citons ses deux principaux résultats en précisant quelques définitions.

DEFINITION 1: Soit K un corps tel que -1 n'est pas un carré dans K. Soit i tel que $i^2 = -1$ et $L = K(i)$.

Notons "-" l'automorphisme de L sur K tel que $\bar{i} = -i$.

Soit $A \in \mu_n(L)$ hermitienne (i.e., ${}^t\bar{A} = A$).

On dira que A est positive si ${}^tXAX$ est positif pour tout $X \in L^n$, donc si et seulement si les mineurs principaux de A sont positifs.

THEOREME 1: Soient K un corps tel que -1 n'est pas un carré dans K, $L = K(i)$, $A \in \mu_n(L)$ hermitienne.

Les trois propriétés suivantes sont équivalents:

(i) $A \geq 0$ pour tout ordre de K

(ii) A est somme de carrés de $K[A]$

(iii) A est somme de carrés de matrices hermitiennes de $\mu_n(L)$.

DEFINITION 2: Soit K ordonnable tel que pour chaque ordre de K, K soit dense dans sa clôture réelle. Soit $K(X_1, \ldots, X_m)$ purement transcendante. Soit $L = K(i)$. On étend l'involution "-" à $L(X_1, \ldots, X_m)$ en posant chaque $\overline{X}_r = X_r$.

Soit $A(X_1, \ldots, X_m) \in \mu_n(L(X_1, \ldots, X_m))$ hermitienne.

Soit un ordre donné sur K. On dira que $A(X_1, \ldots, X_m)$ est définie positive si $A(x_1, \ldots, x_m)$ est positive (au sens de la définition 1) pour tout $(x_1, \ldots, x_m)$ où $A(X_1, \ldots, X_m)$ est définie.

THEOREME 2: Soit K un corps ordonnable tel que pour tout ordre K soit dense dans sa clôture réelle. Soit $L = K(i)$. Soit $A(X_1, \ldots, X_m) \in M_n(L(X_1, \ldots, X_m))$ une matrice hermitienne pour l'involution ci-dessus de $L(X_1, \ldots, X_m)$. Alors les trois propriétés suivantes sont équivalentes:

(i) $A(X_1, \ldots, X_m)$ est définie positive pour tout ordre de K

(ii) $A(X_1, \ldots, X_m)$ est somme de carrés dans $K(X_1, \ldots, X_m)[A(X_1, \ldots, X_m)]$

(iii) $A(X_1, \ldots, X_m)$ est somme de carrés de matrices hermitiennes de $\mu_n(L(X_1, \ldots, X_m))$.

Par des méthodes semblables à celles utilisées pour les matrices symétriques nous pouvons alors obtenir:

THEOREME 3: Soit R un corps ordonné maximal

$$pf(R(X_1, \ldots, X_m)) \leq pf\ \mu_n(R(X_1, \ldots, X_m, i))) \leq 2^m.$$

BIBLIOGRAPHY

1. E. Artin, Uber die Zerlegung definiter Funktionen in Quadrate Abh. Math. Sem. Hamburg 5 (1927), 100-115.

2. N. Bourbaki, Algébre, Chapitre 6, p. 39.

3. J. W. S. Cassels, "On the representation of rational functions as sums of squares," Acta Arithmetica IX (1964), 79-82.

4. J. W. S. Cassels, W. J. Ellison, A. Pfister, On sums of squares and on elliptic curves over function fields, Journal of number theory 3 (1971), 125-150.

5. R. Ciampi, Characterization of a class of matrices as sums of squares. Linear algebra and its applications. Vol. 3 (1970), 45-50.

6. M. R. Christie, Positive definite rational functions of two variables which are not the sum of three squares, Journal of number theory 8 (1976), 224-232.

7. D. Ž. Djoković, Positive semi definite matrices as sums of squares. Preprint. Université de Waterloo, Canada.

8. D. W. Dubois, Note on Artin's solution of Hilbert's 17th problem, Bull. of Amer. Math. Soc. 73 (1967), 540-541.

9. D. W. Dubois and G. Efroymson, Algebraic theory of real varieties, Studies and essays presented to Yu-Why Chen on his sixtieth birthday. October 1970.

10. D. W. Dubois, A Nullstellensatz for ordered fields, Archiv fur Mathematik 8 (1969), 111-114.

11. L. J. Gerstein, A new proof of a theorem of Cassels and Pfister, Proceedings of the American Mathematical Society 41 (1973), 327-329.

12. D. Gondard, Sur le 17ème problème de Hilbert. Thèse 3ème cycle (1973), ORSAY.

13. D. Gondard et P. Ribenboim, Sur le 17ème problème de Hilbert C.R.A.S. 277 (1973), 303-305.

14. D. Gondard, Etude et applications de la théorie des corps ordonnables, DEA Paris 6 (1970).

15. D. Gondard, Sur le 17ème problème de Hilbert dans $\mathbb{R}(V)$ C.R.A.S. Paris 276 (1973), 587-590.

16. D. Gondard et P. Ribenboim, "Fonctions définies positives sur les variétés réelles," Bull. Sci. Math. 2ème série 98 (1974), 39-47.

17. D. Gondard et P. Ribenboim, "Le 17ème problème de Hilbert pour les matrices," Bull. Sci. Math. 2ème série 98 (1974), 49-56.

18. D. Gondard, 17ème problème de Hilbert pour les matrices hermitiennes. Preprint. Brest, 1976.

19. D. Hilbert, Mathematische probleme, Göttinger Nach. (1900), 284-285.

20. D. Hilbert, Uber ternäre definite Formen, Acta. Math. 17 (1893), 169-197.

21. J. S. Hsia et R. P. Johnson, On the representation in sums of squares for definite functions in one variable over an algebraic number field, Am. J. Math. 96 (1974), 448-453.

22. N. Jacobson, Abstract Algebra, Tome III. Van Nostrand, 1964.

23. G. Kreisel et J. L. Krivine, Elements de Logique Mathematique. Théorie des Modèles. Dunod, Paris (1967).

24. E. Landau, Uber die Darstellung definiter Funktionen durch Quadrate, Math. Ann. 62 (1906), 272-285.

25. S. Lang, Algebra. Addison-Wesley (1965).

26. T. S. Motzkin, The arithmetic-geometric inequality. Inequalities. O. Shisha, Academic Press, New York (1967).

27. A. Pfister, Zur Darstellung definiter Funktionen als Summe von Quadraten, Invent. Math. 4 (1967), 229-237.

28. A. Pfister, Multiplikative quadratische Formen, Arch. Math. 16 (1965), 363-370.

29. A. Pfister, Quadratic forms over fields, Proceedings of symposia in pure mathematics, Amer. Math. Soc. 20 (1969), 150-160.

30. Y. Pourchet, Sur la représentation en sommes de carrés des polynômes à une indéterminée sur un corps nombres algébriques, Acta Arithmetica 19 (1971), 89-104.

31. P. Ribenboim, L'arithmétique des Corps. Hermann (1972), Paris.

32. P. Ribenboim, Le théorème des zéros pour les corps ordonnés, Séminaire d'algèbre et théorie des nombres Dubreil-Pisot (1970-71), Exposé 17.

33. A. Robinson, On ordered fields and definite functions, Math. Ann. 130 (1955), 257-271.

34. A. Robinson, Further Remarks on ordered fields and definite functions, Math. Ann. 130 (1956), 405-409.

35. A. Robinson, Introduction to Model Theory and to the Metamathematics of Algebra. North Holland (1963).

36. R. M. Robinson, Some definite polynomials which are not sums of squares of real polynomials, Academie Nationale Russe Novosibirsk (1973).

37. E. Witt, Zerlegung reeller algebraischer Functionen in Quadrate, J. reine angew. Math. 171 (1934), 4-11.

DEPARTEMENT DE MATHEMATIQUES
U.E.R. 47
UNIVERSITE DE PARIS VI
4, PLACE JUSSIEU
75230 PARIS CEDEX 05

Contemporary Mathematics
Volume 8, 1982

CHAINS OF HIGHER LEVEL ORDERINGS

Jonathan Harman

Table of Contents

ACKNOWLEDGMENTS. I would like to thank my thesis advisor, Tsit-Yuen Lam, for his help and encouragement. I would also like to acknowledge my indebtedness to Alex Rosenberg who first introduced me to Becker's work.

INTRODUCTION

This dissertation presents results concerning higher level orderings of fields. A higher level ordering is a generalization, due to E. Becker (B1, B2), of the usual notion of an order. If K is an ordered field, then the non-zero positive elements of K form an additively closed subgroup $P^\cdot$ of index 2 in the multiplicative group $K^\cdot$ of K. A higher level ordering is also an additively closed subgroup $P^\cdot$ of $K^\cdot$, but one requires only that $K^\cdot/P^\cdot$ be a finite cyclic group. We define and study certain infinite sequences, called chains, of higher level orderings.

In Section 1 we define the notion of a chain and give methods for its construction. A chain is an infinite sequence $(P_i)_{i=0}^{\infty}$ of (higher level) orderings such that:

(1) P_0 and P_1 are distinct orders;
(2) P_i for $i \geq 2$ is an ordering with $|K^\cdot/P_i^\cdot| = 2^i$;
(3) for $i \geq 1$, $P_i \cup -P_i = (P_{i-1} \cap P_0) \cup -(P_{i-1} \cap P_0)$.

1980 Mathematics Subject Classification 12J15, 10C04, 06F25, 13J25.

0271-4132/81/0000-0680/$09.50

Thus a chain relates orderings of different 2-power levels. We show that any ordering P for which $|K^{\cdot}/P^{\cdot}| = 2^n$, $n \geq 2$, is in at least one chain.

Becker has shown that an ordering of K determines a valuation of K with an archimedean ordered residue class field. We show that the orderings in a chain must all determine the same valuation and the same archimedean order of the residue class field. This allows us to apply to chains many of the valuation-theoretic methods used by Becker.

The facts about chains developed in Section 1 are used in Section 2 to get results concerning n-pythagorean fields, i.e., real fields for which $K^n + K^n \subset K^n$. We show that if K is 4-pythagorean then K is 2^n-pythagorean for each n. The set $M(K)$ of places of K into $\mathbb{R}$ can be made into a topological space. We show that the condition

$$(*) \quad \text{for all } a \text{ in } K,\ a^2 \text{ is in } \Sigma K^4 \text{ implies } a \text{ is in } \pm\Sigma K^2$$

is equivalent to $M(K)$ being connected. A 4-pythagorean field satisfies this condition. Therefore if K is 4-pythagorean and $|M(K)| < \infty$ then $|M(K)| = 1$ and hence K^{2^n} is a fan for each n. This extends a result of Jacob (J1). We also use the condition $(*)$ to show that $M(K)$ is connected if and only if $M(K(t))$ is connected.

A real closed field has an order which does not extend to any finite extension field. We define chain closed fields analogously. In Section 3 a theory of extensions of chains is developed which is used in Section 4 to characterize chain closed fields. A field R with chain (P_i) is chain closed if and only if R has no odd extensions and $P_0 \cap P_1 = R^2$. Chain closed fields turn out to be the same fields as the real closures of higher level orderings definedby Becker in (B1). We reprove and extend a theorem of Becker's concerning the isomorphism classes of such fields.

Section 5 contains generalizations of results in Section 2 to the case of n-pythagorean fields for n an even positive integer. The main result is: If A is a non-empty set of primes and K is 2p-pythagorean for each p in A then K is 2m-pythagorean for any m divisible only by primes in A.

SECTION 1. CHAINS OF ORDERINGS

In (B2, p. 11) Becker has called a subset P of a field K an <u>ordering</u> if:

(1) $P^{\cdot} = P - 0$ is a subgroup of $K^{\cdot} = K - 0$;

(2) $P^{\cdot} + P^{\cdot} \subset P^{\cdot}$;

(3) $K^{\cdot}/P^{\cdot}$ is a cyclic group.

An ordering P is said to be <u>of higher level</u> if $|K^{\cdot}/P^{\cdot}|$ is finite. It is <u>of level n</u> if $|K^{\cdot}/P^{\cdot}|$ divides n and <u>of exact level n</u> if $n = |K^{\cdot}/P^{\cdot}|$. An ordering of exact level 2 is just the positive cone of a usual order on K. To distinguish them we will call orderings of exact level 2 <u>orders</u>. In this section we will only consider orderings whose exact level is a power of 2.

DEFINITION 1.1. Let K be a field. An infinite sequence $(P_i)_{i=0}^{\infty}$ is a <u>chain</u> of K if:

(1) P_0 and P_1 are different orders;

(2) for $i \geq 2$, P_i is an ordering of exact level 2^i;

(3) for $i \geq 1$, $P_i \cup -P_i = (P_{i-1} \cap P_0) \cup -(P_{i-1} \cap P_0)$.

EXAMPLE 1.2. The field $\mathbb{R}((t))$ contains the chain (P_i) where

$$P_0 = \{\sum_{j=m}^{\infty} a_j t^j : a_m > 0\},$$

and for $i \geq 1$

$$P_i = \{\sum_{j=m}^{\infty} a_j t^j : a_m > 0 \quad \text{and} \quad m \equiv 0 \ (2^i)$$

$$\text{or} \quad a_m < 0 \quad \text{and} \quad m \equiv 2^{i-1} \ (2^i)\}.$$

In fact these are the only orderings of $\mathbb{R}((t))$ of 2-power level.

Our results about chains make heavy use of valuation theory and character theory. First we give some background for the valuation theory. Let v be a valuation with valuation ring A, (multiplicative) value group Γ, maximal ideal I, unit group U, and residue class field k. To differentiate these objects for more than one valuation we will write A_v, A_w, Γ_v, Γ_w, etc. If k is a real field then v is a <u>real</u> valuation. Our references for facts about real valuations are (Br, P). An ordering P is said to be <u>compatible</u> with v if $1 + I \subset P$. In that case the <u>pushdown</u> of P, $\overline{P} = (P \cap A) + I$ is an ordering of k. Becker (B2, Satz 2.2, p. 15) has shown that for P an ordering of higher level there is a unique valuation which we shall call <u>$v(P)$</u> which is compatible with P so that $\overline{P}$ is an archimedean order of k. Let $v = v(P)$. The set of valuation rings of valuations compatible with P is linearly ordered by inclusion. The minimal valuation ring in this chain is A_v. If w is a valuation and $A_w \supset A_v$ then $1 + I_w \subset 1 + I_v$. Therefore P is compatible with w. Suppose P_1 is another ordering compatible with w and that $\overline{P_1} = \overline{P}$ in k_w. Then P_1 is compatible with v and $v = v(P_1)$.

We will show in Theorem 1.8 that for any chain (P_i) the valuations $v(P_i)$ all coincide. Thus there is a valuation compatible with each P_i so that $\overline{P}_i$ is an archimedean order. In fact the orders $\overline{P}_i$ also coincide.

Let C_{2^n} denote the multiplicative cyclic group of order 2^n. For a multiplicative group G let $X_{2^n}(G)$ be the group of characters of G into C_{2^n}, i.e., the group $\mathrm{Hom}(G, C_{2^n})$. A subset P of the field K is an ordering of level 2^n if and only if $P^{\cdot}$ is additively closed and is the kernel of some χ in $X_{2^n}(K^{\cdot})$. The character χ is uniquely determined only in the case of orders. In general χ_1 and χ_2 will determine the same ordering if and only if there is an automorphism τ of C_{2^n} so that $\tau \circ \chi_1 = \chi_2$. It is useful to talk about the exact level of a character. By this we mean the order of its image.

The following theorem describes the set of orderings of level 2^n which are compatible with a valuation v.

THEOREM 1.3. (HR, p. 6, 7) Let v be a real valuation on K and n an integer ≥ 1. Then there exists an isomorphism

$$\Theta : X_{2^n}(K^{\cdot}/(1 + I)) \simeq X_{2^n}(k^{\cdot}) \times X_{2^n}(\Gamma)$$

so that if $\Theta(\chi) = (\pi,\mu)$ then $\overline{\ker \chi} = \ker \pi$. Moreover χ determines an ordering of K if and only if π determines an ordering of k.

For example, suppose P is an ordering of exact level 2^n and $v = v(P)$. Let χ be a character on $K^{\cdot}$ with $P^{\cdot} = \ker \chi$. Then if $\Theta(\chi) = (\pi,\mu)$, $\ker \pi$ dertermines an archimedean order, $\overline{P}$, on k. Therefore $\mathrm{im}(\pi) = C_2$. If $n > 1$ then $\mathrm{im}(\mu)$ must be C_{2^n}. In general the exact level of P is the maximum of the exact levels of π and μ.

We now use Theorem 1.3 to construct chains.

COROLLARY 1.4. If P is an ordering of exact level 2^m for $m \geq 2$ then there is a chain (P_i) with $P = P_m$.

PROOF. First apply the theorem with $v = v(P)$ and $n = m$. Let χ be the character determining P. Then $\Theta(\chi) = (\pi,\mu)$ for some characters π on $k^{\cdot}$ and μ on Γ. The character π is uniquely determined and has exact level 2. Then μ must have exact level 2^m. Thus $\mu^2, \mu^4, \ldots, \mu^{2^m}$ are characters of exact level $2^{m-1}, 2^{m-2}, \ldots, 1$. For $i = 0, \ldots, m$ let

$$\chi_i = \Theta^{-1}(\pi, \mu^{2^{m-i}}).$$

Let P_i be the ordering of K determined by $\ker \chi_i$. Then P_0 and P_1 are different orders (HR, Corollary 1.15, p. 10), P_i is an ordering of exact level 2^i, and $P_m = P$. For $i \geq 1$

$$\chi_i^2 = \Theta^{-1}(\pi, \mu^{2^{m-i+1}}) = \Theta^{-1}(\pi, \mu^{2^{m-i+1}})\Theta^{-1}(\pi, 1) = \chi_{i-1}\,\chi_0.$$

Since

$$\ker \chi_i^2 = P_i^{\cdot} \cup -P_i^{\cdot} \quad \text{and} \quad \ker(\chi_{i-1}\chi_0) = (P_{i-1}^{\cdot} \cap P_0^{\cdot}) \cup -(P_{i-1} \cap P_0^{\cdot})$$

we see that $P_0, \ldots, P_m$ satisfy condition (3) in Definition 1.1. We finish the construction of the chain by induction. Suppose we have already constructed $P_0, \ldots, P_{n-1}$ where $n-1 \geq m$ so that each P_i is compatible with v, each P_i has the same pushdown as P, and the P_i satisfy condition (3). Then $\Theta(\chi_0) = (\pi, \mu_0)$ where μ_0 is a character of Γ of exact level 2 or 1 and $\Theta(\chi_{n-1}) = (\pi, \mu_{n-1})$ where μ_{n-1} is a character of Γ of exact level 2^{n-1}. The character $\mu_{n-1}\,\mu_0$ will also have exact level 2^{n-1}. Since Γ is torsion free the map $\Gamma/\Gamma^{2^{n-1}} \longrightarrow \Gamma/\Gamma^{2^n}$ given by $\gamma \longrightarrow \gamma^2$ is injective. It follows that the map $X_{2^n}(\Gamma) \longrightarrow X_{2^{n-1}}(\Gamma)$ given by $\mu \longrightarrow \mu^2$ is surjective. Thus we can find a character μ_n necessarily of exact level 2^n so that $\mu_n^2 = \mu_{n-1}\,\mu_0$. Let $\chi_n = \Theta^{-1}(\pi, \mu_n)$. Then χ_n determines an ordering P_n of K of exact level 2^n. Since

$$\chi_n^2 = \Theta^{-1}(\pi^2, \mu_n^2) = \Theta^{-1}(\pi, \mu_{n-1})\,\Theta^{-1}(\pi, \mu_0) = \chi_{n-1}\,\chi_0,$$

P_n will satisfy

$$P_n \cup -P_n = (P_{n-1} \cap P_0) \cup -(P_{n-1} \cap P_0).$$

COROLLARY 1.5. Let P_0 and P_1 be different orders of K which are both compatible with a valuation v and have the same pushdown in k. Then there are orderings P_i for $i \geq 2$ so that $(P_0, P_1, P_2, \ldots)$ is a chain.

PROOF. Choose $n = 2$. Let χ_0 and χ_1 be characters of $K^{\cdot}$ determining P_0 and P_1. Then $\Theta(\chi_0) = (\pi, \mu_0)$ and $\Theta(\chi_1) = (\pi, \mu_1)$ where π determines an order of k and μ_0 and μ_1 are characters of Γ of exact level 1 or 2. Since $P_0 \neq P_1$, $\mu_0 \neq \mu_1$. Therefore $\mu_0\,\mu_1$ must have exact level 2. Choose μ_2 to be a character of Γ of exact level 4 such that $\mu_2^2 = \mu_0\,\mu_1$. Let P_2 be the ordering determined by $\Theta^{-1}(\pi, \mu_2)$. Then P_2 has exact level 4 and $P_2 \cup -P_2 = (P_1 \cap P_0) \cup -(P_1 \cap P_0)$. Finish the construction of the chain as in Corollary 1.4.

The chains constructed above and the chain in Example 1.2 are similar in that they each have a valuation compatible with all their orderings and all the pushdowns coincide in the residue class field. The next theorem shows that this is true for any chain.

Let $C_{2^\infty} = \mathbb{Z}(2^\infty) = \varinjlim \mathbb{Z}/2^n\mathbb{Z}$. This group contains a unique subgroup isomorphic to C_{2^n} for each n.

LEMMA 1.6. Let $\{P_i\}_{i=0}^{\infty}$ be a set of orderings of 2-power level satisfying (3) in Definition 1.1. Then there exist characters $\{\chi_i\}_{i=0}^{\infty}$ of $K^{\cdot}$ into C_{2^∞} such that $P_i^{\cdot} = \ker \chi_i$ and $\chi_i^2 = \chi_{i-1}\,\chi_0$.

PROOF. Let χ_0 be any character with $\ker \chi_0 = P_0^{\cdot}$. Assuming that we have already constructed $\chi_0, \ldots, \chi_n$ here is how to get χ_{n+1}: Take any character χ with $\ker \chi = P_{n+1}^{\cdot}$. The condition

$$P_{n+1} \cup -P_{n+1} = (P_n \cap P_0) \cup -(P_n \cap P_0)$$

shows that $\ker \chi^2 = \ker (\chi_n\,\chi_0)$. If $C_{2^m} = \operatorname{im}(\chi_n\,\chi_0)$ then there is an automorphism τ_0 of C_{2^m} so that $\tau_0 \circ \chi^2 = \chi_n\,\chi_0$. Extend τ_0 to an automorphism τ of $C_{2^{m+1}}$ with $\tau|C_{2^m} = \tau_0$. Then $(\tau \circ \chi)^2 = \tau_0 \circ \chi^2 = \chi_n\,\chi_0$. So set $\chi_{n+1} = \tau \circ \chi$.

LEMMA 1.7. Let $\{P_i\}_{i=0}^{\infty}$ be as above and also assume P_0 is an order. Then either $P_i = P_0$ for all i or there is a d such that $P_0 = P_1 = \ldots = P_d$ and $(P_{d+i})_{i=0}^{\infty}$ is a chain.

PROOF. Assume there is an i so that $P_i \neq P_0$. Let m be the least such i and let $d = m - 1$. Choose characters $\{\chi_i\}$ as above. Since $\chi_m^2 = \chi_{m-1}\chi_0 = 1$, P_m must be an order. For $n > d + 1$ the relation $\chi_n^2 = \chi_{n-1}\chi_0$ shows that if P_{n-1} is an ordering of exact level 2^{n-1} then P_n is an ordering of exact level 2^n. Thus $(P_{d+i})_{i=0}^{\infty}$ is a chain.

Let Y be a set of orders of a field K. For each P in Y let χ_P be the character of $K^{\cdot}$ with $P^{\cdot} = \ker \chi_P$. Then Y is a fan if for every P_1, P_2, P_3 in Y, $\chi_{P_1}\chi_{P_2}\chi_{P_3} = \chi_{P_4}$ for some P_4 in Y. A fan is called trivial if it consists of one or two orders. Suppose v is a valuation compatible with each P in Y. Then Theorem 1.3 shows that $\{\overline{P} : P$ is in $Y\}$ forms a fan of k. In (Br, Theorem 2.7, p. 157) it is shown that for any non-trivial fan Y there is a valuation compatible with each P in Y such that the induced fan of the residue class field is trivial.

THEOREM 1.8. Let (P_i) be a chain of K. Then there exists a valuation v compatible with every P_i so that the pushdowns $\overline{P}_i$ all coincide and equal an archimedean order of k.

PROOF. The first step is to prove:

(1) For each $m \geq 2$ there is a valuation w_m compatible with P_0, P_1, and P_m so that $\overline{P}_0 = \overline{P}_1$ in k_{w_m}.

Let $v = v(P_m)$. Let $\{\chi_i\}$ be a set of characters for (P_i) as in Lemma 1.6. The relation $\chi_i^2 = \chi_{i-1}\chi_0$ shows that for all i, $\chi_i^{2^{i-1}} = \chi_1\chi_0$. In Theorem 1.3 take $n = m$. So $\chi_m = \Theta^{-1}(\pi,\mu)$ where π determines an order on k_v and μ is a character of Γ_v of exact level 2^m. Let Q_0 be the order of K determined by $\Theta^{-1}(\pi,1)$ and let Q_1 be that determined by $\Theta^{-1}(\pi,\mu^{2^{m-1}})$. Since

$$\chi_0\chi_1 = \chi_m^{2^{m-1}} = \Theta^{-1}(\pi,1)\,\Theta^{-1}(\pi,\mu^{2^{m-1}})$$

the orders P_0, P_1, Q_0, Q_1 form a fan. If (say) $P_0 = Q_0$ and $P_1 = Q_1$ we are done since then we can take $w_m = v$. So we assume the four orders P_0, P_1, Q_0, Q_1 are distinct. Then there is a valuation w compatible with these orders such that their pushdowns form a trivial fan in k_w. Since Q_0 pushes down to an archimedean order in k_v, A_v is the unique minimal valuation ring compatible with Q_0. Thus any valuation w compatible with Q_0 must have $A_w \supset A_v$ and w will also be compatible with P_m. Therefore if $\overline{P}_0 = \overline{P}_1$ in k_w we can take $w_m = w$. If not then $\overline{P}_0 = \overline{Q}_0$ (say) and $\overline{P}_1 = \overline{Q}_1$. In this case it follows that P_0 and P_1 are compatible with v and in k_v, $\overline{P}_0 = \overline{P}_1 = \overline{Q}_0 = \overline{Q}_1$. So here we take $w_m = v$.

(2) There is a valuation w compatible with every P_i so that $\overline{P}_0 = \overline{P}_1$ in k_w.

Since each w_m is compatible with P_0 the valuation rings A_{w_m} are linearly ordered by inclusion. Let w be the valuation determined by $\bigcup_{m=2}^{\infty} A_{w_m}$. Then w is compatible with each P_i. Since $\overline{P}_0 = \overline{P}_1$ in each k_{w_m}, $\overline{P}_0 = \overline{P}_1$ in k_w.

(3) For each $j \geq 1$ there is a valuation v_j with $A_{v_j} \subset A_{v_{j-1}}$ so that v_j is compatible with every P_i and $\overline{P}_0 = \dots = \overline{P}_j$ in k_{v_j}.

Let $v_1 = w$. Assume we have constructed $v_1, \dots, v_{j-1}$. If $\overline{P}_0 = \overline{P}_j$ in $k_{v_{j-1}}$ then take $v_j = v_{j-1}$. If this is not the case then by Lemma 1.7 $(\overline{P}_{j-1+i})_{i=0}^{\infty}$ is a chain of $k_{v_{j-1}}$. By step (2) there is a valuation w_0 on $k_{v_{j-1}}$ so that $\overline{P}_{j-1}$ and $\overline{P}_j$ have the same pushdown in k_{w_0}. This valuation on $k_{v_{j-1}}$ determines a valuation on K. Let v_j be this valuation.

To finish the proof of the theorem let v_0 be the valuation determined by $\bigcap_{j=1}^{\infty} A_{v_j}$. Then v_0 is compatible with every P_i and the pushdowns $\overline{P}_i$ coincide. If this order is not archimedean then there is a valuation on k_{v_0} compatible with the order and so that its pushdown is archimedean. Let v be the valuation on K determined by that valuation on K_{v_0}.

We will write $v(P_i)$ for the valuation constructed in Theorem 1.8. In a personal communication A. Prestel has indicated an alternate proof of Theorem 1.8 which makes use of the theory of V-topologies.

SECTION 2. 2^n-PYTHAGOREAN FIELDS

A field K is called pythagorean if $K^2 + K^2 \subset K^2$. Real fields which satisfy $K^m + K^m \subset K^m$ for m other than 2 have been studied in (B1, J1, J2). In this section we will study real fields in which $K^m + K^m \subset K^m$ for m a power of 2. In Section 5 we will deal with the case when m is an arbitrary even integer.

DEFINITION 2.1. Let

$$\underline{Q_m(K)} = \{\text{finite sums } \Sigma a_i^m : a_i \text{ are in } K\}.$$

We say K is <u>m-pythagorean</u> if $Q_m(K) = K^m$.

Becker has shown (B2, Satz 2.18, p. 23) that for K real, m even then

$$Q_m(K) = \cap \{\text{orderings of level } m \text{ of } K\}.$$

When m is a 2-power we can use chains to sharpen this result.

PROPOSITION 2.2. Let K be a real field. Then for $n \geq 1$ $Q_{2^n}(K) = \cap$ ({orders of K} $\cup$ {orderings of K of exact level 2^n}).

PROOF. Suppose a is not in $Q_{2^n}(K)$ but a is in every order of K. We want to show that there is an ordering of exact level 2^n which doesn't contain a. By Becker's result there is an ordering P of exact level 2^m, $1 < m \leq n$ such that a is not in P. By Corollary 1.4 there is a chain (P_i) with $P = P_m$. Since a is in P_0 but a is not in P_m the condition

$$P_{m+1} \cup -P_{m+1} = (P_m \cap P_0) \cup -(P_m \cap P_0)$$

shows that a is not in P_{m+1}. By repeating this argument we eventually get that a is not in P_n.

Becker has shown in (B1, Theorem 2.6, p. 63) that a real field which is 2^n-pythagorean is also 2^m-pythagorean for $1 \leq m \leq n$. This follows easily from his result (B1, Theorem 25, p. 62) that if a is a sum of 2^n-th powers then a^{2^m} is a sum of 2^{m+n}-th powers. The next theorem is about real fields which satisfy a sort of converse to this.

THEOREM 2.3. Let K be a real field for which (*) for all a in K, a^2 is in $Q_4(K)$ implies a is in $\pm Q_2(K)$. Then for

$n \geq 1$, $m > n$

$$a^{2^n} \text{ is in } Q_{2^m}(K) \text{ implies } a \text{ is in } \pm Q_{2^{m-n}}(K).$$

PROOF. We will prove the case $n = 1$. The case for general n then follows by induction. Suppose a^2 is in $Q_{2^m}(K)$ for $m \geq 3$. Since $Q_{2^m}(K) \subset Q_4(K)$ the hypothesis (*) implies that a is in $\pm Q_2(K)$. By replacing a by $-a$ if necessary we can assume a is in $Q_2(K)$. In order to show that a is in $Q_{2^{m-1}}(K)$ by Proposition 2.2 it suffices to show that a is in every ordering of exact level 2^{m-1}. Let P be such an ordering. By Corollary 1.4 there is a chain (P_i) so that $P = P_{m-1}$. Since a^2 is in $Q_{2^m}(K)$, a^2 is in P_m. So

$$a \text{ is in } P_m \cup -P_m = (P_{m-1} \cap P_0) \cup -(P_{m-1} \cap P_0).$$

Since a is in P_0 this shows that a is in P_{m-1}.

COROLLARY 2.4. The following are equivalent for a real field K:

(1) K is 2^n-pythagorean for some $n \geq 2$;

(2) K is 4-pythagorean;

(3) K is pythagorean and satisfies (*);

(4) K is 2^n-pythagorean for all n.

PROOF. (1) implies (2): This is Becker's result mentioned above.

(2) implies (3): Since K is 4-pythagorean K is pythagorean. Suppose a^2 is in $Q_4(K) = K^4$. Then there is a b in K so that $a^2 = b^4$. Since K is real we must have $a = \pm b^2$.

(3) implies (4): We show that if K is 2^m-pythagorean for $m \geq 1$ then K is 2^{m+1}-pythagorean. Let a be in $Q_{2^{m+1}}(K)$. Since $Q_{2^{m+1}}(K) \subset Q_{2^m}(K) = K^{2^m}$ there is a b in K so that $a = b^{2^m}$. Since K satisfies (*) and b^{2^m} is in $Q_{2^{m+1}}(K)$, by Theorem 2.3 b is in $\pm Q_2(K) = \pm K^2$. Thus a is in $K^{2^{m+1}}$.

(4) implies (1): This is obviously true.

DEFINITION 2.5. A real field K is called <u>2^n-strictly pythagorean</u> if for every character χ in $X_{2^n}(K^\cdot)$, $\chi(-1) \neq 1$ implies $\ker \chi$ is additively closed. We will call 2-strictly pythagorean fields <u>strictly pythagorean</u>. Such fields have also been called superpythagorean (EL).

Let K be a 2^n-strictly pythagorean field. Then K is 2^n-pythagorean (B1, Theorem 6, p. 10). Becker has shown that K is 2^n-strictly pythagorean for all n if and only if K is 4-strictly pythagorean. Corollary 2.4 is thus the analogue of this result for a weaker condition. In fact Becker shows in (B1, Corollary 2, p. 68) that K is 2^n-strictly pythagorean for all n if and only if K is strictly pythagorean and 4-pythagorean. Although a pythagorean field need not be strictly pythagorean, W. Jacob has shown (J1, Theorem 4) that if $|K^{\cdot}/K^{\cdot 2}| < \infty$ then K is 4-pythagorean implies K is 4-strictly pythagorean. We will extend Jacob's result to a wider class of fields later in this section. First we need some terminology and results from the theory of ordered fields. A reference for this is (BM).

Let K be a real field, $\underline{X(K)}$ its space of orders given the usual topology. Let $\underline{M(K)}$ be the set of $\mathbb{R}$-places of K. These are the places of K into a subfield of $\mathbb{R}$. Since an archimedean order of K gives rise to an isomorphism of K with some subfield of $\mathbb{R}$ each $\mathbb{R}$-place of K corresponds to either (1) an archimedean order of K or (2) a real valuation v of K and an archimedean order of the residue class field of v. Thus if P is an order of K the pair $(v(P),\overline{P})$ determines an $\mathbb{R}$-place of K which is compatible with P. Note that if P is archimedean then $v(P)$ is the trivial valuation. Since all $\mathbb{R}$-places of K arise in this way there is a surjective map $\psi: X(K) \longrightarrow M(K)$ given by $P \longrightarrow$ the place determined by $(v(P),\overline{P})$.

Each a in K determines an evaluation map $M(K) \longrightarrow \mathbb{R} \cup \{\infty\}$ given by $\lambda \longrightarrow \lambda(a)$. Consider $\mathbb{R} \cup \{\infty\}$ as the one point compactification of $\mathbb{R}$ and give $M(K)$ the coarsest topology which makes all the evaluation maps continuous. Then ψ is continuous (BM, Lemma 2.3). We call a subset $X_1 \subset X(K)$ <u>full</u> if $\psi^{-1}(\psi(X_1)) = X_1$. For a in K let

$$\underline{H(a)} = \{P \text{ in } X(K) : a \text{ is in } P\}.$$

This is always a clopen subset of $X(K)$.

THEOREM 2.6. $H(a)$ is full if and only if a^2 is in $Q_4(K)$.

PROOF. Suppose $H(a)$ is full. Let P be an ordering of exact level 4. There is a chain (P_i) with $P = P_2$. By Theorem 1.8 $\psi(P_0 = \psi(P_1)$. Since $H(a)$ is full, a is in P_0 if and only if a is in P_1. Therefore

$$a \text{ is in } (P_0 \cap P_1) \cup -(P_0 \cap P_1) = P_2 \cup -P_2.$$

So a^2 is in P_2. This proves that a^2 is in $Q_4(K)$. Conversely assume a^2 is in $Q_4(K)$. Let P_0 be an order in $H(a)$ and let P_1 be a different order such that $\psi(P_1) = \psi(P_0)$. By Corollary 1.5 there are orderings P_i for $i \geq 2$ so that (P_i) is a chain. Since a^2 is in P_2,

$$a \quad \text{is in} \quad P_2 \cup -P_2 = (P_0 \cap P_1) \cup -(P_0 \cap P_1).$$

Since a is in P_0 we must have a in P_1. Thus P_1 is in $H(a)$.

THEOREM 2.7. Suppose $X(K) = X_1 \cup X_2$ where the X_i are full clopen disjoint sets. Then there is an a in K so that $X_1 = H(a)$.

PROOF. Define a function $f : X(K) \longrightarrow \{\pm 1\}$ by setting $f(P) = 1$ if P is in X_1, $f(P) = -1$ if is in X_2. Give $\{\pm 1\}$ the discrete topology. Since the X_i are clopen f is continuous. Let $\{P_0, P_1, P_2, P_3\}$ be a four element fan in $X(K)$. There is a valuation compatible with all the P_i so that their pushdowns form a trivial fan in the residue class field. Therefore either all four orders determine the same $\mathbb{R}$-place or two determine one and two determine another. In either case since the X_i are full,

$$\sum_{i=0}^{3} f(P_i) \equiv 0 \quad (4).$$

By (M, Corollary 2, p. 8) there is an element a of K so that $f(P) = 1$ if and only if a is in P. Thus $X_1 = H(a)$.

COROLLARY 2.8. A real field K satisfies (*) of Theorem 2.3 if and only if $M(K)$ is connected.

PROOF. Assume K does not satisfy (*). Let a be in K, a^2 in $Q_4(K)$ but a not in $\pm Q_2(K)$. Therefore $H(a)$ is full and $H(a) \neq \emptyset$ or $X(K)$. Let $D_1 = \psi(H(a))$, $D_2 = \psi(H(-a))$. Then D_1 and D_2 are non-empty disjoint clopen sets and $M(K) = D_1 \cup D_2$. This shows that $M(K)$ is not connected. Conversely if $M(K)$ is not connected there exist D_1 and D_2 as above. Let $X_1 = \psi^{-1}(D_1)$, $X_2 = \psi^{-1}(D_2)$. By Theorem 2.7 there is an a in K so that $X_1 = H(a)$. Since X_1 is full a^2 is in $Q_4(K)$. Since $X_1 \neq \emptyset$ or $X(K)$, a is not in $\pm Q_2(K)$. Therefore K does not satisfy (*).

COROLLARY 2.9. If K is 2^n-pythagorean for all n and $|M(K)| < \infty$ then $|M(K)| = 1$ and K is 2^n-strictly pythagorean for all n.

PROOF. By Corollary 2.4 K satisfies (*). Therefore $M(K)$ is connected. Since $M(K)$ is hausdorff (BM, Lemma 2.4) and $|M(K)| < \infty$ we must have $|M(K)| = 1$. Therefore $X(K)$ is a fan (BM, Section 4) and K is strictly pythagorean. Now use the result of Becker quoted after Definition 2.5.

Next we want to give some examples of fields K with $M(K)$ connected. Of course any uniquely ordered field or more generally any field which has a unique $\mathbb{R}$-place will provide an example. In fact $M(K)$ can be infinite and connected. This will follow from Theorem 2.12.

LEMMA 2.10. Let L be an extension of the real field K. If every order of K extends to L then $M(L)$ connected implies $M(K)$ is connected.

PROOF. We will show that K satisfies (*). Since every order of K extends to L, $Q_2(L) \cap K = Q_2(K)$. Let a^2 be in $Q_4(K)$. Then a is in $\pm Q_2(L) \cap K = \pm Q_2(K)$.

PROPOSITION 2.11. Let K be a real field. Then $M(K)$ is connected if and only if $M(K((t)))$ is connected.

PROOF. If $M(K((t)))$ is connected then by Lemma 2.10 so is $M(K)$. Assume $M(K)$ is connected. Let $L = K((t))$. One sees easily that

$$(L) = \{\sum_{i=m}^{\infty} a_i t^i : 0 \neq a_m \text{ is in } \quad (K) \text{ and } m \equiv 0 \ (2^n)\}.$$

If $x = \sum_{i=m}^{\infty} a_i t^i$, $a_m \neq 0$ then x^2 is in $Q_4(L)$ if and only if a_m^2 is in $Q_4(K)$ and $2m \equiv 0 \ (4)$. Since K satisfies (*) this holds if and only if a_m is in $\pm Q_2(K)$ and $m \equiv 0 \ (2)$. Thus x is in $\pm Q_2(L)$. Therefore L satisfies (*) and so $M(L)$ is connected.

This proposition gives the examples $\mathbb{R}((t))$, $\mathbb{Q}((t_1))((t_2))$, etc. Of course these fields have only one $\mathbb{R}$-place.

THEOREM 2.12. Let K be a real field. Then $M(K)$ is connected if and only if $M(K(t))$ is connected.

PROOF. Again Lemma 2.10 furnishes one implication. Suppose $M(K)$ is connected. Let $M(K(t)) = D_1 \cup D_2$ where the D_i are disjoint clopen sets. Let $X_1 = \psi^{-1}(D_1)$, $X_2 = \psi^{-1}(D_2)$. By Theorem 2.7 we can assume

$X_1 = H(x)$ for some x in $K(t)$. We can also assume that $x = af(t)$ where a is in K and $f(t)$ is a monic square free polynomial. We show first that f cannot have a root in any real closure of K. Suppose it does have such a root α. Then $K(\alpha)$ is a real field. Let $p(t)$ be the minimum polynomial of α over K. Let v be the valuation on $K(t)$ determined by $p(t)$. Then $K(\alpha)$ is the residue class field of v. Each order P of $K(\alpha)$ is the pushdown of two orders, P_1 and P_2 of $K(t)$. These orders are determined as follows: If $g(t)$ is in $K(t)$ write $g(t) = p(t)^r h(t)$ where r is an integer, $h(t)$ is in $K(t)$ and $h(\alpha) \neq 0, \infty$. Then

$g(t)$ is in P_1 if and only if $h(\alpha)$ is in P,

$g(t)$ is in P_2 if and only if $(-1)^r h(\alpha)$ is in P.

Write $f(t) = p(t)u(t)$. Then $u(\alpha) \neq 0$. Therefore

$af(t)$ is in P_1 if and only if $au(\alpha)$ is in P,

$af(t)$ is in P_2 if and only if $au(\alpha)$ is in $-P$.

Since $\psi(P_1) = \psi(P_2)$ it is clear that $H(af(t))$ is not full. Thus f has no roots in real closures of K. Let P be an order of $K(t)$ and let R be a real closure of $K(t)$ with respect to P. Let R' be a real closure of K inside of R. Since f is monic and has no roots in R', f is in $Q_2(R'(t))$. Since $Q_2(R'(t)) \subset R^2$ this implies that f is in P. Since this is true for all P we get that f is in $Q_2(K(t))$. Therefore $H(af(t)) = H(a)$. Since $H(a)$ is full a^2 is in $Q_4(K(t))$. Write

$$a^2 = \sum_{i=1}^{m} g_i(t)^4/h_i(t)^4$$

where the $g_i(t)$ and $h_i(t)$ are polynomials. If b is not a root of any $h_i(t)$ then

$$a^2 = \sum_{i=1}^{m} g_i(b)^4/h_i(b)^4$$

shows that a^2 is in $Q_4(K)$. Since K satisfies (*), a is in $\pm Q_2(K) \subset \pm Q_2(K(t))$. Therefore $H(x) = \emptyset$ or $X(K(t))$. Thus $D_1 = \emptyset$ or $M(K(t))$. This shows that $M(K(t))$ is connected.

This theorem gives us the examples $\mathbb{R}(t_1, \ldots, t_n)$ and $\mathbb{Q}(t_1, \ldots, t_n)$. These fields have an infinite number of $\mathbb{R}$-places. The theorem also provides an interesting fact about polynomials: If $g(t_1, \ldots, t_n)$ is in $\mathbb{Q}[t_1, \ldots, t_n]$ and g^2 is a sum of 2^n-th powers in $\mathbb{Q}(t_1, \ldots, t_n)$ then g is $\pm$ a sum of 2^{n-1}-th powers in $\mathbb{Q}(t_1, \ldots, t_n)$.

SECTION 3. EXTENSION THEORY

In this section we develop a theory of extensions of chains. The results will be used in Section 4 to characterize certain fields called chain closed fields. The theory of extensions of higher level orderings has been studied in (B1, B2, and HR). The latter reference uses the theory of group characters to understand the behavior of extensions of orderings. We will do the same here for chains.

DEFINITION 3.1. Let L be an extension of the field K. Let (P_i) be a chain of K, (P_i^L) a chain of L. Then (P_i^L) is an <u>extension of (P_i) of degree d</u> if

$$P_0^L \cap K = \ldots = P_d^L \cap K = P_0$$

and for $i \geq 1$

$$P_{d+i}^L \cap K = P_i.$$

An extension of degree 0 will be called a <u>faithful</u> extension.

We will also use the word faithful in connection with extensions of higher level orderings. An extension P^L of P is <u>faithful</u> if P^L has the same exact level as P. Thus if (P_i^L) is a faithful extension of (P_i) then for each i, P_i^L is a faithful extension of P_i. If (P_i^L) is an extension of degree $d > 0$ then P_0^L and P_1^L are faithful extensions of P_0 but none of the other extensions is faithful.

A main tool for our extension theory will be the following theorem. For this theorem assume that L is an extension of K. Let v_L be a real valuation of L with value group Γ_L, residue class field k_L, etc. Let v be the restriction of v_L to K. By Theorem 1.3 for each n there are isomorphisms

$$\Theta : \chi_{2^n}(K^{\cdot}/(1+I)) \simeq \chi_{2^n}(k^{\cdot}) \times \chi_{2^n}(\Gamma)$$

and

$$\Theta_L : \chi_{2^n}(L^{\cdot}/(1+I_L)) \simeq \chi_{2^n}(k_L^{\cdot}) \times \chi_{2^n}(\Gamma_L)$$

The various inclusion maps $K^{\cdot}/(1+I) \subset L^{\cdot}/(1+I_L)$, $k^{\cdot} \subset k_L^{\cdot}$, $\Gamma \subset \Gamma_L$ give rise to maps of the character groups. These maps will be denoted i^*. The point of the theorem is to relate the maps Θ, Θ_L to the i^* maps.

THEOREM 3.2. (HR, Corollary 2.8, p. 13) In the above situation there is a homomorphism

$$\phi^* : \chi_{2^n}(k_L^\cdot) \longrightarrow \chi_{2^n}(\Gamma)$$

so that the following diagram commutes:

$$\begin{array}{ccc} \chi_{2^n}(L^\cdot/(1+I_L)) & \overset{\Theta_L}{\simeq} & \chi_{2^n}(k_L^\cdot) \times \chi_{2^n}(\Gamma_L) \\ \downarrow i^* & & \downarrow \psi^* \\ \chi_{2^n}(K^\cdot/(1+I)) & \overset{\Theta}{\simeq} & \chi_{2^n}(k^\cdot) \times \chi_{2^n}(\Gamma) \end{array}$$

where $\psi^*(\pi_L,\mu_L) = (i^*(\pi_L), i^*(\mu_L)\cdot\phi^*(\pi_L))$.

We want to apply Theorem 3.2 to extensions of a chain (P_i) of K. To do this we need to know more about sets of characters associated with a chain. We say $\{\chi_i\}_{i=0}^{\infty}$ is a <u>set of characters</u> for the chain (P_i) if these characters satisfy the conditions of Lemma 1.6, i.e., if $\ker \chi_i = P_i^\cdot$ and $\chi_i^2 = \chi_{i-1}\,\chi_0$ for $i \geq 1$.

LEMMA 3.3. Let $\{\chi_i\}$, $\{\lambda_i\}$ be characters for the chain (P_i). Then there is an automorphism τ of C_{2^∞} so that for all i, $\tau \circ \chi_i = \lambda_i$.

PROOF. Since $\ker \chi_i = P_i^\cdot = \ker \lambda_i$ there is an automorphism τ_i of C_{2^i} so that $\tau_i \circ \chi_i = \lambda_i$. We need to show that $\tau_i | C_{2^{i-1}} = \tau_{i-1}$. If so then $\tau = \varinjlim \tau_i$ is the desired automorphism. For $i \geq 1$ we have

$$\tau_i \circ \chi_i^2 = (\tau_i \circ \chi_i)^2 = \lambda_{i-1}\,\lambda_0 = \tau_{i-1} \circ (\chi_{i-1}\,\chi_0).$$

Since $\chi_i^2 = \chi_{i-1}\,\chi_0$ it follows that $\tau_i = \tau_{i-1}$ on $C_{2^{i-1}}$.

LEMMA 3.4. Let L be an extension of K, (P_i^L) an extension of degree d of the chain (P_i) of K. Let $\{\chi_i\}$ be characters for (P_i). Then there are characters $\{\chi_i^L\}$ for (P_i^L) so that

$$\chi_0^L | K = \ldots = \chi_d^L | K = \chi_0$$

and for $i \geq 1$

$$\chi_{d+i}^L | K = \chi_i.$$

PROOF. Choose characters $\{\lambda_i\}$ for (P_i^L). Clearly $\{\lambda_{d+i}|K\}_{i=0}^{\infty}$ gives a set of characters for (P_i). By Lemma 3.3 there is an automorphism τ of C_{2^∞} so that $\tau \circ \lambda_{d+i}|K = \chi_i$. Let $\chi_i^L = \tau \circ \lambda_i$. Then $\{\chi_i^L\}$ is a set of characters for (P_i^L) with the desired properties.

Our main theorem about extensions of chains will be Theorem 3.7. Its proof depends on character theory and on some results in (HR, Section 2). We begin with two more lemmas.

LEMMA 3.5. Let H be a subgroup of the abelian group G such that G/H is finite. Given r, then for all sufficiently large n the image of the map

$$\chi_{2^n}(G) \longrightarrow \chi_{2^n}(H)$$

contains $\chi_{2^r}(H)$.

PROOF. (This proof is based on the suggestion of a referee for HR.) Let 2^a be the exponent of the 2-sylow subgroup of G/H. Let $n \geq a + r$. If g^{2^n} is in $G^{2^n} \cap H$ then since g is in the 2-sylow subgroup of G/H, g^{2^a} is in H. Let μ be in $\chi_{2^r}(H)$. The exact level of μ is $\geq 2^r$.

Since $n \geq a + r$, $\mu(g^{2^n}) = 1$. Thus

$$\chi_{2^r}(H) \subset \chi_{2^n}(H/(G^{2^n} \cap H)).$$

Since $H/(G^{2^n} \cap H) \subset G/G^{2^n}$ the character map

$$\chi_{2^n}(G) \longrightarrow \chi_{2^n}(H/(G^{2^n} \cap H))$$

is surjective. Therefore $\chi_{2^r}(H)$ lies inside the image of $\chi_{2^n}(G)$.

Let L be a finite extension of K. We assume (P_i) is a chain of K and that P_0 extends to an order P_0^L of L. Let $v_L = v(P_0^L)$ and let $v = v_L|K$. Then $v = v(P_i)$. Choose a set of characters $\{\chi_i\}$ for the chain (P_i). Let χ_0^L be the character of $L^{\cdot}$ determined by the order P_0^L. This character determines a character π_L of $k_L^{\cdot}$. We call λ a <u>local extension</u> of χ_i if λ is an extension of χ_i to $L^{\cdot}$ such that $1 + I_L \subset \ker \lambda$ and

λ induces π_L on $k_L^{\cdot}$. Let 2^a be the exponent of the 2-sylow subgroup of Γ_L/Γ.

LEMMA 3.6. In the above situation the following holds:

(1) Each χ_i has a local extension which determines an ordering of L extending P_i.

(2) If λ is a local extension of χ_m then for $1 \le i \le m$

$$\lambda^{2^i} \chi_0^L$$

is a local extension of χ_{m-i}.

(3) There exists a positive integer $d \le a$ so that for all $m > a$ all local extensions of χ_m have exact level 2^{m+d}.

(4) If λ and ν are both local extensions of χ_m, $m \ge 1$, then

$$\lambda^{2^a} = \nu^{2^a}.$$

(5) If λ_m is a local extension of χ_m and λ_{m+i} is a local extension of χ_{m+i}, $i \ge 1$, then

$$\lambda_{m+i}^{2^{a+i}} = \lambda_m^{2^a}.$$

PROOF. Let $i \ge 1$ and choose $n \ge a + i$. In the notation of Theorem 3.2 let $\Theta(\chi_i) = (\pi,\mu)$ where π is a character of $k^{\cdot}$ and μ is in $\chi_{2^n}(\Gamma)$. Then $\pi = i^*(\pi_L)$ and μ has exact level $\le 2^i$. (It can be $< 2^i$ only if $i = 1$ and μ is the trivial character.) The exact level of $\phi^*(\pi_L)$ is ≤ 2. Therefore $\mu\cdot\phi^*(\pi_L)$ has exact level $\le 2^i$. By Lemma 3.5 there is a μ_L in $\chi_{2^n}(\Gamma_L)$ so that $i^*(\mu_L) = \mu\cdot\phi^*(\pi_L)$. Then Theorem 3.2 gives that $\Theta_L^{-1}(\pi_L,\mu_L)$ is a local extension of χ_i which by Theorem 1.3 determines an ordering of L extending P_i. This proves (1). If λ is a local extension of χ_m then

$$(\lambda^{2^i} \chi_0^L)|K = \chi_m^{2^i} \chi_0 = \chi_{m-i}.$$

This proves (2). If $m > a$ then by (HR, Corollary 2.24, p. 22) all local extensions of χ_m have exact level 2^{m+d} for some d. By (HR, Proposition 2.20, p. 20) $d \le a$. Suppose $m' > m$ and λ' is a local extension of $\chi_{m'}$. Then $\lambda'^{2^{m'-m}} \chi_0^L$ is a local extension of χ_m by (2) and so has exact level 2^{m+d}. Since χ_0^L has exact level 2 it follows that λ' has exact level $2^{m'+d}$. This proves (3). Let λ and ν be as in (4). Let $n \geq m + d$. We have

$$\Theta_L(\lambda) = (\pi_L, \mu_1), \quad \Theta_L(\nu) = (\pi_L, \mu_2).$$

Since these are both local extensions of the same character, by Theorem 3.2

$$i^*(\mu_1) \cdot \phi^*(\pi_L) = i^*(\mu_2) \cdot \phi^*(\pi_L).$$

Thus $i^*(\mu_1) = i^*(\mu_2)$. This implies that μ_1/μ_2 is in $\chi_2 n(\Gamma_L/\Gamma)$. Since $\chi_2 n(\Gamma_L/\Gamma)$ has exponent 2^a we see that $\mu_1^{2^a} = \mu_2^{2^a}$. Therefore $\lambda^{2^a} = \nu^{2^a}$. This proves (4). Finally suppose we have λ_m and λ_{m+i} as in (5). By (2) $\lambda_{m+i}^{2^i} \chi_0^L$ is a local extension of χ_m. Clearly (5) now follows from (4).

THEOREM 3.7. Let L be a finite extension of K, (P_i) a chain of K. Then for each extension of P_0 to an order of L there is a unique extension of (P_i) beginning with that order. If d is the degree of the extension then $2^d \leq [L:K]$.

PROOF. We will use the notation established before Lemma 3.6. Let P_0^L be the extension of P_0 to L. For each m, $m > a$, let λ_m be a local extension of χ_m. By Lemma 3.6 (3) λ_m has exact level 2^{m+d}. For each $i \geq 1$ let

$$\chi_i^L = \lambda_{a+i}^{2^{a+d}} \chi_0^L.$$

We claim that $\{\chi_i^L\}$ is a set of characters for a chain extending (P_i). If $d \geq 1$ and $i \geq d$ then

$$\lambda_{a+i}^{2^{a+d}} |K = \chi_{a+i}^{2^{a+d}} = 1.$$

Thus $\chi_0^L, \ldots, \chi_d^L$ are extensions of χ_0. By (2) for $i \geq 1$, χ_{d+i}^L is an extension of χ_i. Now we only need to show that $(\chi_i^L)^2 = \chi_{i-1}^L \chi_0^L$. By construction

$$(\chi_i^L)^2 = \lambda_{a+i}^{2^{a+d+1}}.$$

By (5) this equals $\chi_{a+i-1}^{2^{a+d}}$ which in turn equals $\chi_{i-1}^L \chi_0^L$.

This proves the existence of a chain extending (P_i) and starting with P_0^L. Next assume (Q_i^L) is an extension of (P_i) of degree d' with $Q_0^L = P_0^L$. Note that then $v(Q_i^L) = v(P_0^L)$. Choose characters $\{\nu_i\}$ for (Q_i^L). By Lemma 3.4 we can assume that the ν_i are extensions of the χ_i.

Therefore $\nu_{i+d'}$ is a local extension of χ_i. By (3) $d = d'$. Let $i \geq 1$ and let $m = i+a$. The chain condition implies $\nu_m^{2^a} = \nu_i \nu_0$. By (4) this equals $(\chi_m^L)^{2^a}$ which equals $\chi_i^L \chi_0^L$. Since $\nu_0 = \chi_0^L$ this gives $\nu_i = \chi_i^L$. This proves uniqueness. Finally by (3) $d \leq a$. So

$$2^d \leq 2^a \leq |\Gamma_L/\Gamma| \leq [L:K].$$

The uniqueness part of Theorem 3.7 also applies to the case when L is an algebraic but not necessarily finite extension of K.

COROLLARY 3.8. Let L be an algebraic extension of K. Let (P_i) be a chain of K, (P_i^L) and (Q_i^L) extensions of (P_i) to L. If $P_0^L = Q_0^L$ then for all i $P_i^L = Q_i^L$.

PROOF. Suppose for some j $P_j^L \neq Q_j^L$. Let a be in P_j^L but not in Q_j^L. Then in $E = K(a)$ we would have a situation violating the uniqueness assertion in Theorem 3.7.

The existence part of Theorem 3.7 doesn't apply to algebraic extensions. Let R_0 be a real closure of (K,P_0). The unique order of R_0 is an extension of P_0 but R_0 has no chains.

LEMMA 3.9. Let $(P_0, P_1, P_2, \ldots)$ be a chain of K. Then $(P_1, P_0, P_2, \ldots)$ is also a chain.

PROOF. Choose characters $\{\chi_i\}$ for the chain $(P_0, P_1, P_2, \ldots)$. These characters satisfy $\chi_i^2 = \chi_0 \chi_{i-1}$. This implies that $\chi_i^{2^{i-1}} = \chi_0 \chi_1$. So

$$\chi_i^2 = \chi_i^{2^{i-1}} \chi_1 \chi_{i-1}, \text{ i.e.}$$

$$(\chi_i^2)^{1-2^{i-2}} = \chi_1 \chi_{i-1}.$$

Since $1-2^{i-2}$ is odd if $i > 2$,

$$\ker(\chi_i^2)^{1-2^{i-2}} = \ker \chi_i^2 = P_i^{\cdot} \cup -P_i^{\cdot}.$$

This shows that $P_i \cup -P_i = (P_{i-1} \cap P_1) \cup -(P_{i-1} \cap P_1)$. So $(P_1, P_0, P_2, \ldots)$ is a chain.

We consider these to be two different chains. So if a field has one chain it has at least two. Interestingly the trivial observation in Lemma 3.9 yields an importantfact about faithful extensions of chains.

COROLLARY 3.10. Let L be an odd extension of K. Then every chain (P_i) of K has at least one faithful extension to L.

PROOF. If $(P_0^L, P_1^L, P_2^L, \ldots)$ is a non-faithful extension of (P_i) then P_0^L and P_1^L are both extensions of P_0. Since $(P_1^L, P_0^L, P_2^L, \ldots)$ is also a chain it gives another extension of (P_i). So there are always an even number of non-faithful extensions of (P_i). The number of orders of L extending P_0 is odd therefore (P_i) must have a faithful extension.

We need results about faithful extensions of chains for use in the next section. Corollary 3.10 handles the case when the extension is odd. The next proposition handles the quadratic case.

PROPOSITION 3.11. Let $L = K(\sqrt{a})$ where a is in $K - K^2$. Then a chain (P_i) of K extends faithfully to L if and only if a is in $P_0 \cap P_1$. If this is the case then there are exactly two extensions both faithful.

PROOF. An order P of K extends faithfully to L if and only if a is in P. Then there are two faithful extensions. One contains $\sqrt{a}$ and the other contains $-\sqrt{a}$. If (P_i) extends faithfully to L then both P_0 and P_1 extend faithfully. Therefore a is in $P_0 \cap P_1$. Conversely let a be in $P_0 \cap P_1$. Since P_0 extends faithfully to L, by Theorem 3.7 (P_i) must extend to L and the degrees of its extensions can equal either 0 or 1. Assume (P_i) has a non-faithful extension (P_i^L) of degree 1. Then P_0^L and P_1^L are orders extending P_0, and P_2^L is an ordering extending P_1. Since a is in $P_1 \subset P_2^L$,

$$\sqrt{a} \text{ is in } P_2^L \cup -P_2^L = (P_1^L \cap P_0^L) \cup -(P_1^L \cap P_0^L).$$

This implies

$$\sqrt{a} \text{ is in } P_1^L \cap P_0^L \quad \text{or} \quad -\sqrt{a} \text{ is in } P_1^L \cap P_0^L,$$

a contradiction. Thus the extensions of (P_i) must be faithful. By Theorem 3.7 there are two of them.

We conclude this section by describing the extensions of (P_i) when L is a finite real galois extension of K. If P is an order of K which extends faithfully to L it is well-known that the action of the galois group on the set of orders of L extending P is both faithful and transitive (Ja, Exercise 2, p. 289). By Theorem 3.7 it is clear that this holds also for chains. Moreover the condition that L is galois over K puts a restriction on the degree of the extension. In fact the degree can be no larger than 1.

THEOREM 3.12. Let L be a finite galois extension of K with galois group G. If (P_i) is a chain of K which extends to L then all extensions are of the same degree d. There are $[L:K]$ extensions all of which are conjugate by elements of G. The degree d can equal 0 or 1. If P_1 has a faithful extension to L then $d = 0$, otherwise $d = 1$.

PROOF. Let (P_i^L) be an extension of (P_i) of degree d. Let $n = [L:K]$. For each g in G, $(g(P_i^L))$ is also an extension of (P_i) of degree d. Since

$$\{g(P_0^L) : g \text{ is in } G\}$$

has order $n = |G|$, the same is true for

$$\{(g(P_i^L)) : g \text{ is in } G\}.$$

By Theorem 3.7 these give all the extensions of (P_i). Finally we must show that $d = 0$ if P_1 has a faithful extension and that $d = 1$ otherwise. Let H be a 2-sylow subgroup of G. Let $M = L^H$. Then M is an odd extension of K and so (P_i) has a faithful extension (P_i^M) to M. Because (P_i) has n extensions to L each extension of (P_i) to M must in turn extend to L. If P_1 has a faithful extension to L then it has n of them. Therefore each faithful extension of P_1 to M must in turn extend faithfully to L. Thus (P_i^M) extends to L and P_1^M has a faithful extension to L if and only if P_1 does. By replacing K with M and (P_i) with (P_i^M) we see that we can assume that G has order 2^m. Then G is solvable so there exists a tower of quadratic extensions

$$K = L_0 \subsetneq L_1 \subsetneq \cdots \subsetneq L_m = L.$$

We finish the proof by induction on m. If $m = 1$ then by Theorem 3.7 d can equal 0 or 1. Proposition 3.11 shows that $d = 0$ if and only if P_1 extends faithfully to L. Now assume $m > 1$. If P_1 extends faithfully to L_1 then the case $m = 1$ shows that (P_i) extends faithfully to L_1. We can then apply induction to L/L_1 to get the desired result. Therefore assume P_1 does not extend faithfully to L_1 (and so P_1 does not extend faithfully to L). Let (P_i') be an extension of (P_i) to L_1. The degree of this extension must be 1 so P_0' and P_1' are faithful extensions of P_0. Since P_0 has 2^m faithful extensions to L, both P_0' and P_1' must extend faithfully to L. By induction (P_i') extends faithfully to L. Therefore (P_i) extends to L with degree $d = 1$.

SECTION 4. CHAIN CLOSURES

In (B1, B2) Becker introduced the concept of a real closure of a higher level ordering P of K. This is an algebraic extension R of K maximal with the property that the ordering P extends to an ordering of R of the same exact level. These fields exhibit many properties which tend to justify their consideration as generalizations of real closures of orders. However there is a difference. If P is an order of K then any two real closures of (K,P) are K-isomorphic. This is not true in general when P is a higher level ordering. We will use chains to remedy this for orderings of 2-power level. A chain imposes enough extra structure on the base field K to insure that its "closures" are K-isomorphic. Yet chain closed fields will turn out to be the same fields as Becker's generalizations. A higher level ordering can be a member of many different chains. This accounts for the lack of a "nice" isomorphism theorem for real closures of higher level orderings. Actually Becker has determined conditions for two of these real closures to be isomorphic. We will show that his conditions amount to the condition that the real closures are chain closures of the same chain.

DEFINITION 4.1. Let K be a field, (P_i) a chain of K. An extension R of K is called a <u>chain closure of $(K,(P_i))$</u> if R is an algebraic extension of K maximal with the property that R contains a faithful extension of (P_i). A field R is <u>chain closed</u> if there is a chain of R which does not extend faithfully to any algebraic extension of R.

A Zorn's Lemma argument shows that chain closures R of $(K,(P_i))$ always exist. Note that by Lemma 3.9 R has at least two chains. If we want to specify the chain that is the faithful extension of (P_i) we will write $(R,(P_i^R))$.

In order to use the extension theory of Section 3 to characterize chain closed fields we first prove a well-known fact about fields which have no odd extensions.

LEMMA 4.2. Let K be a field which has no odd extensions inside its algebraic closure. Then every finite non-trivial extension of K contains a quadratic extension.

PROOF. Let L be a finite extension of K. Let N be the normal closure of L over K with galois group G. Let H be a 2-sylow subgroup of G. If $G \neq H$ then K would have an odd extension. Therefore G is a 2-group. Now let V be the galois group of N over L. Then V is contained in some subgroup of index 2 in G. Thus we get a quadratic extension of K inside L.

THEOREM 4.3. Let R be a field with a chain (P_i). Then R is chain closed if and only if $P_0 \cap P_1 = R^2$ and R has no odd extensions in its algebraic closure.

PROOF. Suppose R is chain closed. Then there is a chain (P_i') of R which does not extend faithfully to any algebraic extension of R. By Corollary 3.10 R has no odd extensions. By Proposition 3.11 $P_0' \cap P_1' = R^2$. Then P_0' and P_1' are the only orders of R so also $P_0 \cap P_1 = R^2$. Now assume $P_0 \cap P_1 = R^2$ and R has no odd extensions. By Lemma 4.2 every finite extension of R contains a quadratic extension. By Proposition 3.11 (P_i) does not extend faithfully to any quadratic extension of R. Thus R is chain closed.

PROPOSITION 4.4. Let R be chain closed. Then R has exactly two chains. Let (P_i) be one of them and set $v = v(P_i)$. Then v is henselian with real closed residue class field.

PROOF. Since R has only two orders it follows from (Bl, Corollary 1, p. 42) that R has only one ordering of exact level 2^n for $n \geq 2$. Therefore

$$(P_0, P_1, P_2, \ldots) \quad \text{and} \quad (P_1, P_0, P_2, \ldots)$$

are the only chains of R. The henselization (R',v') of (R,v) contains unique faithful extensions P_0', P_1' of the orders P_0, P_1 (P, Lemma 8.2, p. 123). Since the residue class field of v' equals k, the residue class field of v, it follows that P_0' and P_1' push down to the same order of k as P_0 and P_1. So by Corollary 1.5 there is a chain (P_i') of R'. Since for $i = 0, 1$, P_i' is a faithful extension of P_i it follows that (P_i') is a faithful extension of (P_i). Therefore $R = R'$. Say P_0 and P_1 push down to the order p of k. Since $P_0 \cap P_1 = R^2$ we must have $p = k^2$. If k had an odd extension then by (E, Theorem 27.1, p. 206) so would R. Thus k is real closed (L, Corollary 2.6, p. 231).

Having characterized chain closures and shown some of their properties the next step is to develop an isomorphism theorem for them. Our technique will make use of the already known isomorphism theorem for real closures of orders. We begin with some valuation theory.

Let R be a real field with henselian valuation v, residue class field k. Since v is henselian every order of R is compatible with v. Let R_0 be a real closure of an order P of R. Suppose L_1 and L_2 are two finite extensions of R inside R_0 which are R-isomorphic. The valuation v has a unique extension to L_i, call it v_i, which is also henselian. Let k_i be the residue class field of v_i. The R-isomorphism $\phi : L_1 \to L_2$ must take A_{v_1} to A_{v_2}. Therefore ϕ induces a k-isomorphism $\overline{\phi} : k_1 \to k_2$. Note that since v has a unique extension to R_0 we can consider k_1 and k_2 to be subfields of the residue class field of this extension.

LEMMA 4.5. In the above situation assume that the ramification of v_i/v is odd for $i = 1,2$. If $\overline{\phi}$ is the identity then so is ϕ.

PROOF. Let $P_1 = R_0^2 \cap L_1$. Let p_1 be the pushdown of P_1 in k_1. Clearly both $\phi(P_1)$ and $R_0^2 \cap L_2$ have pushdown p_1 in $k_2 = k_1$. Since the ramification is odd there is a unique order of L_2 which extends P and pushes down to p_1 (Bl, Lemma 12, p. 46 or HR, Remark 2.16 (ii), p. 17). Thus $\phi(P_1) = R_0^2 \cap L_2$. By (L, Theorem 2.8, p. 232) ϕ extends to an automorphism of R_0. The only R-automorphism of R_0 is the identity.

THEOREM 4.6. Let (P_i) be a chain of K. Let R_0 be a real closure of (K,P_0). Then R_0 contains a unique chain closure of $(K,(P_i))$.

PROOF. Let $(E,(P_i^E))$ be a chain closure of $(K,(P_i))$. Let E_0 be a real closure of (E,P_0^E). There is a k-isomorphism $E_0 \simeq R_0$. Then the image of E is a chain closure of $(K,(P_i))$ inside R_0. To prove uniqueness we assume that $(E,(P_i^E))$ and $(F,(P_i^F))$ are two chain closures of $(K,(P_i))$ inside R_0. Let $R = E \cap F$. We want to show that R is chain closed. Then $R = E = F$. Since P_0^E is the only order of E extending P_0 we must have $P_0^E = R_0^2 \cap E$. Similarly $P_0^F = R_0^2 \cap F$. The chains (P_i^E) and (P_i^F) induce chains of R which are faithful extensions of the chain (P_i). Since $P_0^E \cap R = R_0^2 \cap R = P_0^F \cap R$, Corollary 3.8 shows that the induced chains coincide. Call this chain (P_i^R). Let a be in $P_0^R \cap P_1^R$. Then a is in $P_0^E \cap P_1^E$ so $\sqrt{a}$ is in E. Similarly $\sqrt{a}$ is in F. So $\sqrt{a}$ is in R. Therefore $P_0^R \cap P_1^R = R^2$. The final step is to show that R has no odd extensions. For this we need to know more about R.

Let $v = v(P_i^R)$. We prove first that v is henselian. Let $v_E = v(P_i^E)$, $v_F = v(P_i^F)$, $v_0 = v(R_0^2)$. Then v_0 is an extension of v_E and of v_F. In turn v_E and v_F are extensions of v. By Proposition 4.4 v_E and v_F are henselian. So also is v_0 (P, Theorem 8.6, p. 125). By (E, Theorem 17.11, p. 131) R_0 contains a unique hensilization of (R,v) as does E and F. Therefore this henselization must lie inside $E \cap F = R$. Thus v is henselian.

Next we claim that the residue class field of v is real closed. Let k be the residue class field. We also have the residue class fields k_E, k_F, and k_0. Consider k, k_E, and k_F as subfields of k_0. By Proposition 4.4 $k_E = k_F = k_0$. Suppose $k \subsetneq k_0$. Then let α be in $k_0 - k$ and let $\bar{f}$ be the minimal polynomial of α over k. We lift $\bar{f}$ to a monic polynomial f over R which has unit coefficients. By Hensel's Lemma there is a root a_1 of f in E so that $\bar{a}_1 = \alpha$. Similarly there is a root a_2 of f in F with $\bar{a}_2 = \alpha$. Let $L_1 = R(a_1)$, $L_2 = R(a_2)$. Let $\phi : L_1 \longrightarrow L^2$ be the R-isomorphism which takes a_1 to a_2. The extension of v to L_i has residue class field $k(\alpha)$. Since $[L_i : R] = \deg(f) = [k(\alpha) : k]$ this extension must be unramified. So Lemma 4.5 applies and we get $L_1 = L_2$. Thus a_1 is in $E \cap F = R$ so $k(\alpha) = k$. This contradicts the choice of α. Therefore $k = k_0$.

Finally we show that R has no odd extensions and so is chain closed. Let $f(x)$ be a monic irreducible polynomial over R with odd degree. Since E and F are chain closed f has a root a_1 in E and a root a_2 in F. Let $L_1 = R(a_1)$, $L_2 = R(a_2)$. Again we have an R-isomorphism $\phi : L_1 \longrightarrow L_2$. Since $k_R = k_E = k_F$, the extension of v_R to L_i must be totally ramified. Since the degree of L_i

over R is odd the ramification is odd. Lemma 4.5 again shows $L_1 = L_2$. So a_1 is in R.

COROLLARY 4.7. Let (P_i) be a chain of K. Then any two chain closures of $(K,(P_i))$ are K-isomorphic.

PROOF. Let $(E,(P_i^E))$ and $(F,(P_i^F))$ be chain closures of $(K,(P_i))$. Let E_0 be a real closure of (E,P_0^E), F_0 a real closure of (F,P_0^F). There is a K-isomorphism $\phi : E_0 \longrightarrow F_0$. By Theorem 4.6 $\phi(E) = F$.

At this point we would like to compare Becker's theory of real closures of higher level orderings with our theory of chain closures. Let P be an ordering of K of exact level 2^m, $m \geq 2$. Let R be an algebraic extension of K maximal with the property that P extends to an ordering P^R of R of the same exact level. Becker calls (R,P^R) a <u>real closure of (K,P)</u>. He shows (B1, Theorem 24, p. 58) that R is pythagorean, has no odd algebraic extensions, has exactly two orders and for every $n \geq 2$ a unique ordering of exact level 2^n. Corollary 1.4 shows that R has a chain. Thus R is chain closed. Conversely given the ordering P of K let (P_i) be a chain of K with $P = P_m$. Let $(R,(P_i^R))$ be a chain closure of $(K,P_i))$. We claim (R,P_m^R) is a real closure of (K,P). Let L be a finite extension of R. Suppose P_m^R extends faithfully to an ordering P^L of L. Let (P_i^L) be a chain of L with $P_m^L = P^L$. Since P_m^L is a faithful extension of P_m^R the chain (P_i^L) must be a faithful extension of the chain it determines on R. Since R is chain closed we must have $L = R$. So (R,P_m^R) is a real closure of (K,P).

COROLLARY 4.8. Let P be an ordering of K of exact level 2^m, $m \geq 2$. Then two (Becker) real closures (R,P^R) and (E,P^E) are K-isomorphic if and only if they determine the same chains of K.

PROOF. By the above R is chain closed and so has two chains $(P_0^R, P_1^R, \ldots)$ and $(P_1^R, P_0^R, \ldots)$. We must have $P^R = P_m^R$. Since this is a faithful extension of P both chains of R are faithful extensions of the chains they determine on K. Call these chains $(P_0, P_1, \ldots)$ and $(P_1, P_0, \ldots)$. If R and E are K-isomorphic then this isomorphism will take the chains of R to the chains of E. Therefore E will determine the same two chains on K. Conversely if both R and E determine the same chains, $(P_0, P_1, P_2, \ldots)$ and $(P_1, P_0, P_2, \ldots)$ of K then $(R,(P_i^R))$ and $(E,(P_i^E))$ are both chain closures of $(K,(P_i))$. By Corollary 4.7 they are K-isomorphic.

LEMMA 4.9. Let (P_i) and (Q_i) be chains of K. If $P_0 = Q_0$ and for all $i \geq 1$

$$P_0 \cap P_1 \cap P_i = Q_0 \cap Q_1 \cap Q_i$$

then $P_i = Q_i$ for all i.

PROOF. We first show that if (P_i) is a chain then for all $i \geq 1$

$$P_0 \cap P_1 \cap P_i = P_0 \cap P_i.$$

Choose characters $\{\chi_i\}$ for the chain. If a is in $P_0 \cap P_i$, $i \geq 2$ then $\chi_0(a) = 1$ and $\chi_i(a) = 1$. Since

$$\chi_1 = \chi_i^{2^{i-1}} \chi_0,$$

$\chi_1(a) = 1$. So a is in $P_0 \cap P_1 \cap P_i$. Now we prove the lemma by induction. Suppose that we have already shown that $P_i = Q_i$ for $i = 0, \ldots, n-1$. Let a be in P_n. Since

$$P_n \cup -P_n = (P_{n-1} \cap P_0) \cup -(P_{n-1} \cap P_0)$$

$$= (Q_{n-1} \cap Q_0) \cup -(Q_{n-1} \cap Q_0) = Q_n \cup -Q_n$$

either a is in Q_n or $-a$ is in Q_n. Also either a is in Q_0 or $-a$ is in Q_0. Suppose a is in Q_0. Then a is in $P_0 \cap P_n = Q_0 \cap Q_n$, i.e., a is in Q_n. Suppose $-a$ is in Q_0. If also $-a$ is in Q_n then $-a$ is in $Q_0 \cap Q_n = P_0 \cap P_n$, a contradiction. Therefore a is in Q_n. Thus $P_n \subset Q_n$ and by the symmetry of the argument also $Q_n \subset P_n$.

COROLLARY 4.10. (B1, Theorem 12, p. 162) Let P be an ordering of exact level 2^m, $m \geq 2$. Two real closures (R, P^R) and (E, P^E) are K-isomorphic if and only if $R^{2^n} \cap K = E^{2^n} \cap K$ for all n.

PROOF. If R and E are K-isomorphic this is clear. Conversely assume $R^{2^n} \cap K = E^{2^n} \cap K$ for all n. We know R has a chain (P_i^R) which is a faithful extension of a chain (P_i) of K. We also know that R is chain closed and so $P_0^R \cap P_1^R = R^2$. Therefore

$$R^2 \cap K = P_0^R \cap P_1^R \cap K = P_0 \cap P_1.$$

Since this equals $E^2 \cap K$ the two orders of E must also be extensions of P_0 and P_1. Call these P_0^E and P_1^E respectively. Then E has a chain (P_i^E) starting with P_0^E. By (B1, p. 62) or by Corollaries 2.8 and 2.4, R and E are 2^n-pythagorean for each n. By Proposition 2.2

$$R^{2^n} = P_0^R \cap P_1^R \cap P_n^R.$$

A similar fact holds for E. Let $Q_i = P_i^E \cap K$. By Lemma 4.9 $P_i = Q_i$ for all i. By Corollary 4.7 R and E are K-isomorphic.

Next we consider chain closed fields which are algebraic over K but which are not necessarily chain closures of chains of K. We get an isomorphism theorem, but in this case it is not always possible to tell in the base field when two such fields are isomorphic.

THEOREM 4.11. Let R be a chain closed field, (P_i) a chain of R, and R_0 a real closure of (R,P_0). Let a be in $P_0 \cap -P_1$. Then every finite extension L of R inside R_0 is of the form $L = R((a)^{1/2^d})$. Such an L is chain closed and if $d \geq 1$ then the two chains of L are extensions of (P_i) of degree d.

PROOF. By $(a)^{1/2^d}$ we mean the unique positive element α of R_0 satisfying $\alpha^{2^d} = a$. The four square classes of R are R^2, $-R^2$, aR^2, $-aR^2$. Clearly $R(\sqrt{a})$ is the unique quadratic extension of R inside R_0. By Lemma 4.2 any finite non-trivial extension L of R inside R_0 must contain $R(\sqrt{a})$. If we can show that $R(\sqrt{a})$ is chain closed then the theorem will follow by induction on the degree of the extension L/R. Therefore let $L = R(\sqrt{a})$. Let $P_0^L = R_0^2 \cap L$. The chain (P_i) extends to a unique chain of L starting with P_0^L. Call this chain (P_i^L). It must be an extension of (P_i) of degree $d = 1$. The proof of Lemma 4.2 shows that all finite extensions of R are 2-power extensions. This must hold for L also. Therefore L has no odd extensions. Since $P_0 = R^2 + aR^2 = R^2 \cup aR^2$ it follows from (DD, p. 150) that L is pythagorean. Since L has only two orders, P_0^L

and P_1^L, we get $P_0^L \cap P_1^L = L^2$. This proves that L is chain closed.

THEOREM 4.12. Let $(R,(P_i^R))$ be a chain closed field algebraic over K. Then (P_i^R) is an extension of degree d of some chain (P_i) of K. Furthermore R contains a unique chain closure of $(K,(P_i))$.

PROOF. Let α be in $P_0^R \cap -P_1^R$. Let $L = K(\alpha)$ and set $P_i^L = P_i^R \cap L$. Since α is in $P_0^L \cap -P_1^L$, (P_i^R) is a faithful extension of the chain (P_i^L). Because L is a finite extension of K, (P_i^L) must be an extension of degree $d < \infty$ of some chain (P_i) of K. Therefore (P_i^R) is an extension of degree d of (P_i). Let R_0 be a real closure of (R,P_0^R). Then R_0 contains a chain closure $(E,(P_i^E))$ of $(K,(P_i))$. By Theorem 4.11 $M = E(\alpha)$ is chain closed. Let (P_i^M) be the chain of M which starts with $P_0^M = R_0^2 \cap M$. Then $(P_i^M \cap L)$ determines a chain on L starting with P_0^L and extending (P_i). By Corollary 3.8 this chain must equal (P_i^L). Since α is in $P_0^L \cap -P_1^L$, (P_i^M) must be a faithful extension of (P_i^L). Thus M is a chain closure of $(L,(P_i^L))$ inside R_0. By uniqueness $M = R$. Therefore R contains E, the unique chain closure of $(K,(P_i))$ inside R_0.

DEFINITION 4.12. Let R be a chain closed field algebraic over K. We define $\underline{d(R/K)}$ to be the integer d in the above theorem.

COROLLARY 4.13. Let R_1 and R_2 be two chain closed fields algebraic over K. Then R_1 and R_2 are K-isomorphic if and only if:

(1) The chains of R_1 determine the same chains on K as those of R_2 and

(2) $d(R_1/K) = d(R_2/K)$.

PROOF. If R_1 and R_2 are K-isomorphic it is clear that (1) and (2) hold. Assume (1) and (2). If $d = 0$ then Corollary 4.7 gives the result. If $d > 0$ then the chains of R_j determine only one chain on K, $j = 1, 2$. Let (P_i) be this chain. By Theorem 4.12 R_j contains a chain closure E_j of (P_i). There is a K-isomorphism $\phi : E_1 \longrightarrow E_2$. Let a be in $P_0 \cap -P_1$. By Theorem 4.11 $R_j = E_j(\alpha_j)$ where $\alpha_j^{2^d} = a$, $j = 1, 2$. Clearly ϕ extends to a K-isomorphism $\phi' : R_1 \longrightarrow R_2$.

SECTION 5. FURTHER RESULTS ON n-PYTHAGOREAN FIELDS

In Section 2 we proved that a 4-pythagorean real field is 2^j-pythagorean for all j. In this section we will generalize this result. To do this we will work with orderings of level n where n is an arbitrary even positive integer, not necessarily a 2-power. The first theorem allows us to apply results about orderings of 2-power level to the more general case. In what follows we will write $U(v(P))$ for the units of the valuation $v(P)$.

THEOREM 5.1. (HR, Corollary 4.7, p. 35) Let n be an even positive integer. A subset T of a field K is an ordering of exact level n if and only if $T = P \cap S$ where P is an ordering of exact level 2^t, $S^{\cdot}$ is a subgroup of $K^{\cdot}$ such that $S^{\cdot} \supset U(v(P))$, $K^{\cdot}/S^{\cdot}$ is cyclic of order s, odd, and $n = 2^t s$.

COROLLARY 5.2. Let T be an ordering of exact level n, even. Let p be an odd prime dividing n. Then there is an ordering T' of exact level n/p such that $T' \supset T$.

PROOF. If $T = P \cap S$ as in Theorem 5.1 then $p | s = |K^{\cdot}/S^{\cdot}|$. So we can find $S' \supset S$ with $|K^{\cdot}/S^{\cdot\prime}| = s/p$. Let $T' = P \cap S'$.

LEMMA 5.3. Let G be a torsion free abelian group, S a subgroup with G/S cyclic of order n. Let p be a prime dividing n. Then there is a subgroup $S' \subset S$ so that G/S' is cyclic of order np.

PROOF. Choose g in G so that gS generates G/S. Let $h = g^n$. We claim h has order p in S/G^{pn}. If not then there is a b in G so that $h = b^{pn}$. Since G is torsion free we would have $g = h^p$. But then $g^{n/p}$ is in $G^n \subset S$ contradicting the choice of g. We next show that the subgroup generated by hG^{pn} is a direct summand of S/G^{pn}. To prove this it suffices to show that this subgroup is pure (K, Theorem 7, p. 18). Since the subgroup has order p we need only show that hG^{pn} cannot equal $s^{p^k}G^{pn}$ for some integer k and some s in S. If this were so then we would have $h = b^p$ for some b in S which implies $g^{n/p}$ is in S, a contradiction. Thus the subgroup generated by hG^{pn} is a direct summand of S/G^{pn}. Let S' be the inverse image in S of a complement to this subgroup. Then G/S' is cyclic of order np generated by gS'.

COROLLARY 5.4. Let T be an ordering of exact level n, even. Let p be an odd prime dividing n. Then there exists an ordering T' of exact level np such that $T' \subset T$.

PROOF. Write $T = P \cap S$ as in Theorem 5.1. Then $p | s = |K^{\cdot}/S^{\cdot}|$. Since $K^{\cdot}/U(v(P))$ is torsion free we can use Lemma 5.3 to find S' so that $U(v(P)) \subset S' \subset S$ and $K^{\cdot}/S'^{\cdot}$ is cyclic of order ps. Set $T' = P \cap S'$.

The next theorem is proved in (B2, Satz 4.1, p. 28). Here we give a different proof using chains.

THEOREM 5.5. Let K be a real field, n an even integer. Then $Q_n(K)^m \subset Q_{nm}(K)$.

PROOF. It suffices to prove $Q_n(K)^p \subset Q_{np}(K)$ where p is a prime. First assume $p = 2$. Let a be in $Q_n(K)$. We want to show a^2 is in $Q_{2n}(K)$. To do this we must show that a^2 is in every ordering of level $2n$ (see the beginning of Section 2). Let T be an ordering of level $2n$. Then T has exact level r, even, where $r|2n$. If $r|n$ then $T \supset Q_n(K)$ so a and a^2 are in T. Therefore we can assume that if 2^{t-1} is the highest power of 2 which divides n then $2^t|r$. Since n is even $t \geq 2$. Write $T = P \cap S$ as in Theorem 5.1. Here P is an ordering of exact level 2^t. There is a chain (P_i) with $P = P_t$. Since $v(P_0) = v(P_t)$ by Theorem 5.1 $P_0 \cap S$ is an ordering of exact level $2s$ where $s = |K^{\cdot}/S^{\cdot}|$. Since $2s|n$, $P_0 \cap S \supset Q_n(K)$. Therefore a is in $P_0 \cap S$. Since $2^{t-1}|n$, a is in P_{t-1}. Thus

$$a \text{ is in } P_t \cup -P_t = (P_0 \cap P_{t-1}) \cup -(P_0 \cap P_{t-1}).$$

So a^2 is in P_t. Also a^2 is in S since a is in S. Therefore a^2 is in $P_t \cap S = T$. Next assume p is an odd prime. Again let a be in $Q_n(K)$. We want to show a^p is in $Q_{pn}(K)$. Let T be an ordering of exact level r, even, where $r|pn$. As above we can assume that if p^t is the largest power of p which divides n then $p^{t+1}|r$. By Corollary 5.2 there is an ordering T' of exact level r/p such that $T' \supset T$. Then $T' \supset Q_n(K)$ so a is in T'. Since $T'^{\cdot}/T^{\cdot}$ has order p we get a^p is in T.

COROLLARY 5.6. If K is 2m-pythagorean then K is 2p-pythagorean for each prime $p|m$.

PROOF. Let a be in $Q_{2p}(K)$. By Theorem 5.5 $a^{m/p}$ is in $Q_{2m}(K) = K^{2m}$. So there is a b in K such that $a^{m/p} = b^{2m}$. Thus $(a/b^{2p})^{m/p} = 1$. Since K is a real $a = \pm b^{2p}$. The sign must be $+$ since a is in $Q_{2p}(K)$. Therefore $Q_{2p}(K) = K^{2p}$.

Our final theorem generalizes Corollary 2.4. It is interesting to compare this with the results in (J2) about strictly m-pythagorean fields.

THEOREM 5.7. Let A be a non-empty set of primes. If K is 2p-pythagorean for each p in A then K is 2m-pythagorean for any m divisible only by primes in A.

PROOF. It suffices to prove the following. Let $m > 1$ be divisible only by primes in A and assume K is 2m-pythagorean. Let p be in A. Then K is 2mp-pythagorean.

Let a be in $Q_{2mp}(K)$. Then a is in $Q_{2m}(K) = K^{2m}$ so there exists a b in K such that $a = b^{2m}$. We claim b^2 is in $Q_{2p}(K)$. Since $Q_{2p}(K) = K^{2p}$ it then follows that a is in K^{2mp}. Suppose b^2 is not in $Q_{2p}(K)$. Then there is an ordering T of exact level $2p$ so that b^2 is not in T. First assume $p = 2$. Then T has exact level 4 so there exists a chain (P_i) with $T = P_2$. From the definition of a chain it follows that b^2 not in P_2 implies $b^{2^{i-1}}$ is not in P_i, $i \geq 2$. Let 2^{t-1} be the highest power of 2 which divides $2m$. So $t \geq 2$. Since $2m/2^{t-1}$ is odd and $b^{2^{t-1}}$ is not in P_t it follows that b^{2m} is not in P_t. But $P_t \supset Q_{4m}(K)$. This contradicts the choice of a. Next assume p does not divide $2m$. Write $T = P \cap S$ as in Theorem 5.1. Then P is an order so b^2 is in P. Since b^2 is not in T we must have b^2 not in S. Furthermore $K^{\cdot}/S^{\cdot}$ has order p so b^{2m} is not in S. Therefore b^{2m} is not in T. Again since $T \supset Q_{2mp}(K)$ this contradicts the choice of a. Finally assume p is odd and p divides m. Write $2m = 2^t p^r s$ where s is odd and p does not divide s. Since $K^{\cdot}/T^{\cdot}$ is cyclic of order $2p$, b^2 is not in T implies $b^{2^t s}$ not in T. Use repeated applications of Corollary 5.4 to find $T' \subset T$ where T' is an ordering of exact level $2p^{r+1}$. Note that $K^{\cdot}/T'^{\cdot}$ is cyclic of order $2p^{r+1}$ and $T^{\cdot}/T^{\cdot\cdot}$ has order p^r. Therefore if c is in k and c^{p^r} is in T' then c is in T. Since $b^{2^t s}$ is not in T we conclude $b^{2^t sp^r} = b^{2m}$ is not in T'. But $T' \supset Q_{2mp}(K)$ so again this contradicts the choice of a.

Recently Becker has found a proof of Theorem 5.7 which is based on (B2, Satz 2.14, p. 21).

BIBLIOGRAPHY

B1 E. Becker, Hereditarily-pythagorean fields and orderings of higher level, Monografias de Matematica 29, Instituto de Matematica pura e aplicada, Rio de Janero, 1978.

B2 E. Becker, Summen n-ter Potenzen in Körpern, J. reine angew. Math., 307/308 (1979), 8-30.

Br L. Bröcker, Characterization of fans and hereditarily pythagorean fields, Math. Zeit, 151 (1976), 149-163.

BM R. Brown and M. Marshall, The reduced theory of quadratic forms, to appear in Rocky Mtn. J. Math.

DD J. Diller and A. Dress, Zur Galoistheorie pythagoreisher Körper, Arch. Math., 16 (1965), 148-152.

EL R. Elman and T. Y. Lam, Quadratic forms over formally real fields and pythagorean fields, Amer. J. Math., 94 (1972), 1155-1194.

E O. Endler, Valuation theory, Universitext, Springer Verlag, Berlin-Heidelberg-New York, 1972.

HR J. Harman and A. Rosenberg, Extensions of orderings of higher level, preprint.

J1 W. Jacob, On the structure of pythagorean fields, to appear in J. Alg.

J2 W. Jacob, Fans, real valuations, and hereditarily-pythagorean fields, preprint.

Ja N. Jacobson, Lectures in abstract algebra, v. 3, van Nostrand, Princeton NJ, 1964.

K I. Kaplansky, Infinite abelian groups, rev. ed., U. of Mich. Press, Ann Arbor MI, 1969.

L T.-Y. Lam, The algebraic theory of quadratic forms, W. A. Benjamin, Reading MA, 1973.

M M. Marshall, The Witt ring of a space of orderings, Trans. Amer. Math. Soc. 258 (1980), 505-522.

P A. Prestel, Lectures on formally real fields, Monografias de Matematica 22, Instituto de Matematica pura e aplicada, Rio de Janeiro, 1975.

TRW SYSTEMS
R5/2021
1 SPACE PARK
REDONDO BEACH, CALIFORNIA

Contemporary Mathematics
Volume 8, 1982

SOME PROPERTIES OF POSITIVE DERIVATIONS ON f-RINGS

Melvin Henriksen and F. A. Smith

1. INTRODUCTION

Throughout A denotes an *f-ring*; that is, a lattice-ordered ring that is a subdirect union of totally ordered rings. We let $\mathcal{D}(A)$ denote the set of derivations $D : A \longrightarrow A$ such that $a \geq 0$ implies $Da \geq 0$, and we call such derivations *positive*. In [CDK], P. Coleville, G. Davis, and K. Keimel initiated a study of positive derivations on f-rings. Their main results are (i) $D \in \mathcal{D}(A)$ and A archimedean imply $D = 0$, and (ii) if A has an identity element 1 and a is the supremum of a set of integral multiples of 1, then $Da = 0$. Their proof of (i) relies heavily on the theory of positive orthomorphisms on archimedean f-rings and gives no insight into the general case. Below, in Theorem 4 and its corollary, we give a direct proof of (i), and in Theorem 10, we generalize (ii). Throughout, we improve on results in [CDK], and we study a variety of topics not considered therein.

2. THE RESULTS

In the sequel, A will always denote an f-ring, and $A^+ = \{a \in A : a \geq 0\}$ its positive cone. If $a \in A$, let $a^+ = a \vee 0$, $a^- = (-a) \vee 0$, and $|a| = a \vee (-a)$. Then $a = a^+ - a^-$, $|a| = a^+ + a^-$, and $a^+a^- = a^-a^+ = a^+ \wedge a^- = 0$. A subset I of A that is a ring ideal and such that $|b| \leq |a|$, and $a \in I$ imply $b \in I$ is called an *ℓ-ideal*. The ℓ-ideals are the kernels of homomorphisms that preserve lattice as well as ring operations [BKW, Chap. 8].

A *derivation* on A is a linear map $D : A \longrightarrow A$ such that if $a,b \in A$, then $D(ab) = aDb + (Da)b$. A derivation D is called *positive* if $D(A^+) \subset A^+$. The family of all positive derivations on A will be denoted by $\mathcal{D}(A)$.

1980 Mathematics Subject Classification. 13N05, 06F25, 12J15, 13J25.

0271-4132/81/0000-0681/$03.50

In any f-ring *rad* A, the set of all nilpotent elements of A, coincides with the intersection of all the prime ℓ-ideals of A, and hence is an ℓ-ideal [BKW, 9.2.6]. If rad A = {0}, then A is said to be *reduced*. In [CDK], it is shown that if A is commutative and $a^n = 0$, then $[Da]^{2n-1} = 0$. We improve this result next. We begin by observing that if $a, b, \in A^+$ then

$$(1) \qquad ab = 0 \quad \text{implies} \quad aDb = (Da)b = 0.$$

1. PROPOSITION. *Suppose* $a \in A$ *and* $D \in \mathcal{D}(A)$. *Then* $a^n = 0$ *implies* $(Da)^n = 0$. *In particular*, $D[\text{rad } A] \subset \text{rad } A$.

PROOF. Since $a^n = 0$ if and only if $|a|^n = 0$, we may assume $a \in A^+$ and $n > 1$. By (1), $a^{n-1}Da = 0$. So $a^{n-2}(aDa) = 0$. Using (1) again yields $0 = a^{n-2}D(aDa) = a^{n-1}D^2a + a^{n-2}(Da)^2$. Since $a \in A^+$, $a^{n-2}(Da)^2 = 0$. Continuing this process yields $(Da)^n = 0$ and hence that $D[\text{rad } A] \subset \text{rad } A$.

The next example will show that the index of nilpotency of Da need not be less than that of a. We note first that if $D \in \mathcal{D}(A)$ and I is an ℓ-ideal of A such that $D(I) \subset I$, then $D_I \in \mathcal{D}(A/I)$, where

$$(2) \qquad D_I(a+I) = Da+I,$$

2. EXAMPLE. Let R denote all rational functions with real coefficients of negative degree. If $r(x) = \frac{p(x)}{q(x)} \in R$, we may assume that $q(x) = x^m + a_1x^{m-1} + \ldots$ has leading coefficient 1, and we let r(x) be positive if the leading coefficient of p(x) is positive. With this order, R is a totally ordered ring. If $r(x) \in R$, let $Dr(x) = -r'(x)$ be the negative of the usual derivative. Then $D \in \mathcal{D}(R)$, as is $(xD) : R \longrightarrow R$, where $(xD)r(x) = xDr(x) = -xr'(x)$. If n is a positive integer, let I_n denote the set of all r(x) in R of degree $\leq -n$. Clearly I_n is an ℓ-ideal of R, and $(xD)(I_n) \subset I_n$. If $R_n = R/I_n$, and $(xD)_n(r(x)+I_n) = xDr(x) + I_n$, then $(xD)_n \in \mathcal{D}(R_n)$, and $(xD)_n(\frac{1}{x} + I_n) = \frac{1}{x} + I_n$ is nilpotent of index n.

If G is an abelian ℓ-group, and $T : G \longrightarrow$ is an order preserving endomorphism of G such that $x \wedge y = 0$ implies $x \wedge Ty = 0$ for x,y in G^+, then T is called a *positive orthomorphism* of G. If A is reduced, then $x \wedge y = 0$ if and only if $xy = 0$ [BKW, 9.3.1].

So each positive derivation on an f-ring is an orthomorphism by (1). The next result appears implicitly in [CDK]. We include a proof for the sake of completeness.

3. PROPOSITION. *If* P *is a minimal prime* ℓ*-ideal of* A, *and* $D \in \mathcal{D}(A)$, *then* $D(P) \subset P$. *In particular,* $D_P \in \mathcal{D}(A/P)$.

PROOF. As is noted in [BKW, 9.3.2 and 12.1.1], if A is reduced, then each positive orthomorphism of $A(+)$ maps a minimal prime subgroup into itself, and P is a minimal prime ℓ-ideal of A if and only if it is a minimal prime subgroup. So $D(P) \subset P$ if A is reduced. In the general case, if we let $I = \text{rad } A$ in (2), we obtain $D(P) \subset P$.

We do not know if $D(P) \subset P$ for any prime ℓ-ideal of P.

Recall that A is said to be *archimedean* if $a \in A^+$ and $\{na : n=1,2,\ldots\}$ bounded above imply $a = 0$. The next theorem is the key to an alternate proof of the fact that a reduced archimedean f-ring admits no nontrivial derivations [CDK].

4. THEOREM. *Suppose* A *is reduced,* $D \in \mathcal{D}(A)$, $a \in A^+$, *and* n *is a positive integer. Then*

(a) $n(a \wedge a^2)Da \leq (a \vee a^2)Da$,

(b) $nDa(a \wedge a^2) \leq Da(a \vee a^2)$, and

(c) $nD(a^2) \leq (a^2Da + (Da)a^2) \vee Da$.

PROOF. Since A is reduced, $\{0\}$ is an intersection of minimal prime ideals and A is a subdirect sum of totally ordered rings A/P such that P is a minimal prime ℓ-ideal. Thus, by Proposition 3, it suffices to verify these identities in case A is totally ordered and has no proper divisors of 0 [BKW, 9.2.5].

Let $x = (na-a^2)^+ Da$. Then $x \in A^+$. We consider two cases:

(i) Suppose $x = 0$. Then $Da = 0$ or $na \leq a^2$. In either case we obtain

$$naDa \leq a^2Da \quad \text{and} \quad n(Da)a \leq (Da)a^2. \tag{3}$$

(ii) Suppose $x > 0$. Then $Da > 0$ and $a^2 < na$. Hence $aDa + (Da)a \leq nDa$. Since A is totally ordered, $aDa \leq (Da)a$ or $(Da)a \leq a(Da)$.

Suppose the former holds. Then

$$2aDa \leq nDa \quad \text{and hence} \quad (na-2a^2)Da \geq 0.$$

But $Da > 0$, so $2a^2 \le na$. By induction, we get $2^k a^2 \le na$ for $k = 0,1,2,\ldots$. If we choose k so large that $n^2 \le 2^k$, we get

$$na^2 \le a. \tag{4}$$

If, instead, $(Da)a \le aDa$, an obvious modification of this latter argument also yields (4). Pre or post multiplying by Da yields

$$na^2 Da \le aDa \quad \text{and} \quad n(Da)a^2 \le (Da)a. \tag{5}$$

Since either (3) or (5) must hold in A/P for any minimal prime ideal P, the conclusions of (a) and (b) hold.

By (4), if $x > 0$, then $nD(a^2) \le D(a)$. If $x = 0$, then adding the inequalities in (3) yields $nD(a^2) \le (a^2 Da + (Da)a^2)$. Hence (c) holds as well.

5. COROLLARY. [CDK] *If* A *is archimedean and* $D \in \mathcal{D}(A)$, *then* $D(A) \subset \text{rad } A$ *and* $D(A^2) = 0$.

PROOF. By (c) of the last theorem and Proposition 3, if $a \in A$, then $D(a^2) \in \text{rad } A$. Since $aDa \le D(a^2)$, $(Da)^2 \le D(aDa) \le D^2(a^2) \in D(\text{rad } A) \subset \text{rad } A$ by Proposition 1. Since each element of rad A is nilpotent, so is Da.

If $a,b \in A$, then $D(ab) = aDb + (Da)b = 0$, since $(\text{rad } A)A = A(\text{rad } A) = 0$ in an archimedean f-ring [BKW, 12.3.11]. Hence $D(A^2) = 0$.

6. PROPOSITION. *Suppose* $e^2 = e \in A$ *and* $D \in \mathcal{D}(A)$.

(a) $(De)^2 = e(De)e = (De)e(De) = 0$.

(b) *If* A *is reduced or has an identity element or* e *is in the center of* A, *then* $De = 0$.

PROOF. Since $e^2 = e$, we have

$$eDe + (De)e = De \tag{6}$$

Multiplying (6) on the left by e yields

$$e(De)e = 0. \tag{7}$$

Applying D to (7), we obtain

$$eD[(De)e] + (De)^2 e = 0 = e(De)^2 + D(eDe)e.$$

Hence

(8) $$e(De)^2 = (De)^2e = 0.$$

Multiplying both sides of (6) on the left by (De) and using (8) yields

(9) $$(De)e(De) = (De)^2.$$

By (7), (8), and (9), we obtain

$$[eDe - (De)e]^2 + (De)e(De) = 0.$$

Hence $(De)^2 = (De)e(De) = 0$, which together with (7), completes the proof of (a).

Clearly $De = 0$ if $\text{rad } A = \{0\}$. If $eDe = (De)e$, then by (6) and (7), $De = 2eDe = 0$. If A has an identity element, then each of its idempotents is in the center of A by [BKW, 9.4.20]. This completes the proof of (b).

The next example shows that the hypotheses of (b) above cannot be omitted.

7. EXAMPLE. *A totally ordered ring with an idempotent* e *and a positive derivation* D *such that* $De \neq 0$.

Let S denote the algebra over the real field $\mathbb{R}$ (with the usual order) with basis $\{e,z\}$, where $e^2 = e$, $ez = z^2 = 0$, and $ze = z$. If $x = \alpha e + \beta z \in S$, let $x > 0$ if $\alpha > 0$ or $\alpha = 0$ and $\beta > 0$. If we let $Dx = zx - xz = \alpha z$, then $D \in \mathcal{D}(S)$, and $De = z \neq 0$.

If $D \in \mathcal{D}(A)$, let $\ker D = \{a \in A : Da = 0\}$. If G is an abelian ℓ-group and $H \subset G$, let $H^{\perp} = \{g \in G : |g| \wedge |h| = 0$ for all $h \in H\}$, and let $H^{\perp\perp} = (H^{\perp})^{\perp}$. Note that $H^{\perp}$ is an ℓ-subgroup of G (that is, H is a subgroup and $|a| \leq |b|$, and $b \in H^{\perp}$ implies $a \in H$). A *band* in G is an ℓ-subgroup H of G such that if $K \subset H$ and $\sup K \in G$, then $\sup K \in H$. If H is a subset of G, the intersection $B(H)$ of all the bands in G containing H is also a band. Moreover, $B(H) \subset H^{\perp\perp}$. See [LZ, Theorem 19.2]. An element e of G such that $\{e\}^{\perp} = 0$ is called a *weak order unit* of G. An element e of an f-ring A such that $ex = 0$ or $xe = 0$ implies $x = 0$ is called *regular*. Note that if $e \in A$ is regular, then e is a weak order unit, and the converse holds if A is reduced.

The following lemma will be useful in what follows.

8. LEMMA. *Suppose* A *is an* f-*ring and* $D \in \mathcal{D}(A)$.

(a) $xDx \wedge (Dx)x \geq 0$ *for every* $x \in A$.

(b) *If* A *is reduced, then* D *is an* ℓ-*endomorphism.*

(c) *If* A *has an identity element* 1, *and* n *is a positive integer, then* $nDx \leq xDx \wedge (Dx)x$ *for every* $x \in A^+$ *and* $D(I) \subset I$ *for every* ℓ-*ideal* I *of* A.

PROOF. (a) holds since this inequality holds whenever A is totally ordered.

(b) holds since if A is reduced, then D is a positive orthomorphism and hence an ℓ-endomorphism [BKW, 12.1].

(c) by Proposition 6(b), $1 \in \ker D$, and by (a) $(x-n1)D(x-n1) \geq 0$. Hence $nDx \leq xDx$. Similarly, $nDx \leq (Dx)x$. Hence $x \in I$ implies $Dx \in I$ since I is an ℓ-ideal.

Next, we provide some examples to show that the hypotheses of (b) and (c) above cannot be omitted.

9. EXAMPLES. (i) Let E denote the direct sum of two copies of the real line $\mathbb{R}$ with trivial multiplication, and let $(r,s) \geq 0$ mean $r \geq s \geq 0$. As is noted in [GJ, 5B], the map $D : E \longrightarrow E$ such that $D(r,s) = (r,0)$ is a positive endomorphism that is not an ℓ-homomorphism. To see the latter, note that $(1,2)^+ = (2,2)$. So $D[(1,2)^+]$ $= (2,0) \neq (1,0) = [D(1,2)]^+$.

(ii) Let R and (xD) be as in Example 2, and let $y = \frac{1}{x}$. Then $n(xD)y = \frac{n}{x}$, while $y(xD)y = x^{-2}$, so the conclusion of (c) fails.

The next theorem summarizes most of what we know about kernels of positive derivations.

10. THEOREM. *Suppose* $D \in \mathcal{D}(A)$, $x \in A$, *and* n *is a positive integer.*

(a) *If* e *is regular, and* $ex \in \ker D$, *then* $x \in \ker D$.

(b) *If* A *is reduced then:*

(i) $x \in \ker D$ *implies* $\{x\}^{\perp\perp} \in \ker D$,

(ii) $x^n \in \ker D$ *implies* $x \in \ker D$,

(iii) $\ker D$ *is a band,*

(iv) $D^n = 0$ *implies* $D = 0$, *and*

(v) $e^2 = e \in A$ *implies* $e \in \ker D$.

(c) *If* A *has an identity element and* $\mathcal{U}(A)$ *is the smallest band containing the units of* A, *then* $\mathcal{U}(A) \subset \ker D$. *In particular,* $\operatorname{rad} A \subset \ker D$. *Also, if* $x^2 \le x$, *then* $x \in \ker D$.

PROOF. (a) By (1), $D(ex) = 0$ implies $eDx = 0$, which, in turn implies $Dx = 0$.

(b) (i) By Lemma 8(b), and [BKW, 3.2.2], $D(\{x\}^{\perp\perp}) \subset D(\{x\}^{\perp})^{\perp} \subset \{(Dx)\}^{\perp\perp} = \{0\}$ since $x \in \ker D$ and A is reduced.

(ii) follows from (i) and the fact that $\{x\}^{\perp\perp}$ is the intersection of all the minimal prime ℓ-ideals that contains x [BKW, 3.4.12].

(iii) As was noted above, the smallest band containing $\ker D$ is contained in $\{(\ker D)\}^{\perp\perp}$ and the latter is contained in $\ker D$ by (i).

(iv) Since x is a difference of positive elements, it suffices to show that $Dx = 0$ whenever $x \in A^+$. The proof will proceed by induction on n. It is obvious when $n = 1$. Assume that $D^n(A) = 0$ implies $D(A) = 0$ whenever A is a reduced f-ring and $n \ge 1$ is an integer. If $0 = D^{n+1}(A) = D^n(D(A))$, then $D^n(D(A)^{\perp\perp}) = 0$ by (i). So $D(D(A)^{\perp\perp}) = 0$ by the induction hypothesis. In particular, $D^2(x^2) = 0$. Since $xDx \le D(x^2)$, $0 = D(xDx) = xD^2x + (Dx)^2$. So $(Dx)^2 = 0 = Dx$ since A is reduced.

(v) is a restatement of Proposition 16(b).

(c) That $\mathcal{U}(A) \subset \ker D$ follows directly from (a) and (b) (iii) above. If $x^n = 0$, then $(1-x)(1 + x + \cdots + x^{n-1}) = 1$, so $1-x$ is a unit and $x = 1 - (1-x) \in \mathcal{U}(A) \subset \ker D$. Finally, if $x^2 \le x$, then $D(x^2) = xDx + (Dx)x \le Dx \le xDx \wedge (Dx)x$ by Lemma 8(c). Hence $xDx = (Dx)x = 0$. Thus $Dx = 0$. This completes the proof of Theorem 10.

11. EXAMPLES AND REMARKS. The assumption that A is reduced in Theorem 10(b) cannot be dropped. For example, if $A = C[0,1]$, the ℓ-group of continuous real-valued functions on $[0,1]$, with trivial multiplication for all $f \in C[0,1]$, we let $Df = f(\frac{1}{2})$, then $D \in \mathcal{D}(A)$, and $\ker D$ fails to be a band [DV, p. 12]. Also, the plane E^2 with the usual coordinatewise addition and trivial multiplication admits positive endomorphisms that are nilpotent. (For example, let $T(a,b) = (0,a)$ for all $(a,b) \in E^2$).

Theorem 10(c) generalizes [CDK, Theorem 7] where it is shown that $\ker D$ contains the supremum of any set of elements bounded above by some integral multiple of the identity element.

As in [P], we let $I_0(A) = \{a \in A : n|a| \leq x$ for some $x \in A^+$ and $n = 1,2,\ldots\}$. Clearly $I_0(A)$ is an ℓ-ideal and $I_0(A) = \{0\}$ if and only if A is archimedian.

12. THEOREM. *Suppose* $D \in \mathcal{D}(A)$.

(a) *If* A *is reduced, then* $D(A^2) \subset I_0(A)$.

(b) *If* A *has an identity element, then* $D(A) \subset I_0(A)$. *If, moreover,* A *is reduced and* $I_0(A) \subset \mathcal{U}(A)$, *then* $D = 0$.

PROOF. (a) follows immediately from Theorem 4 and the fact that $ab \leq (a \vee b)^2$ whenever $a,b \in A^+$.

(b) That $D(A) \subset I_0(A)$ is a restatement of Lemma 10(c). If $I_0(A) \subset \mathcal{U}(A)$, then by Theorem 10(c), $D^2(A) \subset D(\mathcal{U}(A)) = \{0\}$. Hence if A is reduced, then $D = 0$ by Theorem 10(b).

13. EXAMPLES AND REMARKS.

(a) The reader may easily verify for the f-ring R of Example 2, $I_0(R) = I_2$, while $(xD)(R) = R$. So the hypothesis in Theorem 12(b) that A has an identity element may not be dropped if we wish to have $D(A) \subset I_0(A)$.

(b) Let S denote the ring of all functions of the form

$$\textstyle\sum_{i=0}^{n} a_i x^{r_i}$$

where a_i is an integer and r_i is a nonnegative rational number, ordered lexicographically, with the coefficient of the largest power of x dominating. Then $I_0(S) = S$, and $\mathcal{U}(A)$ is the set of constant polynomials. So, the condition of Theorem 12(b) fails. Despite this, $D \in \mathcal{D}(S)$ implies $D = 0$.

For if $D \in \mathcal{D}(S)$, then $D(x) = D((x^{1/2})^2 = 2x^{1/2}D((x^{1/4})^2)$ $= 4x^{3/4}D((x^{1/8})^2) = \cdots = 2^n\, x^{1-1/2n}\, D(x^{1/2n})$. Hence $2^n|D(x)$ for $n = 0,1,2,\ldots$. Since the coefficients of any element of S are integers, it follows that $D(x) = 0$. A similar argument will show that $x^r \in \ker D$ whenever r is a nonnegative rational number. It follows that $D = 0$.

We do not, however, know of any such example that is an algebraa over an ordered field. If S^* is the result of allowing the coefficients of the elements of S to be arbitrary rational numbers, and we let $D(x^r) = rx^r$ for any positive rational number r, then D is a positive derivation. To see why, map x^r to e^{rx} and note that S^* is isomorphic as an ordered ring to a subring of the ring of exponential polynomials, and the usual derivative on the latter maps the image of S^* into itself.

Our last result applies more general theorems and techniques of Herstein [H_1] [H_2] to the context of positive derivations.

14. THEOREM. *Suppose* A *is reduced and* $D \in \mathcal{D}(A)$.

(a) *If* $D \neq 0$, *then the ring* S *generated by* $\{Da : a \in A\}$ *contains a nonzero ideal of* A.

(b) *If* S *is commutative, then* S *is contained in the center of* A.

(c) *If* $z \in A$ *commutes with every element of* S, $(az-za) \in \ker D$ *for every* $a \in A$. *If, in addition,* A *is totally ordered and* $D \neq 0$, *then* z *is in the center of* A.

PROOF. (a) It is shown in [H_1] that the conclusion holds for any derivation on any ring if $D^3 \neq 0$. Since A is reduced, $D^3 \neq 0$ implies $D \neq 0$ by Theorem 10(b).

(b) Suppose $a \in S$ and $x \in A$. Then

$$0 = (Da)D(ax) - D(ax)(Da) = Da[aDx + (Da)x] - [aDx + (Da)x]Da = Da[(Da)x - x(Da)].$$

By [H_3, Lemma 1.1.4], Da is in the center of A.

(c) The second statement is shown in [H_2], and the first follows immediately from the second and Theorem 10(b).

BIBLIOGRAPHY

[BKW] A. Bigard, K. Keimel and S. Wolfenstein, *Groupes et Anneaux Reticulés*, Lecture Notes in Mathematics 608, Springer-Verlag, New York, 1977.

[CDK] P. Colville, G. Davis and K. Keimel, *Positive derivations on f-rings*, J. Austral Math. Soc. 23 (1977), 371-375.

[DV] E. De Jonge and A. Van Rooij, *Introduction to Riesz Spaces*, Mathematical Center Tracts 78, Amsterdam, 1977.

[GJ] L. Gillman and M. Jerison, *Rings of Continuous Functions*, D. Van Nostrand Co., Princeton, New Jersey, 1960.

[H1] I. Herstein, *A note on derivations*, Canad. Math. Bull. 21 (1978), 369-370.

[H2] ___________, *A note on derivations II*, ibid., 22 (1979), 509-511.

[H3] ___________, *Rings with Involution*, University of Chicago Press, Chicago, Illinois, 1976.

[LZ] W. Luxemburg and A. Zaanen, *Riesz Spaces I*, North Holland Publ. Co., Amsterdam, 1971.

[P] J. Pairó, *Yosida-Fukamiya's theorem for* f-*rings*, pre-print.

HARVEY MUDD COLLEGE
CLAREMONT, CALIFORNIA 91711

KENT STATE UNIVERSITY
KENT, OHIO 44242

PITZER COLLEGE
CLAREMONT, CALIFORNIA 91711

Contemporary Mathematics
Volume 8, 1982

QUADRATIC FORMS OVER FIELDS WITH FINITELY MANY ORDERINGS

Jonathan Lee Merzel

CHAPTER 1

INTRODUCTION

This paper examines certain topics in the study of fields with preorderings. A preordering of a field F is a subset T of F such that:

(i) $T + T \subseteq T$
(ii) $T \cdot T \subseteq T$
(iii) $F^2 \subseteq T$
(iv) $T \cap -T = \{0\}$.

We will deal primarily with the case where $F^\cdot/T^\cdot$ is a finite (multiplicative) group. In this case, there are finitely many orderings (i.e., maximal preorderings) containing T. We write X/T for the set of all orderings containing T.

Two extreme cases have been studied in the literature: the SAP case, in which $|X/T| = n$ if $|F^\cdot/T^\cdot| = 2^n$; and the fan case, in which $|X/T| = 2^{n-1}$ if $|F^\cdot/T^\cdot| = 2^n$. The philosophy of this paper is that one expects some sort of smooth transition in the behavior of invariants of preordered fields between these cases. In Chapter 2, we study the number of archimedean orderings in X/T. Numbers of equivalence classes under equivalence relations defined by Bröcker and Marshall are studied in Chapter 3; similarly for an equivalence relation of Brown in Chapter 5. Chapter 4 is devoted to the stability index introduced by Bröcker. The index of Berman's $A(F)$ in $F^\cdot$ is examined in Chapter 6, and the structure of the square class graph is inductively characterized in Chapter 7.

$\mathbb{N}$, $\mathbb{Z}$, $\mathbb{Q}$, and $\mathbb{R}$ are used, as usual, for the sets of natural, integer, rational, and real numbers respectively. F and K are

---

1980 Mathematics Subject Classification. 10C04, 12D15, 13K05.

0271-4132/81/0000-0801/$12.25

generally reserved for (formally) real fields, and T for preorderings. F^2 denotes the set of all squares in F, the set of all sums of squares in F being written ΣF^2.

An ordering will usually be thought of as a positive cone, i.e., a preordering P satisfying the additional property that $P \cup -P = F$. The associated signature map is written σ_P, and the associated linear order on F is written $<_P$.

"Valuation" refers here to a Krull valuation v, with A_v, I_v, Γ_v, and F_v used to designate the corresponding valuation ring, maximal ideal, value group, and residue class field respectively. An ordering P and a Krull valuation v are said to be compatible if $1 + I_v \subseteq P$; a preordering T and a Krull valuation v are said to be strongly compatible if $1 + I_v \subseteq T$.

An important equivalence relation introduced by Bröcker has been used frequently: We write $P_1 \sim P_2$ if and only if $P_1 = P_2$, or P_1 and P_2 induce the same order topology on F.

Two results are used so frequently that it seems wise to mention them here:

THEOREM (Baer-Krull; see Chapter 6 of (19)). Let (F,T) be a preordered field, v a nontrivial valuation strongly compatible with T. Then $\overline{T}$ (meaning $\overline{T \cap A_v}$; bar denotes passage to F_v) is a preordering on F_v, and:

(i) $F^\cdot/T^\cdot \simeq F_v^\cdot/\overline{T}^\cdot \times \Gamma_v/v(T^\cdot)$ (group isomorphism)

(ii) $X(F)/T \simeq X(F_v)/\overline{T} \times (\Gamma_v/v(T^\cdot))^*$ (homeomorphism)

Here, $*$ denotes dual $(\mathrm{Hom}(\cdot, \{\pm 1\}))$. Also, we remark that these bijections preserve the action of orderings on elements of F.

THE V-TOPOLOGY THEOREM (A. Stone; see (22)). Let $U_1, \ldots, U_k$ be nonempty subsets of a field open with respect to distinct V-topologies (in particular, with respect to distinct order topologies). Then

$$\bigcap_{i=1}^{k} U_i \neq \phi.$$

A preordering T on a field F is called a fan if for every $a \in F^\cdot$, $a \notin -T$, we have $T + aT \subseteq T \cup aT$. (This terminology was introduced by Becker and Köpping in (2).) Following Craven, we call $A \subseteq X(F)$ a box if $A = X/T$ for some fan T; if $|X/T| = 2^n$, we may call A a 2^n-box.

Most notations and terminology follow Lam (12), Lam (13), and Prestel (19). Bröcker's 1977 paper (5) provided most of the key ideas and constructions used herein. For an interesting axiomatic approach to this subject, and background for Chapter 3, the reader is referred to Marshall's papers (14), (15), (16), (17).

CHAPTER 2

REMARKS ON $\mathcal{O}(n)$

In his paper (5), L. Bröcker defined a function $\mathcal{O}$ from the set of positive integers $\mathbb{N}$ to the power set of $\mathbb{N}$ as follows.

DEFINITION 2.1. (i) $\mathcal{O}(1) = \{1\}$.

(ii) For any $n > 1$, an integer a is an element of $\mathcal{O}(n)$ if and only if there exists a decomposition of n,

$$n = \sum_{i=1}^{r} (p_i + q_i),$$

and there exist integers $a_i \in \mathcal{O}(q_i)$ $(1 \leqq i \leqq r)$ with $p_i \in \mathbb{N} \cup \{0\}$, $q_i \in \{1, 2, \ldots, n-1\}$, and

$$a = \sum_{i=1}^{r} 2^{p_i} a_i.$$

This definition of $\mathcal{O}(n)$ is somewhat unwieldy, though convenient in some respects. An easier characterization is also found in (5).

OBSERVATION 2.2 (<u>Bröcker</u>). For $n \geqq 2$,

$$\mathcal{O}(n) = 2\mathcal{O}(n-1) \cup \bigcup_{i=1}^{[n/2]} (\mathcal{O}(i) + \mathcal{O}(n-i)).$$

The raison d'être of $\mathcal{O}$ is that it exhibits the possible numbers of orderings that fields may admit. See Bröcker (5) for details. It is not surprising that further investigation of combinatorial properties of $\mathcal{O}$ should yield more information about the set of orderings of a field. It is difficult to give enough credit to B. Reznick for opening many lines of investigation into $\mathcal{O}$.

The purpose of this chapter is to develop some interesting properties of the function $\mathcal{O}$. No apology is made for presentation of a few results which are left unused, since many of the useful results predate their applications.

An important observation on $\mathcal{O}(n)$ is found in the thesis (3) of L. Berman.

PROPOSITION 2.3 (<u>Berman</u>). For $n \geqq 2$,

$$\mathcal{O}(n) = 2\mathcal{O}(n-1) \cup (1 + \mathcal{O}(n-1)).$$

Instead of proving this result, we use Observation 2.2 to prove a slight strengthening (Proposition 2.6). Then we use 2.3 to prove the much stronger 2.7. It may be objected that Berman's result may be as easy a path as any to Bröcker's observation; those of this turn of mind may proceed directly from Berman's result to Proposition 2.7, and deduce Proposition 2.6 as a corollary.

DEFINITION 2.4. $k \in \mathcal{O}(n)$ is called <u>decomposable</u> if there exist $a,b,r,s \in \mathbb{N}$ such that $n = a + b$, $k = r + s$, $r \in \mathcal{O}(a)$ and $s \in \mathcal{O}(b)$. $k \in \mathcal{O}(n)$ is <u>indecomposable</u> otherwise.

The preceding contains an abuse of language; one should actually refer to an ordered pair (k,n) with $k \in \mathcal{O}(n)$ as decomposable or indecomposable, but no harm will be done. One should remark that Observation 2.2 implies that for $k \in \mathcal{O}(n)$, $n > 1$, either $k \in \mathcal{O}(n)$ is decomposable or $k/2 \in \mathcal{O}(n-1)$ (possibly both).

DEFINITION 2.5. For $n \geqq 2$, $k \in \mathcal{O}(n)$ is <u>1-decomposable</u> if $k-1 \in \mathcal{O}(n-1)$.

It is clear from the definitions that $k \in \mathcal{O}(n)$ 1-decomposable implies that $k \in \mathcal{O}(n)$ is decomposable. The converse, viewed in conjunction with Observation 2.2, provides a slight strengthening of Berman's result.

PROPOSITION 2.6. $k \in \mathcal{O}(n)$ is decomposable implies that $k \in \mathcal{O}(n)$ is 1-decomposable.

PROOF. We use induction on n, the case $n = 1$ holding vacuously. Write $k = r + s$, $r \in \mathcal{O}(a)$, $s \in \mathcal{O}(b)$, $a + b = n$. By induction, either $r - 1 \in \mathcal{O}(a-1)$ or $r/2 \in \mathcal{O}(a-1)$. In the latter case, iterate the same consideration until one in any case obtains $r = 2^i r_0$ with r $r_0 - 1 \in \mathcal{O}(a-i-1)$. Similarly, $s = 2^j s_0$ with $s_0 - 1 \in \mathcal{O}(b-j-1)$. Say (by symmetry) $i \leqq j$. Then $r/2^i - 1 \in \mathcal{O}(a-i-1)$ and $s/2^i \in \mathcal{O}(b-i)$. Adding and using 2.2, $r/2^i - 1 + s/2^i = k/2^i - 1 \in \mathcal{O}(n-2i-1)$. Using 2.2 again, we may alternately double and add 1 (for a total of $2i$ operations, each incrementing the argument of $\mathcal{O}$ by 1) to produce $k - 1 \in \mathcal{O}(n-1)$. Q.E.D.

Clearly Proposition 2.3 now follows. We may now produce a substantial strengthening. The following proof is a slight modification of an argument worked out by T. Y. Lam.

PROPOSITION 2.7. For $1 < a \leq b$,

$$\mathcal{O}(a) + \mathcal{O}(b) \subseteq \mathcal{O}(a-1) + \mathcal{O}(b+1).$$

PROOF. We proceed by induction on b. The case $b = 2$ is easily checked, noting that $\mathcal{O}(1) = \{1\}$, $\mathcal{O}(2) = \{2\}$ and $\mathcal{O}(3) = \{3,4\}$. For the general case, let $x \in \mathcal{O}(a)$, $y \in \mathcal{O}(b)$. In view of Proposition 2.3 we may separate three cases:

CASE 1. $x - 1 \in \mathcal{O}(a-1)$. Then $x - 1 + y \in \mathcal{O}(a-1) + \mathcal{O}(b)$, so

$$x + y = x - 1 + y + 1 \in \mathcal{O}(a-1) + (1 + \mathcal{O}(b+1)) \subseteq \mathcal{O}(a-1) + \mathcal{O}(b+1).$$

CASE 2. $y - 1 \in \mathcal{O}(b-1)$. If $b - 1 \geqq a$, the inductive hypothesis may be invoked to give

$$x + (y - 1) \in \mathcal{O}(a) + \mathcal{O}(b-1) \subseteq \mathcal{O}(a-1) + \mathcal{O}(b).$$

Then adding one,

$$x + y \in \mathcal{O}(a-1) + (\mathcal{O}(b) + 1) \subseteq \mathcal{O}(a-1) + \mathcal{O}(b+1).$$

If $b = a$, we may write $y - 1 \in \mathcal{O}(a-1)$, $x \in \mathcal{O}(b)$; the rest is just Case 1 with x and y interchanged.

CASE 3. $x/2 \in \mathcal{O}(a-1)$ and $y/2 \in \mathcal{O}(b-1)$. If $a - 1 > 1$, induction yields

$$x/2 + y/2 \in \mathcal{O}(a-1) + \mathcal{O}(b-1) \subseteq \mathcal{O}(a-2) + \mathcal{O}(b);$$

so

$$x + y \in 2\mathcal{O}(a-2) + 2\mathcal{O}(b) \subseteq \mathcal{O}(a-1) + \mathcal{O}(b+1).$$

If $a = 2$, then

$$x = 2 \quad \text{and} \quad x + y = 2 + y = 1 + (y + 1) \in \mathcal{O}(a-1) + \mathcal{O}(b+1).$$

Q.E.D.

On the following page, an $\mathcal{O}(n)$ table is presented. It is convenient to arrange the table in such a way that space is left where "missing" entries would be. To effect this, we label both rows and columns, using n as a row index and r as a column index. The nth row consists of the elements of $\mathcal{O}(n)$.

$O(n)$ TABLE

n/r	0	1	2	3	4	5	6	7	8	9	10	11	12	13	14	15	16	17	18	19	20	21	23	23	24	25	26	27	28	29	30
1	1																														
2	2																														
3	3	4																													
4	4	5	6		8																										
5	5	6	7	8	9	10		12				16																			
6	6	7	8	9	10	11	12	13	14		16	17	18		20				24								32				
*7	7	8	9	10	11	12	13	14	15	16	17	18	19	20	21	22		24	25	26		28				32	33	34		36	
*8	8	9	10	11	12	13	14	15	16	17	18	19	20	21	22	23	24	25	26	27	28	29	30		32	33	34	35	36	37	38
*9	9	10	11	12	13	14	15	16	17	18	19	20	21	22	23	24	25	26	27	28	29	30	31	32	33	34	35	36	37	38	39
.																															
.																															
.																															

*designates truncated rows.

An element $k \in O(n)$ appears in the rth column, where $r = k - n$. One easily sees by induction that the smallest entry of $O(n)$ is n (and the largest is 2^{n-1}), so that it is natural to begin numbering columns with $r = 0$.

Some facts about $O(n)$ are suggested by the appearance of this table. For example, the property $1 + O(n) \subseteq (n+1)$ is reflected in the lack of "holes" in columns of the $O(n)$ table. One may also notice the lack of holes in lines of (r versus n) slope -1; this is the content of the following proposition, which is due to B. Reznick (20).

PROPOSITION 2.8. If $k \in O(n)$ and $k > n$, then $k \in O(n+1)$.

PROOF. The cases $n = 1$ and $n = 2$ are checked directly from the table. (Both are vacuous anyway.) For $n > 2$ we use induction.

CASE 1. $k - 1 \in O(n-1)$. Then $k > n$ implies $k - 1 > n - 1$, and induction applies to give $k - 1 \in O(n)$. So

$$k = (k - 1) + 1 \in O(n) + 1 \subseteq O(n+1).$$

CASE 2. $k/2 \in O(n-1)$. If $k/2 > n - 1$, induction yields $k/2 \in O(n)$, whence

$$k = (k/2)\cdot 2 \in 2O(n) \subseteq O(n+1).$$

If $k/2 = n - 1$, we must show $k = 2n - 2 \in O(n+1)$. However, $n - 2 \in O(n-2)$ implies $2(n-2) \in O(n-1)$; twice adding 1 yields $2n - 2 \in O(n+1)$. Q.E.D.

In some later connections, it will be necessary to consider more general types of decompositions.

DEFINITION 2.9. We say that $k \in O(n)$ <u>decomposes into t parts</u> ($t \geq 1$) if there are expressions

$$(*) \qquad k = \sum_{i=1}^{t} \alpha_i, \quad n = \sum_{i=1}^{t} \beta_i$$

with $\alpha_i \in O(\beta_i)$ for $i = 1, \ldots, t$. We refer to such expressions as <u>(t-fold) decompositions</u> of $k \in O(n)$.

PROPOSITION 2.10. If $k \in O(n)$ decomposes into $t \geq 2$ parts then $k - 1 \in O(n-1)$ decomposes into $t - 1$ parts.

PROOF. Write a decomposition for $k \in O(n)$ as in (*). Rewrite $\alpha_1 + \alpha_2 = 1 + (\alpha_1 + \alpha_2 - 1)$; by Proposition 2.6, we have

$\alpha_1 + \alpha_2 - 1 \in \mathcal{O}(\beta_1 + \beta_2 - 1)$. So we have

$$k - 1 = (\alpha_1 + \alpha_2 - 1) + \sum_{1=3}^{t} \alpha_i, \quad n - 1 = (\beta_1 + \beta_2 - 1) + \sum_{i=3}^{t} \beta_i$$

where $\alpha_1 + \alpha_2 - 1 \in \mathcal{O}(\beta_1 + \beta_2 - 1)$ and $\alpha_i \in \mathcal{O}(\beta_i)$ for $i \geqq 3$. This provides a decomposition of $k - 1 \in \mathcal{O}(n-1)$ into $t-1$ parts (and also shows that $k - 1 \in \mathcal{O}(n-1)$). Q.E.D.

Two more propositions have occasionally proved useful, especially for computation of examples.

PROPOSITION 2.11. $2k \in \mathcal{O}(n)$, $k \geqq n - 1$ implies $k \in \mathcal{O}(n-1)$.

PROOF. If $k \notin \mathcal{O}(n-1)$, then $2k - 1 \in \mathcal{O}(n-1)$; but $2k - 1$ is odd, so we must have $2k - 2 \in \mathcal{O}(n-2)$. Certainly if $k \geqq n - 1$ we have $k - 1 \geqq n - 3$, so by induction (one must check the cases $n = 1, 2$ and 3 directly) on n, $k - 1 \in \mathcal{O}(n-3)$. Again $k - 1 \geqq n - 2$, so by Proposition 2.8, $k - 1 \in \mathcal{O}(n-2)$, and so $k \in \mathcal{O}(n-1)$.

PROPOSITION 2.12. If $k \in \mathcal{O}(n)$ is indecomposable, then $2k \in \mathcal{O}(n+1)$ is indecomposable, providing $n > 1$.

PROOF. If $2k \in \mathcal{O}(n+1)$ were decomposable, then by 2.6, $2k - 1 \in \mathcal{O}(n)$; $2k - 1$ odd now gives $2k - 2 \in \mathcal{O}(n-1)$. Since $k \in \mathcal{O}(n)$, certainly $k - 1 \geqq n - 2$, so by Proposition 2.11, $k - 1 \in \mathcal{O}(n-2)$. (So this argument fails for $n = 2$, but that case checks easily.) Again, $k - 1 \geqq n - 1$ so by Proposition 2.8, $k - 1 \in \mathcal{O}(n-1)$, and $k \in \mathcal{O}(n)$ would be decomposable. Q.E.D.

Two important functions have been introduced by Reznick.

DEFINITIONS 2.13. For $r \geqq 1$:

$$L(r) := \min \{n : n + r \in \mathcal{O}(n)\}.$$
$$\lambda(r) := r + L(r).$$

The tabular interpretation of $L(r)$ is the number of the first row in which an entry is found in column r. $\lambda(r)$ is then the "leading entry" of column r. For example, a glance at the table should suffice to see that $L(7) = 5$ and $\lambda(7) = 12$. The importance of $L(r)$ is easily explained: By virtue of the fact that no "holes" appear in columns of the $\mathcal{O}(n)$ table, the entire table may be reconstructed from the function $L(r)$. That is, L carries all tne information of $\mathcal{O}$. (So, of course, does λ.)

It will be of interest later to know that λ is monotone. For this we develop a sequence of easy results.

LEMMA 2.14. $\lambda(r)$ is an even integer for $r \geqq 1$.

PROOF. $\lambda(r)$ is the "leading entry" of column r. If $\lambda(r)$ were odd, $\mathcal{O}(n) = 2\mathcal{O}(n-1) \cup (1 + \mathcal{O}(n-1))$ would imply $\lambda(r) - 1 \in \mathcal{O}(n-1)$, where $n = L(r)$. But then $n - 1 + r = \lambda(r) - 1 \in \mathcal{O}(n-1)$, yielding $L(r) \leqq n - 1$, a contradiction. Q.E.D.

LEMMA 2.15. For $n \geqq 2$, $k \in \mathcal{O}(n)$ implies $k + 2 \in \mathcal{O}(n+1)$.

PROOF. Induction on n; the case $n = 2$ is checked easily. Let $k \in \mathcal{O}(n)$, $n \geqq 3$.

CASE 1. $k - 1 \in \mathcal{O}(n-1)$. Induction gives $k + 1 \in \mathcal{O}(n)$, and so $k + 2 \in \mathcal{O}(n+1)$.

CASE 2. $k/2 \in \mathcal{O}(n-1)$. Then $k/2 + 1 \in \mathcal{O}(n)$, and (doubling) $k + 2 \in \mathcal{O}(n+1)$. Q.E.D.

The preceding lemma will be generalized in Proposition 2.20.

PROPOSITION 2.16 (Reznick). For $r \geqq 1$,

$$L(r+1) = L(r) \pm 1.$$

PROOF. First $\lambda(r+1) \in \mathcal{O}(L(r+1))$, so by Proposition 2.8 $\lambda(r+1) \in \mathcal{O}(L(r+1)+1)$. Since $r = \lambda(r+1) - (L(r+1) + 1)$, we have $L(r) \leqq L(r+1) + 1$, i.e., $L(r+1) \geqq L(r) - 1$. On the other hand, $\lambda(r) \in \mathcal{O}(L(r))$ implies by Lemma 2.15 that $\lambda(r) + 2 \in \mathcal{O}(L(r)+1)$. Since $(\lambda(r) + 2) - (L(r) + 1) = r + 1$, we may assert that $L(r+1) \leqq L(r) + 1$. Finally, $L(r) = L(r+1)$ is impossible, since by Lemma 2.14 $r \equiv L(r) \pmod 2$. Q.E.D.

COROLLARY 2.17. λ is monotone; in fact,

$$\lambda(r+1) = \lambda(r) \quad \text{or} \quad \lambda(r) + 2.$$

PROOF. Add $r + 1$ to both sides in Proposition 2.16. Q.E.D.

In view of Corollary 2.17, one can easily see that for any positive even integer $k \geqq 4$, $\lambda^{-1}\{k\}$ is a finite set of consecutive integers. Also, determination of $|\lambda^{-1}\{k\}|$ for each such k determines the function λ. Reznick has obtained a simple formula for $|\lambda^{-1}\{k\}|$

involving the binary expansion of k. (See Appendix to Chapter 2.) Two more easy facts about L and λ respectively are given in the following proposition.

PROPOSITION 2.18. For $a, b \geqq 1$,

$$L(a+b) \leqq L(a) + L(b) - 2 \quad \text{and} \quad \lambda(a+b) \leqq \lambda(a) + \lambda(b) - 2.$$

PROOF. $L(a+b) \leqq L(a) + L(b) - 1$ is immediate from Proposition 2.6. Adding $a + b$ to both sides we get $\lambda(a+b) \leqq \lambda(a) + \lambda(b) - 1$. But by Lemma 2.14, we may strengthen this to $\lambda(a+b) \leqq \lambda(a) + \lambda(b) - 2$. Finally, subtracting $a + b$ we get $L(a+b) \leqq L(a) + L(b) - 2$. Q.E.D.

It is possible to characterize decomposability in terms of L or λ: indecomposables are found at "tops of columns".

PROPOSITION 2.19. $k \in \mathcal{O}(n)$ is indecomposable if and only if $n = L(k-n)$ (if and only if $k = \lambda(k-n)$).

PROOF. We have $L(k-n) \leqq n$ since $k \in \mathcal{O}(n)$. If $L(k-n) < n$ then $n = L(k-n) + (n - L(k-n))$, $k = \lambda(k-n) + (n - L(k-n))$ is a decomposition of $k \in \mathcal{O}(n)$. If $L(k-n) = n$, then clearly $k - 1 \notin \mathcal{O}(n-1)$, so $k \in \mathcal{O}(n)$ is 1-indecomposable, thus indecomposable by Proposition 2.6. Q.E.D.

Before moving on to results of another type, let us produce the promised generalization of Lemma 2.15.

PROPOSITION 2.20. $\mathcal{O}(n) + 2^t \subseteq \mathcal{O}(n+1)$ if $0 \leqq t \leqq [n/2]$.

PROOF. The case $t = 0$ is contained in Observation 2.2, while $t = 1$ is Lemma 2.15. So we will consider only $t \geqq 2$. Proceed by induction on n, small cases being checked easily. Let $x \in \mathcal{O}(n)$.

CASE 1. n is odd.

<u>Subcase 1</u>. $x - 1 \in \mathcal{O}(n-1)$. By induction, $x - 1 + 2^t \in \mathcal{O}(n)$, so $x + 2^t \in \mathcal{O}(n+1)$.

<u>Subcase 2</u>. $x/2 \in \mathcal{O}(n-1)$. By induction, $x/2 + 2^{t-1} \in \mathcal{O}(n)$, so $x + 2^t \in \mathcal{O}(n+1)$.

CASE 2. n is even.

<u>Subcase 1</u>. $x/2 \in \mathcal{O}(n-1)$. Then $x/2 + 2^{t-1} \in \mathcal{O}(n)$ by induction, so $x + 2^t \in \mathcal{O}(n+1)$.

<u>Subcase 2</u>. $x - 1 \in O(n-1)$ and $x - 2 \in O(n-2)$. Then $x - 2 + 2^{t-1} \in O(n-1)$ by induction, so

$$x - 2 + 2^t = x - 2 + 2^{t-1} + 2^{t-1} \in O(n)$$

by induction, and $x + 2^t \in O(n+1)$ by Lemma 2.15.

<u>Subcase 3</u>. $x - 1 \in O(n-1)$ and $(x - 1)/2 \in O(n-2)$. Then $(x-1)/2 + 2^{t-1} \in O(n-1)$ by induction, so $x - 1 + 2^t \in O(n)$ and $x + 2^t \in O(n+1)$. Q.E.D.

The previous results on L and λ provide a "vertical" description of $O(n)$. We now develop "horizontal" results, that is, information on location and size of gaps, as defined below.

DEFINITION 2.21. If $k_1, k_2 \in O(n)$ with $k_1 < k_2$, and if $k_1 < k < k_2$ implies $k \notin O(n)$, we say k_1 and k_2 <u>span a gap in $O(n)$, of length $k_2 - k_1$</u>. Note that a gap of length 1 represents no "missing" values; we may call such a gap trivial, or loosely say that there is "no gap".

PROPOSITION 2.22. All gaps have length a power of 2.

PROOF. Use induction on n. Small values of n are easily checked. Let k_1, k_2 span a gap of length greater than 1. Then $k_1 - 1 \notin O(n-1)$, for otherwise by Lemma 2.15 we would have $k_1 + 1 \in O(n)$, and $k_1 < k_1 + 1 < k_2$. Also, $k_2 - 1 \notin O(n-1)$, for otherwise by Proposition 2.8, $k_2 - 1 \in O(n)$, and again $k_1 < k_2 - 1 < k_2$. Thus $k_1/2, k_2/2 \in O(n-1)$, and clearly these span a gap in $O(n-1)$. By induction, $k_2/2 - k_1/2$ is a power of 2, so the same is true of $k_2 - k_1$. Q.E.D.

It may be of interest to identify the first nontrivial gap in $O(n)$.

DEFINITION 2.23. $\Phi(n) := \max \{k \in \mathbb{N} : i \in O(n) \text{ for } i = n, \ldots, k\}$.

LEMMA 2.24. Let $n \in \mathbb{N}$ be fixed. If $k_1, k_2 \in O(n)$ span a gap of length greater than 2 and k_1 is minimal for all such pairs k_1, k_2 then $\Phi(n+1) = k_1 + 2$.

PROOF. $k_1 \in O(n)$ must be indecomposable, for otherwise $k_1 - 1 \in O(n-1)$ and $k_1 + 1 \in O(n)$. Now if $k_1 + 3 \in O(n+1)$, then k_1 even (by indecomposability) implies that $k_1 + 2 \in O(n)$, contradicting $k_2 - k_1 > 2$. Thus $k_1 + 2 \in O(n+1)$ is the "initial point" of a nontrivial gap in $(n+1)$. Suppose there is a nontrivial gap in $O(n+1)$ spanned by c_1 and c_2 with $c_1 < k_1 + 2$. Since $c_1 + 1 \notin \mathcal{O}(n+1)$, we see that $c_1 - 1$ and $c_1 \notin \mathcal{O}(n)$. This contradicts minimality of k_1, so $k_1 + 2$ is the smallest "initial point" of a nontrivial gap in $\mathcal{O}(n+1)$, and $\Phi(n+1) = k_1 + 2$. Q.E.D.

LEMMA 2.5. If $k_1, k_2 \in \mathcal{O}(n)$ span a gap of length greater than 2 and k_1 is minimal as before, then $\Phi(n-1) = k_1/2$.

PROOF. $k_1, k_2 \in \mathcal{O}(n)$ must be indecomposable, for $k_1 - 1 \in \mathcal{O}(n-1)$ would imply $k_1 + 1 \in \mathcal{O}(n)$ while $k_2 - 1 \in \mathcal{O}(n-1)$ would imply $k_2 - 1 \in \mathcal{O}(n)$. Thus $k_1/2, k_2/2 \in \mathcal{O}(n-1)$, and they clearly span a gap in $\mathcal{O}(n-1)$. Now suppose $c_1, c_2 \in \mathcal{O}(n-1)$ span a gap with $c_1 < k_1/2$. The same argument as for k_1, k_2 shows that $c_1, c_2 \in \mathcal{O}(n-1)$ are indecomposable. We claim that $2c_1 \in \mathcal{O}(n)$ would initiate a gap of length greater than 2 in $\mathcal{O}(n)$, contradicting minimality of k_1. First, suppose $2c_1 + 1 \in \mathcal{O}(n)$. Then $2c_1 \in \mathcal{O}(n-1)$, and so either $c_1 \in \mathcal{O}(n-2)$ or $2c_1 - 1 \in \mathcal{O}(n-2)$. If $c_1 \in \mathcal{O}(n-2)$ then $c_1 + 1 \in \mathcal{O}(n-1)$ contradicting c_1, c_2 spanning a gap. If $2c_1 - 1 \in \mathcal{O}(n-2)$ then $2c_1 - 1 \in \mathcal{O}(n-1)$. But c_1 indecomposable in $\mathcal{O}(n-1)$ implies by Proposition 2.12 that $2c_1 \in \mathcal{O}(n)$ is indecomposable. Thus $2c_1 + 1 \notin \mathcal{O}(n)$. Finally suppose $2c_1 + 2 \in \mathcal{O}(n)$. Since $c_1 + 1 \notin \mathcal{O}(n-1)$, we must have $2c_1 + 1 \in \mathcal{O}(n-1)$. But then $2c_1 + 1 \in \mathcal{O}(n)$, contradicting what we have just shown. This shows that $2c_1 \in \mathcal{O}(n)$ initiates a gap of length greater than 2, a contradiction as planned. Q.E.D.

COROLLARY 2.26. $\Phi(n) =$

(i) $2^{n/2+1} - 2$, n even.

(ii) $2^{(n+1)/2} + 2^{(n-1)/2} - 2$, n odd.

PROOF. Early cases for n are checked easily. A remark of Bröcker's in (Br 2) is that $2^{n-1} - 2^{n-3}$, 2^{n-1} span a gap in $\mathcal{O}(n)$, so that for $n \geqq 5$ gaps of length greater than 2 exist. So assembling Lemmas 2.24 and 2.25, we may say that $\Phi(n+2) = 2\Phi(n) + 2$. The rest is simple induction. Q.E.D.

APPENDIX TO CHAPTER 2

Some further investigations into $\mathcal{O}$ have been made by B. Reznick. In this appendix we develop some of these results. In particular, we get a count on $\lambda^{-1}\{k\}$ for even $k \geq 4$. We start with one of Reznick's definitions which has proved invaluable.

DEFINITION 2.A.1. For $k \in \mathbb{N}$,

$$\mu(k) := \min \{n \in \mathbb{N} : k \in \mathcal{O}(n)\}.$$

So $\mu(k)$ denotes the first row in which k appears. Note that $\mu(k)$ is always defined, since $k \in \mathcal{O}(k)$.

OBSERVATION 2.A.2. k even $\geqq 4$ implies $k \in \mathcal{O}(\mu(k))$ is indecomposable.

PROOF. Suppose not. By Proposition 2.6, we then have $k - 1 \in \mathcal{O}(\mu(k)-1)$; since $k - 1$ is odd, $k - 2 \in \mathcal{O}(\mu(k)-2)$. Now if $\mu(k) - 2 = 1$, then $\mu(k) = 3$ and necessarily $k = 4$. This case checks since $3 \notin \mathcal{O}(2)$. Otherwise, Lemma 2.15 applies, and $k \in \mathcal{O}(\mu(k)-1)$, contradicting the definition of μ. Q.E.D.

μ may be fairly easily computed, as shown in the next proposition, and its consequence Proposition 2.A.4.

PROPOSITION 2.A.3 (<u>Reznick</u>). For $k > 1$,

$$\mu(k) = \begin{cases} 1 + \mu(k-1), & k \text{ odd} \\ 1 + \mu(k/2), & k \text{ even}. \end{cases}$$

PROOF. Suppose $k > 1$ is odd. Then $k - 1 \in \mathcal{O}(\mu(k)-1)$, so $\mu(k-1) \leqq \mu(k) - 1$. On the other hand $k \in \mathcal{O}(\mu(k-1)+1)$, so $\mu(k) \leqq \mu(k-1) + 1$.

Now suppose k is even. By Observation 2.A.2, we have $k/2 \in \mathcal{O}(\mu(k)-1)$, so $\mu(k/2) \leqq \mu(k) - 1$. But clearly $k \in \mathcal{O}(\mu(k/2)+1)$, so $\mu(k) \leqq \mu(k/2) + 1$. Q.E.D.

PROPOSITION 2.A.4 (<u>Reznick</u>). For $k \in \mathbb{N}$, write $k = \varepsilon_t \cdots \varepsilon_{0(2)}$ for the base 2 representation of k, with $\varepsilon_t = 1$. Then

$$\mu(k) = t + \sum_{i=0}^{t} \varepsilon_i.$$

PROOF. Using Proposition 2.A.3, this is a very easy mathematical induction exercise. Q.E.D.

Now we make the connection between μ and λ. Let $k \geqq 4$ be even. We have observed that $\lambda^{-1}\{k\}$ is a finite set of consecutive integers.

PROPOSITION 2.A.5. With k as above, $\max (\lambda^{-1}\{k\}) = k - \mu(k)$.

PROOF. $k \in \mathcal{O}(\mu(k))$ is indecomposable by Observation 2.A.2, so $\lambda(k-\mu(k)) = k$. However, if $r \in \lambda^{-1}\{k\}$ then $k \in \mathcal{O}(k-r)$, so $\mu(k) \leqq k - r$ and $r \leqq k - \mu(k)$. Q.E.D.

PROPOSITION 2.A.6. With k as above, $\min (\lambda^{-1}\{k\}) = k - \mu(k-2) - 1$.

PROOF. First note that $k \in \mathcal{O}(\mu(k-2)+1)$ by Lemma 2.15. To get $\lambda(k-\mu(k-2)-1) = k$, we must show that $k \in \mathcal{O}(\mu(k-2)+1)$ is indecomposable (see Proposition 2.19). But if we had $k - 1 \in \mathcal{O}(\mu(k-2))$, then $k - 1$ odd implies $k - 2$ is in $\mathcal{O}(\mu(k-2)-1)$, a contradiction.

Now let r be given with $\lambda(r) = k$, $r < k - \mu(k-2) - 1$ if possible. Since $k-2 \in \mathcal{O}(\mu(k-2))$ and $\mu(k-2) \leqq k - r - 2$, we may say by Proposition 2.8 that $k - 2 \in \mathcal{O}(k-r-2)$, provided that $k - 2 \geqq k - r - 2$, which is always true. Thus $k - 2 \in \mathcal{O}(k-r-2)$, and so $\lambda(r) \leqq k - 2$, a contradiction. Q.E.D.

COROLLARY 2.A.7. $|\lambda^{-1}\{k\}| = \mu(k-2) - \mu(k) + 2$.

PROOF. Direct from Propositions 2.A.5 and 2.A.6, and the fact that $\lambda^{-1}\{k\}$ is a set of consecutive integers. Q.E.D.

We can now apply the formula of Proposition 2.A.4 to give $|\lambda^{-1}\{k\}|$ more explicitly.

PROPOSITION 2.A.8 (Reznick). Let $k \geqq 4$ be an even integer, and write $k = 2^s \cdot u$, u odd. If $u = 1$, $|\lambda^{-1}\{k\}| = s - 1$. If $u > 1$, $|\lambda^{-1}\{k\}| = s$.

PROOF. If $u = 1$, the base 2 representation of $k - 2$ is $\varepsilon_s \cdots \varepsilon_0\ (2)$ with $\varepsilon_s = 1$ and $\varepsilon_i = 0$ for $i \neq s$. The base 2 representation for $k - 2$ would then be $\varepsilon'_{s-1} \cdots \varepsilon'_0\ (2)$ with $\varepsilon'_i = 1$ for $i \neq 0$ and $\varepsilon'_0 = 0$. So $\mu(k) = s + 1$, $\mu(k-2) = 2s - 2$, and $|\lambda^{-1}\{k\}| - s - 1$.

If $u \neq 1$, write the base 2 representation of k as $\varepsilon_t \cdots \varepsilon_s \varepsilon_{s-1} \cdots \varepsilon_0\ (2)$, where $\varepsilon_t = \varepsilon_s = 1$ and $\varepsilon_i = 0$ for i satisfying $0 \leqq i \leqq s - 1$. Then the representation for $k - 2$ is $\varepsilon_t \cdots \varepsilon_{s+1} \varepsilon'_s \cdots \varepsilon'_1 \varepsilon_0\ (2)$ where $\varepsilon'_s = 0$ and $\varepsilon'_i = 1$ for $1 \leqq i \leqq s - 1$. So we may compute:

$$\mu(k) = t + \sum_{i=0}^{t} \varepsilon_i, \quad \mu(k-2) = t + \sum_{i=s+1}^{t} \varepsilon_i + \sum_{i=1}^{s-1} \varepsilon'_i = t + \sum_{i=0}^{t} \varepsilon_i + s - 2,$$

and so $|\lambda^{-1}\{k\}| = s$. Q.E.D.

Reznick has also proved some beautiful results about $|\mathcal{O}(n)|$. We prove the main result, connecting $|\mathcal{O}(n)|$ with the Fibonacci sequence.

DEFINITION 2.A.9. For $n \geqq 2$, let $NEW(n) = \mathcal{O}(n) \setminus \mathcal{O}(n-1)$ and set $NEW(1) = \{1\}$.

PROPOSITION 2.A.10. For $n \geqq 3$,

$$NEW(n) = (2 \cdot NEW(n-1)) \dot{\cup} (1 + 2 \cdot NEW(n-2)).$$

PROOF. Let $k \in NEW(n)$ be even. We can't have $k - 1 \in \mathcal{O}(n-1)$, for if so, $k - 1$ odd implies $k - 2 \in \mathcal{O}(n-2)$, and then $k \in \mathcal{O}(n-1)$ by Lemma 2.15. (Actually, one must check $n = 3$ separately as Lemma 2.15 does not apply if $n - 2 = 1$.) Thus $k/2 \in \mathcal{O}(n-1)$. Clearly $k/2 \notin \mathcal{O}(n-2)$ for otherwise $k \in \mathcal{O}(n-1)$. Therefore $k/2 \in NEW(n-1)$.

Now let $k \in NEW(n)$ be odd. Then $k - 1 \in \mathcal{O}(n-1)$; but certainly $k - 1 \notin \mathcal{O}(n-2)$, for then $k \in \mathcal{O}(n-1)$. Thus $k - 1 \in NEW(n-1)$. By the first part of this argument, $(k - 1)/2 \in NEW(n-2)$, provided $n - 1 \geqq 3$. The case $n = 3$ being easy to check directly, we have in any case $(k - 1)/2 \in NEW(n-2)$.

Thus far, we have shown $NEW(n)$ is contained in $2 \cdot NEW(n-1) \cup 1 + 2 \cdot NEW(n-2)$. Let $k \in NEW(n-1)$. Then $2k \in \mathcal{O}(n)$. But if $2k \in \mathcal{O}(n-1)$, then $k \in \mathcal{O}(n-2)$ by Proposition 2.11. Therefore $2k \in NEW(n)$. Finally, take $k \in NEW(n-2)$. Then $2k + 1 \in \mathcal{O}(n)$. But if $2k + 1 \in \mathcal{O}(n-1)$, then $2k \in \mathcal{O}(n-2)$, and we would have $k \in \mathcal{O}(n-3)$ again by Proposition 2.11. Therefore $2k + 1 \in NEW(n)$, finishing the reverse inclusion.

Q.E.D.

COROLLARY 2.A.11 (<u>Reznick</u>). $|NEW(n)| = F_n$ (the nth Fibonacci number) and $|\mathcal{O}(n)| = F_{n+2} - n$.

PROOF. $|NEW(n)| = F_n$ directly from Proposition 2.A.10 using $NEW(1) = \{1\}$ and $NEW(2) = \{2\}$.

By Proposition 2.8 we may write $\mathcal{O}(n) = \mathcal{O}(n-1) \cup NEW(n) \setminus \{n-1\}$. So $|\mathcal{O}(n)| = |\mathcal{O}(n-1)| + F_n - 1$. If we assume inductively that $|\mathcal{O}(n-1)| = F_{n+1} - (n-1)$, then $|\mathcal{O}(n)| = F_{n+1} - (n-1) + F_n - 1 = F_{n+2} - n$ as desired. The induction begins with $|\mathcal{O}(1)| = 1 = F_3 - 1$.

Q.E.D.

CHAPTER 3

ARCHIMEDEAN AND NONARCHIMEDEAN ORDERINGS

Let F be a field, T a preordering on F with $|F^{\cdot}/T^{\cdot}| = 2^n < \infty$. It is well known that examples exist (among number fields, for example) where $|X/T| = n$ and where all orderings in X/T are archimedean. At the other extreme, suppose that $|X/T| = 2^{n-1}$, $n > 2$, so that T is a nontrivial fan. Then a result of Bröcker (see (4)) shows that some nontrivial valuation is strongly compatible with T. Since archimedean orderings are never compatible with (Krull) valuations, all orderings in X/T must be nonarchimedean. In this chapter, we effect a transition between these cases.

We set $X_{non}(F) = \{P \in X(F) : P$ is nonarchimedean$\}$ and $X_{arch}(F) = \{P \in X(F) : P$ is archimedean$\}$.

DEFINITION 3.1. For $n \in \mathbb{N}$, $k \in \mathcal{O}(n)$ let $\underline{f(n,k)}$ be the minimum value attained for $|X_{non}(F)/T|$ as (F,T) varies over preordered fields with $|F^{\cdot}/T^{\cdot}| = 2^n$ and $|X/T| = k$.

We can restate the above remarks: $f(n,) = 0$; $f(n,2^{n-1}) = 2^{n-1}$. In Corollary 3.6 we produce f explicitly, and in Corollary 3.7 we record the possibilities for $|X_{non}(F)/T|$.

Let us fix the notations: F is a field, T a preordering on F with $|F^{\cdot}/T^{\cdot}| = 2^n$.

The following proposition is implicitly contained in Bröcker's paper (5).

PROPOSITION 3.2. Let $X/T = \dot{\bigcup}_{i=1,\ldots,r} X_i$ with each X_i closed under $\sim$ (Bröcker's equivalence relation; see Chapter 1). Let $T_i = \cap X_i$ (that is, $T_i = \bigcap_{P \in X_i} P$), $|F^{\cdot}/T_i^{\cdot}| = 2^{m_i}$. Then:

(i) $F^{\cdot}/T^{\cdot} \xrightarrow[\text{nat}]{} \prod_{i=1}^{r} F^{\cdot}/T_i^{\cdot}$ is an isomorphism.

(ii) $n = \sum_{i=1}^{r} m_i$.

(iii) $|X/T| = \sum_{i=1}^{r} |X/T_i|$.

(iv) $|X/T_i| \in \mathcal{O}(m_i)$.

PROOF. (iv): This is part of the Hauptsatz in (5).

(iii): We need to show only that X_i is saturated, i.e., that $X_i = X/T_i$. Clearly $X_i \subseteq X/T_i$. Let $P \in X/T_i$. Suppose that the topology induced by P on F is distinct from all those induced by various $Q \in X_i$. Then the V-Topology Approximation Theorem would apply to show that $\phi \neq -P^{\cdot} \cap (\bigcap_{Q \in X_i} Q^{\cdot}) = -P^{\cdot} \cap T_i^{\cdot}$, contradicting $P \in X/T_i$. Thus $P \sim Q$ for some $Q \in X_i$, and so $P \in X_i$ as X_i is closed under $\sim$.

(ii): This will follow from (i).

(i): Prescribe $a_1, \ldots, a_r \in F^{\cdot}$. For any i $(1 \leqq i \leqq r)$, we may write $T_i^{\cdot} = T_{i,1}^{\cdot} \cap \cdots \cap T_{i,k_i}^{\cdot}$, where each $T_{i,j}$ is an intersection of a Bröcker class. Then

$$a_i T_i^{\cdot} = a_i T_{i,1}^{\cdot} \cap \cdots \cap a_i T_{i,k_i}^{\cdot}, \quad \text{and} \quad \bigcap_{i=1}^{r} a_i T_i^{\cdot} = \bigcap_{i,j} a_i T_{i,j}^{\cdot}.$$

Now any $T_{i,j}^{\cdot}$ is an intersection of a finite number of sets open with respect to a common order topology, and so $T_{i,j}^{\cdot}$ is open in that topology. The V-Topology Theorem again applies to give $\bigcap_{i=1}^{r} a_i T_i^{\cdot} \neq \phi$. Q.E.D.

COROLLARY 3.3. Let

$$T_{non} = \bigcap_{P \in X_{non}(F)/T} P \quad \text{and} \quad T_{arch} = \bigcap_{P \in X_{arch}(F)/T} P.$$

Write $|F^{\cdot}/T_{non}^{\cdot}| = 2^b$, $|F^{\cdot}/T_{arch}^{\cdot}| = 2^a$. Then $c := |X_{non}(F)/T| \in \mathcal{O}(b)$, $a = |X_{arch}(F)/T|$, $a + c = |X/T|$, and $a + b = n$.

PROOF. Apply Proposition 3.2, recalling that archimedean orderings form singleton $\sim$ classes. (Note that $X_{non}(F)/T = X/T_{non}$ and $X_{arch}(F)/T = X/T_{arch}$ from the proof of (iii) of 3.2.) Q.E.D.

We now have a necessary condition on $|X_{non}(F)/T|$ and $|X_{arch}(F)/T|$: If $|X/T| = k$, $|X_{non}(F)/T| = c$ and $|X_{arch}(F)/T| = a$ then we must have $c \in \mathcal{O}(b)$ where $a + b = n$. The next theorem shows that no other conditions are required. The proof uses ideas similar to those in (5). We precede the theorem with a lemma from valuation theory.

LEMMA 3.4. Let (F,v) be a valued field, Λ an ordered group containing Γ_v, and $\lambda \in \Lambda$ suchthat the sum $\Gamma_v + \mathbb{Z}\cdot\lambda$ is direct. Then v may be extended to a valuation w on the simple transcendental extension $F(x)$ of F such that $w(x) = \lambda$, $F(x)_w = F_v$ and Γ_w is the subgroup $\Gamma_v + \mathbb{Z}\cdot\lambda$ of Λ (with induced ordering).

PROOF. All of this is contained in Proposition 5, Chapter B of (21) except for the independence assumption on λ and Γ_v and the conclusion that $F(x)_w = F_v$. Let

$$f = \sum_{i=0}^{m} a_i x^i / \sum_{i=0}^{n} b_i x^i$$

be a unit of A_w. For $i \neq j$ we have $w(a_i x^i) \neq w(a_j x^j)$ unless $a_i = a_j = 0$ by our independence assumption, and similarly for monomials in the denominator. Among monomials appearing in the numberator, let $a_i x^i$ have minimum value; similarly find $b_j x^j$. By our remarks and the fact that our fraction is a unit

$$w(a_i x^i) = w(\sum_{k=1}^{m} a_k x^k) = w(\sum_{k=1}^{n} b_k x^k) = w(b_j x^j).$$

By independence, $i = j$ and $v(a_i) = v(b_i)$. It is now clear, by factoring $a_i x^i$ out of the numerator and $b_i x^i$ out of the denominator, that $f = (a_i/b_i)\cdot u$, where $u \in 1 + I_w$. Thus $F_v = F(x)_w$. Q.E.D.

THEOREM 3.5. Let $k, b, c, n \in \mathbb{N}$, $a \in \mathbb{N} \cup \{0\}$ with $n = a + b$, $k = a + c \in \mathcal{O}(n)$, and $c \in \mathcal{O}(b)$. Then there exists a countable pythagorean field F such that:

(i) $|F^{\cdot}/F^{\cdot 2}| = 2^n$.

(ii) F has exactly a archimedean orderings.

(iii) F has exactly c nonarchimedean orderings.

PROOF. We induct on n. Standard examples abound for $n = 1$.

CASE 1. $c - 1 \in \mathcal{O}(b-1)$. Inductively produce a countable pythagorean field F_0 with $|F_0^{\cdot}/F_0^{\cdot 2}| = 2^b$ having one archimedean and $c - 1$ nonarchimedean orderings. (If $a = 0$ and $b = n$, the original construction in (5) yields a satisfactory F.) Let $K = F_0(x, x^{1/2}, x^{1/4}, x^{1/8}, \ldots)$ and let v be the valuation on K associated withsthe place at 0 from K to F_0. Note that Γ_v is 2-divisible and v is rank 1. Within a fixed algebraic closure of K, let (H,w) be a henselization of (K,v), and let $R_1, \ldots, R_a$ be real closures of K with respect to any a archimedean orderings on K. (There are sufficiently many archimedean orderings on K. F_0 may be viewed as a countable subfield of $\mathbb{R}$ via its archimedean ordering; specializing x to any positive real transcendental over F_0 yields an archimedean ordering on K.) We claim that $F = R_1 \cap R_2 \cap \dots \cap R_a \cap H$ is the required field.

F is obviously countable and pythagorean. As $F^{\cdot 2} = R_1^{\cdot 2} \cap \dots \cap R_a^{\cdot 2} \cap H^{\cdot 2}$, we have an injection

$$\pi : F^{\cdot}/F^{\cdot 2} \longrightarrow R_1^{\cdot}/R_1^{\cdot 2} \times \cdots \times R_a^{\cdot}/R_a^{\cdot 2} \times H^{\cdot}/H^{\cdot 2}.$$

We check that π is also surjective. Choose α_i (= ±1 WLOG) $\in R_i^{\cdot}$, $i = 1, \dots, a$ and choose $\beta \in H^{\cdot}$. Each $F \cap \alpha_i R_i^{\cdot 2}$ is just a "strictly" positive or negative cone of an archimedean ordering on F. (H,w) is a henselization of $(F,w|_F)$; so if $\{P_k\}$ is the set of orderings on F compatible with $w|_F$, each P_k induces the same topology on F, P_k lifts to some $Q_k \in X(H)$, and $\bigcap_k Q_k^{\cdot 2} = H^{\cdot 2}$. So $\beta H^{\cdot 2} \cap F = \bigcap_k \beta P_k^{\cdot}$ is an open set in the topology of $w|_F$; as $F^{\cdot}/F^{\cdot 2} \longrightarrow H^{\cdot}/H^{\cdot 2}$ is surjective, this set is also nonempty. Therefore, the V-Topology Approximation Theorem applies, and π is surjective.

Since $\Gamma_v = \Gamma_w$ is 2-divisible, we may conclude that $|H^{\cdot}/H^{\cdot 2}| = |F_0^{\cdot}/F_0^{\cdot 2}| = 2^b$, and $|X(H)| = |X(F_0)| = |\{P \in X(K) : P$ is compatible with $v\}| = c$. The fact that π is an isomorphism gives us $|F^{\cdot}/F^{\cdot 2}| = 2^a 2^b = 2^n$.

We have a archimedean and c nonarchimedean orderings already acco nted for on F (distinct since they restrict to distinct orderings on K). Suppose there is a further ordering P on F; unless P induces the same topology as some ordering already accounted for on F, the V-Topology Theorem would apply once more to give surjectivity of

$$\pi^* : F^{\cdot}/F^{\cdot 2} \longrightarrow R_1^{\cdot}/R_1^{\cdot 2} \times \cdots \times R_a^{\cdot}/R_a^{\cdot 2} \times H^{\cdot}/H^{\cdot 2} \times F^{\cdot}/P^{\cdot}$$

which is clearly impossible. Thus P would have to induce the same topology as some Q among the $k = a + c$ orderings known. But if $P \neq Q$, then Q is nonarchimedean, and P and Q are compatible with some valuation. However, Q is compatible with the rank 1 valuation $w|_F$; it follows that P is also compatible with $w|_F$, and so must be the restriction to F of some ordering on H. That is, P is accounted for.

CASE 2. $c - 1 \in \mathcal{O}(b-1)$. Repeatedly applying Proposition 2.3, we may write $c = 2^s \cdot t$ with $t \in \mathcal{O}(b-s)$, and $t - 1 \in \mathcal{O}(b-s-1)$. Inductively produce a field F_0 which is countable and pythagorean with $|F_0^{\cdot}/F_0^{\cdot 2}| = 2^{b-s}$, $|X_{non}(F_0)| = t - 1$ and $|X_{arch}(F_0)| = 1$. Let $K = F_0(x_1, \dots, x_s)$.

Iterated use of Lemma 3.4 gives us a place from K to $F_0 \cup \{\infty\}$ such that the associated value group Γ is the subgroup of $\mathbb{R}$ generated by the first s primes. Once again, K admits many archimedean orderings. Note that $|\Gamma/2\Gamma| = 2^s$ so that

$$|H^{\cdot}/H^{\cdot 2}| = |F_0^{\cdot}/F_0^{\cdot 2}|\cdot|\Gamma/2\Gamma| = 2^{b-s}\cdot 2^s = 2^b$$

and

$$|X(H)| = |X(F_0)|\cdot|(\Gamma/2\Gamma)^* = t\cdot 2^s = c.$$

(We are using $v(H^{\cdot 2}) = 2\Gamma$.) Also, the valuation used here is again rank rank 1 ($\Gamma < \mathbf{R}$). Now we may proceed exactly as in Case 1. Q.E.D.

COROLLARY 3.6. $f(n,k) = \lambda(k-n)$. (For λ, see 2.13.)

PROOF. Let $k \in \mathcal{O}(n)$, and set $r = k - n$. Then we may write

$$n = (n - L(r)) + L(r), \quad k = (n - L(r)) + (k - n + L(r)) = (n - L(r)) + \lambda(r);$$

since $\lambda(r) \in \mathcal{O}(L(r))$, we may take $a = n - L(r)$, $b = L(r)$, $c = \lambda(r)$ in Theorem 3.5. Therefore $f(n,k) \leqq \lambda(r)$.

On the other hand, suppose (F,T) is a preordered field such that $|F^{\cdot}/T^{\cdot}| = 2^n$ and $|X/T| = k$. Let $a = |X_{arch}/T|$, $c = |X_{non}/T|$. By Proposition 3.2, $c \in \mathcal{O}(b)$, where $b = n - a$. Since $c - (n - a)$ $= (a + c) - n = k - n = r$, we see that $\lambda(r) \leqq c$. Taking min over all (F,T) with $|F^{\cdot}/T^{\cdot}| = 2^n$, $|X/T| = k$ we get $\lambda(r) \leqq f(n,k)$. Q.E.D.

COROLLARY 3.7. The precise set of possibilities for $|X_{non}(F)/T|$ as (F,T) ranges through preordered fields with $|F^{\cdot}/T^{\cdot}| = 2^n$ and $|X/T| = k$ is $\{\lambda(r),\lambda(r)+1,\ldots,k\}$ where $r = k - n$, and we temporarily set $\lambda(0) = 0$.

PROOF. Corollary 3.6 gives $\lambda(r)$ as a lower bound, while k is clearly an upper bound. Now let $t \in \mathbf{N}$, $0 \leqq t \leqq k - \lambda(r) = n - L(r)$. Write $n = (n - L(r) - t) + (L(r) + t)$, $k = (n - L(r) - t) + (\lambda(r) + t)$. We may apply Theorem 3.5 to get a pair (F,T) as above with precisely $\lambda(r) + t$ nonarchimedean orderings, provided $\lambda(r) + t \in \mathcal{O}(L(r) + t)$. But this is clear from the definition of $\mathcal{O}(n)$ and the fact that $\lambda(r) \in \mathcal{O}(L(r))$. Q.E.D.

CHAPTER 4

BRÖCKER CLASSES AND MARSHALL CLASSES

In this chapter, we investigate the possible numbers of Bröcker classes for fields. We also study an equivalence relation of M. Marshall's, and relate it to that of Bröcker.

We take our beginning point from Proposition 3.2 which gives us, upon choosing the X_i's to be Bröcker classes, that the partition of $X(F)$ corresponding to $\sim$ induces a decomposition (Definition 2.9) of $k \in \mathcal{O}(n)$, where as usual $k = |X/T|$ and $|F^{\cdot}/T^{\cdot}| = 2^n$. One can ask for a converse; this is provided by the following theorem, which is extremely close to the Hauptsatz in (5).

THEOREM 4.1. Let $k \in \mathcal{O}(n)$ have a decomposition

$$k = \sum_{i=1}^{t} k_i, \quad n = \sum_{i=1}^{t} n_i, \quad k_i \in \mathcal{O}(n_i).$$

Then there is a field F with exactly t Bröcker classes $X_1, \ldots, X_t$ such that $|F^{\cdot}/T_i^{\cdot}| = 2^{n_i}$ (where $T_i = \bigcap_{P \in X_i} P$) and $|X/T_i| = |X_i| = k_i$ for $i = 1, \ldots, t$.

PROOF. We mimic and use explicitly Bröcker's original construction. For each i, $1 \leqq i \leqq t$, we may find a pythagorean field F_i of finite transcendence degree over $\mathbb{Q}$ such that $|F_i^{\cdot}/F_i^{\cdot 2}| = 2^{n_i}$ and $|X(F_i)| = k_i$. This is possible by the Hauptsatz in (5). We would like to "adjust" these fields so that all have the same transcendence degree over $\mathbb{Q}$. To do this, we cite the trick appearing in (5) for raising transcendence degree by 1 without perturbing the critical data $|F_i^{\cdot}/F_i^{\cdot 2}| = 2^{n_i}$ and $|X(F_i)| = k_i$: Replace F_i with a henselization of $F_i(x, x^{1/2}, x^{1/4}, x^{1/8}, \ldots)$ with respect to the place at 0. Without loss of generality, then, we may assume that all F_i's are algebraic extensions of a rational function field $\mathbb{Q}(x_1, \ldots, x_s)$. Endow $\mathbb{Q}(x_1, \ldots, x_s)(y)$ (y an indeterminate) with the degree valuation v; then $\Gamma_v = \mathbb{Z}$ and $\mathbb{Q}(x_1, \ldots, x_s)(y)_v = \mathbb{Q}(x_1, \ldots, x_s)$. By a standard theorem in valuation theory (see, e.g., Corollary 27.2 in (9)) there is an algebraic extension K of $\mathbb{Q}(x_1,\ldots,x_s)(y)$ admitting distinct extensions $v_1,\ldots,v_t$ of v suc that $K_{v_i} = F_i$ and $\Gamma_{v_i} = \mathbb{Q}$ for $i = 1,\ldots,t$. Within a fixed algebraic closure of K, let (H_i,w_i) be a henselization of (K,v_i) for $i = 1,\ldots,t$. We claim that $F = \bigcap_{i=1}^{t} H_i$ fulfills the desired conditions. Note that (H_i,w_i) is a herselization of $(F,w_i|F)$.

Since $\Gamma_{w_i} = \mathbb{Q}$ is 2-divisible, we have $|H_i^{\cdot}/H_i^{\cdot 2}| = |F_i^{\cdot}/F_i^{\cdot 2}| = 2^{n_i}$ and $|X(H_i)| = |X(F_i)| = k_i$. The argument which was used in the proof of Theorem 3.5, mutatis mutandis, shows that $F^{\cdot}/F^{\cdot 2} \longrightarrow H_1^{\cdot}/H_1^{\cdot 2} \times \ldots \times H_t^{\cdot}/H_t^{\cdot 2}$ is surjective, and that each ordering on F is induced by some ordering on some H_i. The orderings induced on F by any particular H_i are all in one Bröcker class. Also, if $P \in X(H_i)$, $Q \in X(H_j)$ with $i \neq j$, then $P \cap F \sim Q \cap F$ is impossible since valuation rings compatible with an ordering form a chain, and P and Q are compatible with distinct rank 1 valuations. (If P, Q were both compatible with w, then $A_w \subseteq A_{w_i} \cap A_{w_j}$ with $i \neq j$ corresponding to P, Q respectively; then P would be compatible with w_i and w_j, which is impossible.)

Finally, notice that if

$$T_i = \bigcap \{P;\ P \in X^{w_i}(F)\}$$

then

$$|F^{\cdot}/T_i^{\cdot}| = |H_i^{\cdot}/H_i^{\cdot 2}| = 2^{n_i}, \quad \text{and} \quad |X/T_i| = |X(H_i)| = k_i.$$

Q.E.D.

In order to discuss the possible numbers of Bröcker classes, we define a function b.

DEFINITION 4.2. Let (F,T) be a preordered field. We define $b(F,T)$ to be the number of $\sim$ classes on X/T.

LEMMA 4.3. If (F,T) is a preordered field with $|F^{\cdot}/T^{\cdot}| = 2^n$, $|X/T| = k \in \mathcal{O}(n)$ indecomposable, then $b(F,T) = 1$.

PROOF. If $b(F,T) > 1$, we would obtain a t-fold (and hence a 2-fold) decomposition of $k \in \mathcal{O}(n)$, $t > 1$, contradicting indecomposability. Q.E.D.

PROPOSITION 4.4. Fix $n \in \mathbb{N}$, $r \in \mathbb{N} \cup \{0\}$ with $n + r \in \mathcal{O}(n)$. Then as (F,T) vary with restrictions $|F^{\cdot}/T^{\cdot}| = 2^n$, $|X/T| = n + r$:

(i) If $r = 0$, $b(F,T)$ will take on all values from 1 to n.

(ii) If $r > 0$, $b(F,T)$ will take on exactly the values 1 through $n - L(r) + 1$.

PROOF. (i): This is clear from Theorem 4.1, and the fact that for every $m \in \mathbb{N}$, $m \in \mathcal{O}(m)$ so that t-fold decompositions of $n \in \mathcal{O}(n)$ can be found for every $t \leqq n$.

(ii): Note that we may write $n + r = \lambda(r) + \sum_{i=1}^{n-L(r)} 1$ with $\lambda(r) \in \mathcal{O}(L(r))$. By grouping 1's, we may produce t-fold decompositions for any $t = 1, \ldots, n - L(r) + 1$ for $n + r \in \mathcal{O}(n)$. Thus $b(F,T)$ takes on each of these values.

To show that $b(F,T) > n - L(r) + 1$ is impossible, we show that $n + r \in \mathcal{O}(n)$ cannot admit a t-fold decomposition with $t > n - L(r) + 1$. The proof is by induction on $n - L(r)$. The case $n - 1(r) = 0$ is Lemma 4.3. (Note $n + r \in \mathcal{O}(n)$ implies $n \geqq L(r)$, so we may initialize the induction with $n - L(r) = 0$.) For the general case, suppose $n + r \in \mathcal{O}(n)$ admits a t-fold decomposition with $t > n - L(r) + 1$. Then by Proposition 2.10 $n + r - 1 \in \mathcal{O}(n-1)$ admits a (t-1)-fold decomposition. But then $t - 1 > (n - 1) - L(r) + 1$ contradicts the inductive hypothesis.

Q.E.D.

We now turn our attention to a relation $\tilde{m}$ defined by M. Marshall in (14). Although Marshall's definition is given in the context of abstract spaces of orderings, we give here the concrete interpretation in the theory of ordered fields.

DEFINITION 4.5. Let (F,T) be a preordered field. For $P_1, P_2 \in X/T$ we write $P_1 \tilde{m} P_2$ if and only if $P_1 = P_2$ or there exists a nontrivial fan $T_0 \supseteq T$ such that $P_1, P_2 \in X/T_0$. (One may always find such a T_0, if it exists, with the property $|X/T_0| = 4$.)

Note that $P \tilde{m} Q$ implies that $P \sim Q$, and that $\tilde{m}$ is reflexive and symmetric. In the context of abstract spaces of orderings, Marshall has proved that $\tilde{m}$ is actually an equivalence relation. We can give a direct proof here for the case of ordered fields. The advantages of doing so despite the prior existence of a more general theorem are:

(1) The relationship between $\sim$ and $\tilde{m}$ is clarified, at least for $|X/T| < \infty$, and

(2) This provides an example of how the theory of finite spaces of orderings can influence the general theory ($|X/T|$ infinite). For this, we prepare a reduction to the finite case.

PROPOSITION 4.6. $\tilde{m}$ is a transitive relation on every $X(F)/T$ if and only if it is a transitive relation on every $X(F)/T$ with $|F^{\cdot}/T^{\cdot}| < \infty$.

PROOF. Let (F,T) be arbitrary, and suppose $P_1, P_2, P_3 \in X/T$ are distinct orderings with $P_1 \tilde{m} P_2 \tilde{m} P_3$. Then there are orderings P_4, P_5, P_6 and P_7 such that $\{P_1,P_2,P_4,P_5\}$ and $\{P_2,P_3,P_6,P_7\}$ are 4-boxes. Let $T_0 = \bigcap_{i=1}^{7} P_i$. Then we have $P_1 \tilde{m} P_2 \tilde{m} P_3$ in $X(F)/T_0$, and $|F^{\cdot}/T_0^{\cdot}| < \infty$. (In fact, $|F^{\cdot}/T_0^{\cdot}| \leqq 2^5$.) So if $\tilde{m}$ is transitive in such cases, $P_1 \tilde{m} P_3$ in $X(F)/T_0$, and then certainly $P_1 \tilde{m} P_3$ in $X(F)/T$. Q.E.D.

Thus, in order to show $\tilde{m}$ to be an equivalence relation, it suffices to consider the case $|X/T| < \infty$. To do this, our strategy will be to define a new equivalence relation R inductively (in terms of its associated partition), and show that $R = \tilde{m}$.

DEFINITION 4.7. Let (F,T) be a preordered field with $|X(F)/T| < \infty$. We define a partition on $X(F)/T$ called the <u>R-partition</u> (with associated equivalence relation <u>R</u>) recursively as follows:

Partition X/T into Bröcker classes, with associated preorderings T_i. Split any doubletons into singletons. Leave alone any X/T_i which have more than 2 elements and have finest compatible valuation v (A_v smallest such that v is strongly compatible with T_i) with $|\Gamma_v/v(T_i)| \neq 1$; but in the case $|\Gamma_v/v(T_i)| = 1$ replace X/T_i with its "carbon copy" $X/\overline{T_i}$, which divides into smaller Bröcker classes. Then iterate.

(A more formal and certainly more opaque definition of R is as follows: If $X(F)/T = \{P\}$ we let $P R P$. In general, $P_1 R P_2$ in X/T if and only if $P_1, P_2 \in X/T_i$ for some i, and for v the finest valuation strongly compatible with T_i we have either

(i) $|\Gamma_v/v(T_i^{\cdot})| > 1$ and, if $|\Gamma_v/v(T_i^{\cdot})| = 2$, $|X/\overline{T_i}| > 1$ <u>or</u>

(ii) $|\Gamma_v/v(T_i^{\cdot})| = 1$ and $\overline{P_1} R \overline{P_2}$ in $X/\overline{T_i}$. Here we use that $X/\overline{T_i}$ must split into more than one Bröcker class, so induction on the size of Bröcker classes can be used.)

LEMMA 4.8. Let (F,T) be a preordered field, v a valuation strongly compatible with T. Then

(i) $A \subseteq X/T$ a box $\longrightarrow \{\overline{P} : P \in A\} \subseteq X/\overline{T}$ is a box.

(ii) $B \subseteq X/\overline{T}$ a box $\longrightarrow \{P : \overline{P} \in B\} \subseteq X/T$ is a box.

PROOF. This is direct from the remarks in Chapter 1 and from the characterization of fans given in (2). Q.E.D.

Before proceeding with the next proposition, observe that from the definition of R, $P_1 R P_2$ implies $P_1 \sim P_2$.

PROPOSITION 4.9. For any preordered field (F,T) with $|F^{\cdot}/T^{\cdot}| < \infty$, we have $\tilde{m} = R$ on X/T.

PROOF. First assume $P_1 R P_2$, $P_1 \neq P_2$; induct on the number of orderings in the Bröcker class X/T_i containing P_1 and P_2. Let v be the finest valuation strongly compatible with T_i. Since $P_1 \neq P_2$, we may use the definition of R to distinguish two cases.

CASE 1. $|\Gamma_v/v(T_i^{\cdot})| \neq 1$.

Subcase 1. $|X/\overline{T_i}| = 1$ or 2. Then Lemma 4.8 shows that X/T_i is a box; thus T_i is a fan containing P_1 and P_2, and cannot be trivial since $P_1 \neq P_2$ and an R class cannot be a doubleton. Therefore $P_1 \tilde{m} P_2$.

Subcase 2. $|X/\overline{T_i}| \geqq 4$. Choose distinct $Q_1, Q_2 \in X/\overline{T_i}$ so that $\{\overline{P_1}, \overline{P_2}\} \subseteq \{Q_1, Q_2\}$. Then Lemma 4.8 shows that $\{P \in X/T_i : \overline{P} = Q_1 \text{ or } Q_2\}$ is a box containing P_1 and P_2, and this set must have at least 4 elements since we are in the case $|\Gamma_v/v(T_i^{\cdot})| \neq 1$. (We could just as easily have included the case $|X/\overline{T_i}| = 2$ in this subcase rather than Subcase 1.) So we have $P_1 \tilde{m} P_2$.

CASE 2. $|\Gamma_v/v(T_i^{\cdot})| = 1$. The recursion for R forces $\overline{P_1} R \overline{P_2}$ in $X/\overline{T_i}$. As $X/\overline{T_i}$ must split into more than one Bröcker class (lest we "refine" v), induction implies that $\overline{P_1} \tilde{m} \overline{P_2}$. But then Lemma 4.8 applied to a 4-box containing $\overline{P_1}$ and $\overline{P_2}$ gives us $P_1 \tilde{m} P_2$. This completes Case 2.

Now let us assume $P_1 \tilde{m} P_2$, $P_1 \neq P_2$. As $P_1 \sim P_2$, let X/T_i be the Bröcker class containing P_1 and P_2, and take v as before. If $|\Gamma_v/v(T_i^{\cdot})| \neq 1$, $P_1 R P_2$ perforce. If $|\Gamma_v/v(T_i^{\cdot})| = 1$, any 4-box in X/T_i containing P_1 and P_2 will "push down" by Lemma 4.8 to a 4-box containing $\overline{P_1}$ and $\overline{P_2}$ in $X/\overline{T_i}$. So $\overline{P_1} \tilde{m} \overline{P_2}$. Induction applies as in Case 2 above, yielding $\overline{P_1} R \overline{P_2}$. The recursive definition of R then assures $P_1 R P_2$. Q.E.D.

COROLLARY 4.10. For all preordered fields (F,T), $\tilde{m}$ is an equivalence relation on X/T.

REMARK 4.11. In Marshall's paper "Classification of Finite Spaces of Orderings" (14), it is shown that when a finite space of orderings is decomposed by $\tilde{m}$, each equivalence class is isomorphic to the space of orderings of some pythagorean field. It follows that decomposition by $\tilde{m}$ leads to decomposition of $k \in \mathcal{O}(n)$ just as for $\sim$; the same upper bound $n - L(r) + 1$ (or n if $r = 0$) can be established for the number of $\tilde{m}$ classes. In fact, this bound is still faithful; e.g., use a field with exactly $\lambda(r)$ nonarchimedean orderings. However, the analog of Theorem 4.1 does not hold. For example, one may easily see that the number of orderings in an $\tilde{m}$-class is either 1 or even.

CHAPTER 5

THE STABILITY INDEX

An important invariant for a preordered field (F,T) has been introduced by Bröcker (4), called the (reduced) stability index.

DEFINITION 5.1. For a preordered field (F,T), we define the (reduced) stability index $st(F,T)$ by

$$st(F,T) = \log_2(\max\{|X/T_0| : T_0 \supseteq T,\ T_0 \text{ a fan}\}).$$

Note that $|X/T_0|$ is a power of 2 for any fan T_0 of finite index, so that $st(F,T)$ is a nonnegative integer if X/T is a finite set.

In this chapter, we establish conditions on the possible values of $st(F,T)$ for $|F^{\cdot}/T^{\cdot}| = 2^n$, $|X/T| = k \in \mathcal{O}(n)$. These conditions are most easily realized and visualized by describing certain regions in the $\mathcal{O}(n)$ Table, i.e., sets of pairs (n,k) with $k \in \mathcal{O}(n)$, outside of which certain stabilty indices cannot occur. As a trivial example, any preordered field (F,T) with $|F^{\cdot}/T^{\cdot}| = 2^n$ and $|X/T| = k > n$ must have some nontrivial fan T_0 containing T. Thus $st(F,T) \geqq 2$ if $k > n$. We find it most convenient to describe the regions by referring to a "location" in the $\mathcal{O}(n)$ Table as a triple (n,r,k) with $k \in \mathcal{O}(n)$ and $r = k - n$. The redundancy permits handy use of all three parameters.

DEFINITION 5.2. For each $a \in \mathbb{N}$, define

$$\underline{B_a} := \{(n,r,k) : k \in \mathcal{O}(n),\ k = n+r,\ r \geqq 2^a-a-1,\ k \leqq 2^{a-1}n-(a-1)2^{a-1},\ n \geqq 2\}.$$

Also set $B_0 = \{(1,0,1)\}$.

REMARK 5.3. B_0 is so chosen since $st(F,T) = 0$ only when $|F^{\cdot}/T^{\cdot}| = 2^1$ and $|X/T| = 1$, corresponding to values $n = 1$, $r = 0$, $k = 1$ for our parameters. For $a \geqq 1$, B_a may be pictured as the region of the $\mathcal{O}(n)$ Table on or between two rays emanating from $(a+1, 2^a-a-1, 2^a)$: One ray is vertical; the other proceeds in the direction of increasing n, incrementing k values by 2^{a-1} as n increases by 1. A table indicating the position of B_a for small a appears on the following page.

REMARK 5.4. $(n,r,k) \in B_a \longrightarrow n \geqq a + 1$. To see this, note that $2^a \in \mathcal{O}(a+1)$ is indecomposable for $a \geqq 2$, so that $\lambda(2^a-a-1) = 2^a$ for $a \geqq 2$. Then the condition $r \geqq 2^a - a - 1$ implies $\lambda(r) \geqq 2^a$ by monotonicity of λ (Corollary 2.17). But then $k \in \mathcal{O}(n)$, $k \geqq \lambda(r)$ implies $n \geqq a+1$, since the largest entry in $\mathcal{O}(n)$ is 2^{n-1}. If $a = 0$, the condition is vacuous, while if $a = 1$, one has the stipulation $n \geqq 2$ in the definition of B_a.

In order to set up successful induction, we must investigate the behavior of the reduced stability index under two circumstances: When X/T is decomposed into Bröcker classes, and when there is a valuation strongly compatible with T.

LEMMA 5.5. Let (F,T) be a preordered field. Write

$$X/T = X_1 \dot{\cup} X_2 \dot{\cup} \cdots \dot{\cup} X_t$$

for the partition of X/T into Bröcker classes. Set T_i equal to the intersection of the orderings in X_i, for $i = 1, \ldots, t$. Then, providing X/T is not SAP, $st(F,T) = \max\{st(F,T_i) : i = 1, \ldots, t\}$.

PROOF. Let $T^* \supseteq T$ be a fan of maximal index in $F^{\cdot}$. Since X/T is not SAP, T^* must be a nontrivial fan. Thus (see (4)) there is a valuation strongly compatible with T^*, and so $X/T^* \subseteq X_i$ for some i. Thus $T^* \supseteq T_i$, showing $st(F,T) \leqq \max\{st(F,T_i) : i = 1, \ldots, t\}$. The reverse inequality is immediate from the definition of $st(F,T)$ since each $T_i \supseteq T$.

Q.E.D.

LEMMA 5.6. Let (F,T) be a preordered field with X/T a single Bröcker class. Let v be the finest valuation (i.e., A_v minimal) strongly compatible with T. Then (writing as usual $\bar{\ }$ for passage to F_v)

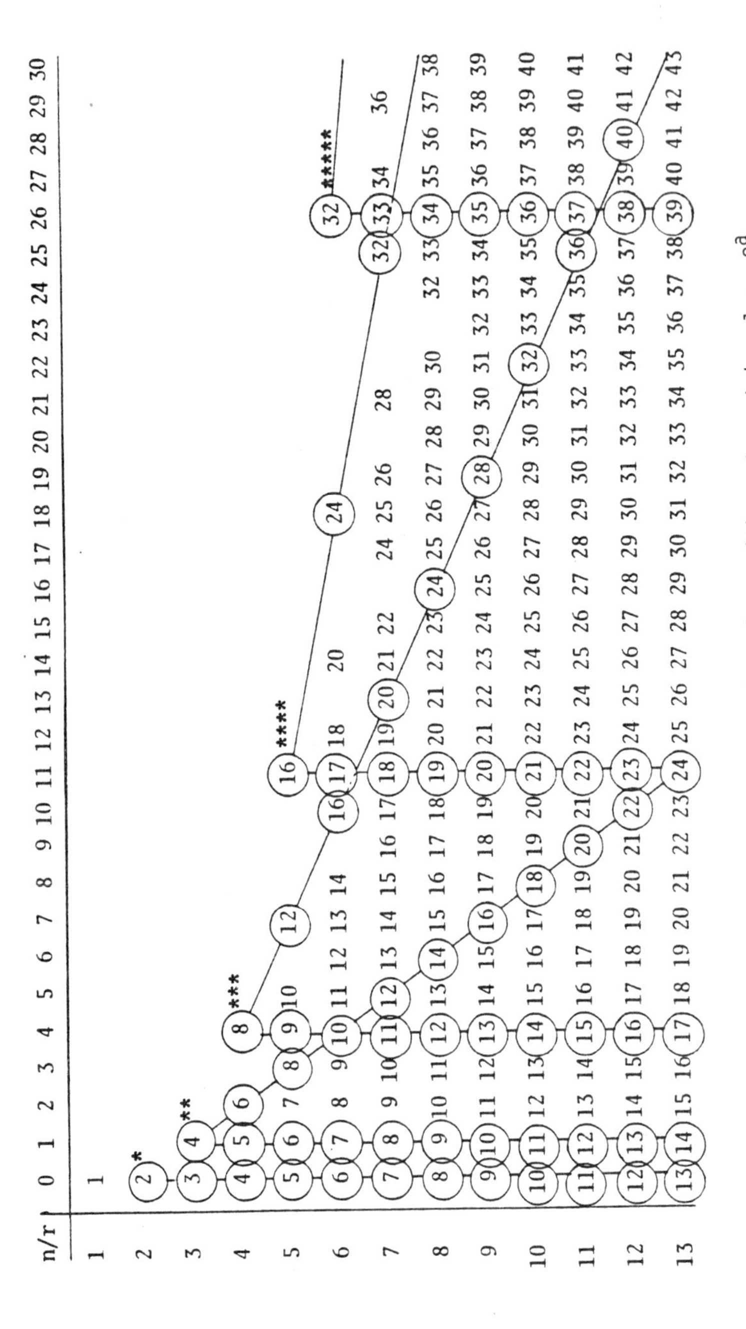

Region B_a is the angle with interior, with vertex at k-value 2^a.

a asterisks designate the vertex of the angle for B_a.

$$st(F,T) = st(\overline{F},\overline{T}) + t, \quad \text{where} \quad |\Gamma_v/v(T^\cdot)| = 2^t.$$

PROOF. Let $\{Q_1,\ldots,Q_{2^m}\} \subseteq X/\overline{T}$ be a box with m maximal, so that $st(\overline{F},\overline{T}) = m$. Then $\{P \in X/T : \overline{P} = Q_i \text{ for some } i\}$ is seen by Lemma 4.8 to be a box; its cardinality is 2^{m+t} by 1.10 (ii). Thus $st(F,T) \geqq m + t$. On the other hand, if $\{P_1,\ldots,P_s\}$ is a box in X/T, $\{\overline{P}_1,\ldots,\overline{P}_s\}$ is a box in $X/\overline{T}$, also by Lemma 4.8. But then $|\{\overline{P}_1,\ldots,\overline{P}_s\}| \leqq 2^m$, and $|\{P_1,\ldots,P_s\}| \leqq 2^{m+t}$ as each ordering in $X/\overline{T}$ has 2^t "lifts" in X/T. Thus $st(F,T) \leqq m + t$. Q.E.D

Now it becomes necessary to check that the sets B_a have been appropriately defined so that induction can function smoothly. Assertion (ii) of the following lemma is tailored for application of Lemma 5.6, while assertion (i), which says that the sets B_a form a sort of "filtration", is designed for application of Lemma 5.5.

LEMMA 5.7. (i) For $\alpha \leqq \beta$, $\beta \neq 0$, we have $B_\alpha + B_\beta \subseteq B_\beta$.

(ii) $(n,r,k) \in B_\alpha$ implies $(n+1,\ n+2r-1,\ 2k) \in B_{\alpha+1}$.

PROOF. (i) If $\alpha = 0$, the proof is easy. So assuming $0 < \alpha \leqq \beta$, we may write, given $(n_1,r_1,k_1) \in B_\alpha$ and $(n_2,r_2,k_2) \in B_\beta$,

$$r_1 \geqq 2^\alpha - \alpha - 1, \quad k_1 \leqq 2^{\alpha-1}n_1 - (\alpha-1)2^{\alpha-1} \quad \text{and}$$

$$r_2 \geqq 2^\beta - \beta - 1, \quad k_2 \leqq 2^{\beta-1}n_2 - (\beta-1)2^{\beta-1}.$$

So

$$r_1 + r_2 \geqq 2^\alpha - \alpha - 1 + 2^\beta - \beta - 1 \geqq 2^\beta - \beta - 1, \quad \text{and}$$

$$\begin{aligned} k_1 + k_2 &\leqq 2^{\alpha-1}n_1 - (\alpha-1)2^{\alpha-1} + 2^{\beta-1}n_2 - (\beta-1)2^{\beta-1} \\ &\leqq 2^{\beta-1}n_1 - (\alpha-1)2^{\alpha-1} + 2^{\beta-1}n_2 - (\beta-1)2^{\beta-1} \\ &= 2^{\beta-1}(n_1+n_2) - (\beta-1)2^{\beta-1} - (\alpha-1)2^{\alpha-1} \\ &\leqq 2^{\beta-1}(n_1+n_2) - (\beta-1)2^{\beta-1} \end{aligned}$$

which are the desired conditions.

Also, $k_1 \in \mathcal{O}(n_1)$ and $k_2 \in \mathcal{O}(n_2)$, so $k_1 + k_2 \in \mathcal{O}(n_1+n_2)$; $k_1 = n_1 + r_1$ and $k_2 = n_2 + r_2$ give us $k_1 + k_2 = (n_1 + n_2) + (r_1 + r_2)$ This verifies (i).

(ii) For $\alpha = 0$, check directly. Otherwise:

$$k \leqq 2^{\alpha-1}n - (\alpha-1)2^{\alpha-1} \longrightarrow 2k \leqq 2^{\alpha}n - (\alpha-1)2^{\alpha} = 2^{\alpha}(n+1) - \alpha 2^{\alpha}.$$

Also,

$$r \geqq 2^{\alpha} - \alpha - 1 \longrightarrow n + 2r - 1 \geqq 2^{\alpha+1} - (\alpha + 1) - 1 + (n - \alpha - 1)$$

$$\geqq 2^{\alpha+1} - (\alpha + 1) - 1$$

since by our preceding remark, $n \geqq \alpha + 1$.

Finally,

$$k \in \mathcal{O}(n) \longrightarrow 2k \in \mathcal{O}(n+1) \quad \text{and}$$

$$k = n + r \longrightarrow 2k = (n + 1) + (n + 2r - 1).$$

Q.E.D.

It is time to state and prove the proposition toward which we have been building.

PROPOSITION 5.8. Let (F,T) be a preordered field with $|F^{\cdot}/T^{\cdot}| = 2^n$, $|X/T| = k$. Then, setting $r = k - n$, we have

$$(n,r,k) \in B_{st(F,T)}.$$

PROOF. We use induction on n.

CASE 1. X/T is SAP.

Then $k = n$. If $n \geqq 2$, the intersection of two orderings will provide a fan T^* of maximal index in $F^{\cdot}$ (among those containing T), namely 4; as $|X/T^*| = 2^1$, $st(F,T) = 1$. One sees that $(n,0,n) \in B_1$ for any $n \geqq 2$. If $n = 1$, then $st(F,T) = 0$, and $(1,0,1) \in B_0$.

CASE 2. X/T is not SAP, and partitions into more than one Bröcker class.

Let us write as in Lemma 5.5

$$X/T = X_1 \dot{\cup} X_2 \dot{\cup} \cdots \dot{\cup} X_t, \quad \text{and} \quad T_i = \bigcap_{P \in X_i} P$$

for $i=1,\ldots,t$. Since X/T is not SAP, Lemma 5.5 applies to give $st(F,T) = \max\{st(F,T_i) : i=1,2,\ldots,t\}$. Let

$$|F^{\cdot}/T_i^{\cdot}| = 2^{n_i}, \quad |X/T_i| = k_i, \quad \text{and} \quad r_i = k_i - n_i.$$

Then with componentwise addition,

$$(n,r,k) = \sum_{i=1}^{t} (n_i,r_i,k_i).$$

As $t > 1$, each $n_i < n$; therefore we may inductively assume that $(n_i,r_i,k_i) \in B_{st(F,T_i)}$ for each $i=1,\ldots,t$. Lemma 5.7 (i) now applies without a hitch to give $(n,r,k) \in B_{st(F,T)}$, since X/T non-SAP implies $\max\{st(F,T_i) : i=1,\ldots t\} \geqq 2$.

CASE 3. X/T is not SAP, and consists of a single Bröcker class.

Let v be the finest valuation strongly compatible with T. As usual, write $^{-}$ for passage to F_v.

If $|\Gamma_v/v(T^{\cdot})| = 1$, then Lemma 5.6 yields $st(F,T) = st(\overline{F},\overline{T})$. Also in this case, $|\overline{F}^{\cdot}/\overline{T}^{\cdot}| = |F^{\cdot}/T^{\cdot}|$ and $|X/\overline{T}| = |X/T|$. $X/\overline{T}$ cannot consist of a single Bröcker class, as existence of a valuation strongly compatible with $\overline{T}$ would force a finer valuation than v still strongly compatible with T. (This is Bröcker's primary tool in (5).) These considerations reduce the problem to the already handled Case 2.

If $|\Gamma_v/v(T^{\cdot})| = 2^t$ with $t > 0$, then Lemma 5.6 yields $st(F,T) = st(\overline{F},\overline{T}) + t$. Inductively, we may assume that

$$(n-t,\ (k/2^t)-n+t,\ k/2^t) \in B_{st(\overline{F},\overline{T})}.$$

Now t applications of Lemma 5.7 (ii) finish the proof. Q.E.D.

REMARK 5.9. The information in Proposition 5.8 should be thought of in a negative sense; that is, (n,r,k) <u>outside</u> B_a forces a stability index unequal to a. One sees on the table just presented that the $\mathcal{O}(n)$ Table is actually broken into trapezoidal regions by the rays bounding the B_a's. By a little analytic geometry, one can reformulate Proposition 5.8 as follows: (With identical notation as in Proposition 5.8, and the requirement that X/T is not SAP).

If for positive integers a and i with $1 \leqq i \leqq a-1$ we have

$$2^a - a - 1 \leqq r < 2^{a+1} - a - 2, \qquad \text{and}$$
$$(r + i2^i)/(2^i - 1) \leqq n < (r + (i-1)2^{i-1})/(2^{i-1} - 1),$$

then

$$i + 1 \leqq st(F,T) \leqq a.$$

REMARK 5.10. The upper and lower bounds for $st(F,T)$ generated in this manner may fail to be faithful. For example, although $(7,12,19) \in B_3$, one may trace the reasoning of Proposition 5.8 to see that no preordered field (F,T) satisfying $|F^{\cdot}/T^{\cdot}| = 2^7$, $|X/T| = 19$ exists with $st(F,T) = 3$. Our reformulation predicts only that $3 \leqq st(F,T) \leqq 4$.

CHAPTER 6

BROWN CLASSES

Besides the equivalence relations placed on X/T by Bröcker and Marshall, there is a rather different equivalence relation defined by R. Brown. This relation has proved a useful construct both in contexts involving or not involving the reduced theory. The main content of this chapter is an upper bound for the number of equivalence classes with respect to this relation.

DEFINITION 6.1. Let (F,T) be a preordered field. We say that $P_1, P_2 \in X/T$ are <u>Brown-equivalent</u> if $A(\mathbb{Q},P_1) = A(\mathbb{Q},P_2)$ and if P_1 and P_2 induce the same (archimedean) ordering on the corresponding residue class field. We call equivalence classes under this relation <u>Brown classes</u>. <u>$Br(F,T)$</u> denotes the number of Brown classes in X/T. (The cardinality of $Br(F,T)$ admits the following interpretation: It is the number of distinct places from F to $\mathbb{R}$ compatible with at least one ordering containing T.).

Observe that if P_1 and P_2 are Brown-equivalent, then there is a valuation compatible with both (unless, possibly, they are equal), and so certainly $P_1 \sim P_2$. Thus the Brown equivalence relation is finer than $\sim$; inductive considerations for bounds on the number of Brown classes should proceed smoothly provided that X/T consists of more than one Bröcker class.

Let us now analyze the case where X/T is a single Bröcker class. Let v be the finest valuation strongly compatible with T. If $P \in X/T$, the valuations compatible with P form a chain with $A(\mathbb{Q},P)$ at the bottom; thus $A(\mathbb{Q},P) \subseteq A_v$. In fact, using $\bar{\ }$ for passage to F_v, $\overline{A(\mathbb{Q},P)} = A(\mathbb{Q},\bar{P})$. The place (into $\mathbb{R}$) corresponding to $A(\mathbb{Q},P)$ then factors through F_v. As a result, there is a 1-1 correspondence between the set of Brown classes in X/T and the set of Brown classes in $X/\bar{T}$.

These remarks motivate the following definition and proposition.

DEFINITION 6.2. Inductively define functions η and ν with common domain $\{(n,k) : k \in \mathcal{O}(n)\}$ as follows.

$\eta(1,1) = \nu(1,1) = 1.$

For $n > 1$:

If $k \in \mathcal{O}(n)$ is indecomposable, $\nu(n,k) = 1$. Otherwise,

$$\nu(n,k) = \max \{ \sum_{i=1}^{t} \eta(n_i,k_i) : n = \sum_{i=1}^{t} n_i,\ k = \sum_{i=1}^{t} k_i \text{ is a decomposition of } k \in \mathcal{O}(n),\ t > 1\}.$$

In any case,

$$\eta(n,k) = \begin{cases} \nu(n,k), & \text{if } k/2 \notin \mathcal{O}(n-1). \\ \max \{\nu(n,k), \eta(n-1, k/2)\}, & \text{if } k/2 \in \mathcal{O}(n-1). \end{cases}$$

PROPOSITION 6.3. For a preordered field (F,T) with $|F^{\cdot}/T^{\cdot}| = 2^n$ and $|X/T| = k$, the number of Brown classes in X/T is less than or equal to $\eta(n,k)$.

PROOF.

CASE 1. X/T has more than one Bröcker class. For the Bröcker partition let us write $X/T = X_1 \dot{\cup} X_2 \dot{\cup} \dots \dot{\cup} X_t$, and set $T_i = \bigcap_{P \in X_i} P$. Inductively, $Br(F,T_i) \leqq \eta(n_i,k_i)$ where $|X/T_i| = k_i$, $|F^{\cdot}/T_i^{\cdot}| = 2^{n_i}$. But since Brown-equivalence is finer than $\sim$,

$$Br(F,T) = \sum_{i=1}^{t} Br(F,T_i) \leqq \sum_{i=1}^{t} \eta(n_i,k_i) \leqq \nu(n,k) \leqq \eta(n,k).$$

CASE 2. X/T is a single Bröcker class. Let v be the finest valuation strongly compatible with T. Use $\overline{}$ for passage to F_v. The preceding remarks show that $Br(F,T) = Br(\overline{F},\overline{T})$.

Subcase 1. $|\Gamma_v/v(T^{\cdot})| = 1$. Then $X/\overline{T}$ consists of more than one Bröcker class, and we are essentially in Case 1.

Subcase 2. $|\Gamma_v/v(T^{\cdot})| = 2^s > 1$. Inductively, $Br(\overline{F},\overline{T}) \leqq \eta(n-s, k/2^s)$. But $Br(F,T) = Br(\overline{F},\overline{T}) \leqq \eta(n-s, k/2^s) \leqq \eta(n-s+1, k/2^{s-1}) \leqq \dots \leqq \eta(n-1, k/2) \leqq \eta(n,k)$. Q.E.D

In order to get more computable bounds, we prove two lemmas which simplify the computation of $\eta(n,k)$.

LEMMA 6.4. $\nu(n,k) = \max\{\eta(n_1,k_1)+\eta(n_2,k_2) : n=n_1+n_2,$
$k = k=k_1+k_2$ is a decomposition of $k\in O(n)\}$

provided that $k \in O(n)$ is decomposable.

PROOF. Realize the maximum in the definition of $\nu(n,k)$ by the decomposition

$$n = \sum_{i=1}^{t} n_i', \qquad k = \sum_{i=1}^{t} k_i',$$

so that

$$\nu(n,k) = \sum_{i=1}^{t} \eta(n_i',k_i').$$

Then

$$\eta(\sum_{i=1}^{t-1} n_i', \sum_{i=1}^{t-1} k_i') \geqq \nu(\sum_{i=1}^{t-1} n_i', \sum_{i=1}^{t-1} k_i') \geqq \sum_{i=1}^{t-1} \eta(n_i', k_i').$$

Thus we obtain

$$\eta(\sum_{i=1}^{t-1} n_i', \sum_{i=1}^{t-1} k_i') + \eta(n_t', k_t') \geqq \sum_{i=1}^{t} \eta(n_i', k_i') = \nu(n,k).$$

But the reverse inequality holds by the definition of ν. So taking

$$n_1 = \sum_{i=1}^{t-1} n_i', \quad k_1 = \sum_{i=1}^{t-1} k_i', \quad n_2 = n_t' \text{ and } k_2 = k_t',$$

we can realize $\nu(n,k)$ by a 2-fold decomposition of $k \in O(n)$. Q.E.D.

LEMMA 6.5. $\nu(n,k) = \eta(n-1,k-1)$ for $k \in O(n)$ decomposable.

PROOF. Realize the maximum in Lemma 6.4 via $n = n_1 + n_2$ and $k = k_1 + k_2$ so that $\nu(n,k) = \eta(n_1,k_1) + \eta(n_2,k_2)$. By induction, we may assume that $\eta(n_i,k_i) = \eta(n_i-1,k_i-1) + 1$ or $\eta(n_i,k_i) = \eta(n_i-1,k_i/2)$ for $i=1,2$. If for example $\eta(n_1,k_1) = 1 + \eta(n_1-1,k_1-1)$, then $\eta(n-1,k-1) \geqq \eta(n_1-1,k_1-1) + \eta(n_2,k_2) = \eta(n_1,k_1) + \eta(n_2,k_2) - 1 = \nu(n,k) -1$. The reverse inequality holds by definition of ν, finishing the proof in this case (or the case $\eta(n_2,k_2) = 1 + \eta(n_2-1,k_2-1)$).

So we may assume that $\eta(n_i,k_i) = \eta(n_i-1,k_i/2)$ for $i=1,2$.

By iteration and symmetry we obtain an integer $j > 0$ such that $\eta(n_i,k_i) = \eta(n_i-j,k_i/2^j)$ for $i=1,2$ and $\eta(n_1-j,k_1/2^j) = 1 + \eta(n_1-j-1,k_1/2^j-1)$. Adding the equations $\eta(n_1,k_1) = \eta(n_1-j-1,k_1/2^j-1)$, $\eta(n_2,k_2) = \eta(n_2-j,k_2/2^j)$ we get

$$\nu(n,k) = 1 + \eta(n_1-j-1,k_1/2^j-1) + \eta(n_2-j,k_2/2^j) \leqq 1 + \eta(n-2j-1,k/2^j-1).$$

Alternately applying the principles $\eta(a,b) \leqq \eta(a+1,2b)$ and $\eta(a,b) \leqq \eta(a+1,b+1)$ for $b \in \mathcal{O}(a)$, we get (after $2j$ operations) $\nu(n,k) \leqq 1 + \eta(n-1,k-1)$. But then as before, equality must hold. Q.E.D.

COROLLARY 6.6. $\eta(n,k) = \max\{\eta(n-1,k/2), \eta(n-1,k-1) + 1\}$ if $k/2 \in \mathcal{O}(n-1)$ and $k - 1 \in \mathcal{O}(n-1)$; otherwise, $\eta(n,k) = \eta(n-1,k/2)$ if $k/2 \in \mathcal{O}(n-1)$, or $\eta(n-1,k-1) + 1$ if $k - 1 \in \mathcal{O}(n-1)$.

PROOF. Immediate from the definition and Lemma 6.5. Q.E.D.

It is now possible to state a simple upper bound for $Br(F,T)$ which behaves in monotone fashion with respect to k.

PROPOSITION 6.7. For a preordered field (F,T) with $|F^{\cdot}/T^{\cdot}| = 2^n$, $|X/T| = k$, if $2^{i-1}(n - 1 + 1) < k \leqq 2^i(n - 1)$ for some $i \in \{0,1,2,\ldots,n-2\}$ then $Br(F,T) \leqq n - i$.

PROOF. By Proposition 6.3, it is sufficient to show that $\eta(n,k) \leqq n - i$. We do this by induction on n.

CASE 1. $\eta(n,k) = 1 + \eta(n-1,k-1)$.

$k > 2^{i-1}(n - i + 1)$ implies $k - 1 > 2^{i-1}(n - i + 1) - 1$ $= 2^{i-1}(n - i) + 2^{i-1} - 1 \geqq 2^{i-1}(n - i)$ (unless $i = 0$) $= 2^{i-1}((n - 1) - i + 1)$. So inductively, $\eta(n-1,k-1) \leqq (n - 1) - i$ and $\eta(n,k) \leqq n - i$ if $i \neq 0$. If $i = 0$, $Br(F,T) \leqq n - i$ is obvious anyway since $k \leqq 2^i(n - i)$ gives us $k = n$.

CASE 2. $\eta(n,k) = \eta(n-1,k/2)$.

$k > 2^{i-1}(n - i + 1)$ implies $k/2 > 2^{i-2}(n - i + 1)$ $= 2^{(i-1)-1}((n - 1) - (i - 1) + 1)$. So inductively $\eta(n-1,k/2)$ $\leqq (n -1) - (i - 1) = n - i$ and hence $\eta(n,k) \leqq n - i$. Q.E.D.

REMARK 6.8. While Proposition 6.7 effects a monotone tendency for $Br(F,T)$ from SAP $(k=n)$ to fan $(k=2^{n-1})$, one loses the faithfulness one could hope for from Proposition 6.3. For example, $12 = 2^{2-1}(7-2+1) < 19$ $\leqq 2^2(7-2) = 20$, but $\eta(7,19) = 4 < 7-2$. Via a trifle more computation, one also obtains $176 = 2^{5-1}(15-5+1) < 319 \leqq 2^5(15-5) = 320$, but $\eta(15,319) = 8 < 15-5$.

CHAPTER 7

THE INDEX OF $A(F)$ IN $F^{\cdot}$

In this chapter and the next, we need the reduced analog of $D_F(<1,x>)$ for $x \in F^{\cdot}$. If (F,T) is a preordered field, we will suppress F and write $D_T<1,x>$ for

$$\{t_1 + t_2x : t_1, t_2 \in T, t_1 + t_2x \neq 0\}.$$

Note that $y \in D_T<1,x>$ if and only if for each $\sigma \in X/T$ such that $\sigma(x) = 1$, $\sigma(y) = 1$ also.

Let (F,T) be a preordered field with $|F^{\cdot}/T^{\cdot}| < \infty$, and write $X_1 = \{P_{11},\ldots,P_{1m_1}\}$, $X_2 = \{P_{21},\ldots,P_{2m_2}\}$, $\ldots$, $X_q = \{P_{q1},\ldots,P_{qm_q}\}$ for the Brocker classes in X/T. An element of $F^{\cdot}/T^{\cdot}$ can be identified with a system

$$\{\varepsilon_{i,j} : i = 1,\ldots,q; \; j = 1,\ldots,m_i; \; \varepsilon_{i,j} = \pm 1\}$$

of coset representatives for the groups $F^{\cdot}/P^{\cdot}_{ij}$. Let us call such a system a <u>sign-prescription</u>. Of course, not every sign-prescription can be "filled" by an element of $F^{\cdot}/T^{\cdot}$; that is, there may not exist $x \in F^{\cdot}$ such that $xP^{\cdot}_{ij} = \varepsilon_{i,j}P^{\cdot}_{i,j}$ for all i,j.

PROPOSITION 7.1. With notation as above, set

$$T_i = \bigcap_{j=1}^{m_i} P_{ij}.$$

Then a prescription $\{\varepsilon_{i,j} : i=1,\ldots,q; \; j=1,\ldots m_i\}$ can be filled by an element of $F^{\cdot}/T^{\cdot}$ if and only if for each $i = 1,\ldots,q$ the prescription $\{\varepsilon_{i,j} : j = 1,\ldots,m_i\}$ can be filled by an element of $F^{\cdot}/T^{\cdot}_i$.

PROOF. This is in essence a restatement of Proposition 3.2 (i). Q.E.D.

The object of study in this chapter is $A(F,T)$, as introduced by Berman (3). (The generalization here to preordered fields (F,T) is not really an innovation.)

DEFINITION 7.2. Let (F,T) be a preordered field. $x \in F^{\cdot}$ is called <u>rigid</u> (<u>with respect to T</u>) if $D_T<1,x> = T^{\cdot} \cup xT^{\cdot}$. We define

$$\underline{A(F,T)} = \{x \in F^{\cdot} : \text{either } x \text{ or } -x \text{ is not rigid wrt } T\}.$$

We agree that -1 is nonrigid even when $|F^\cdot/T^\cdot| = 2$, so we always have $1 \in A(F,T)$. We consider x as belonging to $F^\cdot/T^\cdot$ when convenient, and then view $A(F,T) \subseteq F^\cdot/T^\cdot$.

LEMMA 7.3. Let $|F^\cdot/T^\cdot| < \infty$, and suppose X/T consists of more than one Bröcker class, and $|X/T| > 2$. Then $F^\cdot = A(F,T)$.

PROOF. Use notation as in the beginning of this chapter. Let $x \in F^\cdot$ be rigid with respect to T. Suppose x corresponds to sign prescription $\{\varepsilon_{i,j}\}_{i,j}$. If x is negative with respect to orderings in distinct X_k and X_m, we can fill the prescription $\{\varepsilon'_{i,j}\}_{i,j}$ where $\varepsilon'_{i,j} = \varepsilon_{i,j}$ for $i \neq k$, and all $\varepsilon'_{k,j} = 1$, by Proposition 7.1; say y fills this prescription. Then $y \in D_T\langle 1,x\rangle$, but $y \notin T^\cdot$ and $y \notin xT^\cdot$. This contradicts the rigidity of x, and shows that rigid x must be negative on orderings in at most one X_i.

Now suppose that $x \notin A(F)$, so that in addition $-x$ is rigid. Then we see from the above that there can be just two Bröcker classes X_1 and X_2, with x positive at all orderings in (say) X_1 and negative at all orderings in X_2. But if $|X_2| > 1$, we can surely find y positive at all orderings in X_1 and positive at some but not all orderings in X_2; such a y yields a contradiction as above. So $|X_2| = 1$; arguing symmetrically with $-x$ yields $|X_1| = 1$. Thus if $x \notin A(F)$ exists, then $b(F,T) = 1$ or $|X/T| = 2$. Q.E.D.

Now suppose X/T is a single Bröcker class. Let v be the finest valuation strongly compatible with T, and use $\overline{}$ to denote passage to F_v. If $|\Gamma_v/v(T^\cdot)| = 1$, we may reduce as usual to the case of more than one Bröcker class, and $A(F,T) = F^\cdot$ (except when $|X/T| = 2$) So suppose $|\Gamma_v/v(T^\cdot)| = 2^s > 1$. We may identify $F^\cdot/T^\cdot$ with $\overline{F}^\cdot/\overline{T}^\cdot \times \Gamma/v(T^\cdot)$ and X/T with $X/\overline{T} \times (\Gamma_v/(T^\cdot))^*$ (see Chapter 1), so that for $(\overline{x},\overline{\gamma}) \in F^\cdot/T^\cdot$ and $(\sigma,\chi) \in X/T$, $(\sigma,\chi)(\overline{x},\overline{\gamma}) = \sigma(\overline{x})\chi(\overline{\gamma}) \in \{\pm 1\}$.

PROPOSITION 7.4. In the context above, all elements of $F^\cdot/T^\cdot$ of the form $(\overline{x},\overline{\gamma})$ with $\overline{\gamma} \neq 0$ are rigid.

PROOF. First suppose that some element $(\overline{y},0) \in D_T\langle 1,(\overline{x},\overline{\gamma})\rangle$. Let $\sigma \in X/\overline{T}$ be arbitrary. Choose $\chi \in (\Gamma_v/v(T^\cdot))^*$ so that $\sigma(\overline{x}) = \chi(\overline{\gamma})$; this may be done since $\overline{\gamma} \neq 0$. Then $(\sigma,\chi)(\overline{x},\overline{\gamma}) = 1$, so also $(\sigma,\chi)(\overline{y},0) = \sigma(\overline{y}) = 1$. As σ was arbitrary, this shows that $\overline{y} = \overline{T}$, and $(\overline{y},0) = T$.

Next suppose $(\overline{y},\overline{\delta}) \in D_T\langle 1,(\overline{x},\overline{\gamma})\rangle$ with $\overline{\delta} \neq 0$. We claim first that $\overline{\delta} = \overline{\gamma}$. If not, fix any $\sigma \in X/\overline{T}$ and choose $\chi \in (\Gamma_v/v(T^\cdot))^*$

such that $\chi(\overline{\gamma}) = \sigma(\overline{x})$ and $\chi(\overline{\delta}) = -\sigma(\overline{y})$; we get the contradiction $(\sigma,\chi)(\overline{x},\overline{\gamma}) = 1$ and $(\sigma,\chi)\ (\overline{y},\overline{\delta}) = -1$. Thus $\overline{\delta} = \overline{\gamma}$.

Now choosing for each $\sigma \in X/\overline{T}$ a $\chi \in (\Gamma_v/v(T^\cdot))^*$ such that $\chi(\overline{\gamma}) = \sigma(\overline{x})$, we obtain $\sigma(\overline{x}) = \sigma(\overline{y})$ for all $\sigma \in X/\overline{T}$; thus $(\overline{x},\overline{\gamma}) = (\overline{y},\overline{\delta})$. Q.E.D.

PROPOSITION 7.5. With the same notations, $(\overline{x},0)$ is rigid with respect to T if and only if $\overline{x}$ is rigid with respect to $\overline{T}$.

PROOF. Arguments exactly as in the previous proposition show that $(\overline{y},\overline{\gamma}) \in D_T<1,(\overline{x},0)>$ implies that $\overline{\gamma} = 0$, and that $(\overline{y},0) \in D_T<1,(\overline{x},0)>$ if and only if $\overline{y} \in D_{\overline{T}}<1,\overline{x}>$. Q.E.D.

COROLLARY 7.6. Let (F,T) be a preordered field with $|F^\cdot/T^\cdot| < \infty$. Then $A(F,T)$ is a subgroup of $F^\cdot$. If X/T consists of a single Bröcker class and if v is the finest valuation strongly compatible with T, then

$$|F^\cdot/A(F,T)| = |\overline{F}^\cdot/A(\overline{F},\overline{T})|\cdot|\Gamma_v/v(T^\cdot)|.$$

PROOF. $A(F,T)$ is already seen to be a group unless X/T is a single Bröcker class with $|\Gamma_v/v(T^\cdot)| > 1$ for v the finest valuation strongly compatible with T. In that case, Propositions 7.4 and 7.5 show that

$$A(F,T) = \{(\overline{x},0) : \overline{x} \in A(\overline{F},\overline{T})\}$$

(noting that $-1 \longleftrightarrow (-1,1)$), so that we get $A(F,T)$ a group by induction. Also, $|A(F,T)| = |A(\overline{F},\overline{T})|$ while $|F^\cdot/T^\cdot| = |\overline{F}^\cdot/\overline{T}^\cdot|\cdot|\Gamma_v/v(T^\cdot)|$, which gives us the final conclusion. Q.E.D.

We remark that $A(F,T)$ is already known to be a group in general (i.e., when $F^\cdot/T^\cdot$ need not be finite).

We can now give bounds for $|F^\cdot/A(F,T)|$ of a similar nature to those for Brown classes (cf. Proposition 6.7).

PROPOSITION 7.7. Let (F,T) be a preordered field with $|F^\cdot/T^\cdot| = 2^n$, $|X/T| = k$. If for some i, $1 \leqq i \leqq n - 2$, $2^{i-1}(n - i + 1) \leqq k < 2^i(n - i)$ then $|F^\cdot/A(F,T)| \leqq 2^{i-1}$. For $k = 2^{n-1}$ we have $|F^\cdot/A(F,T)| = 2^{n-1}$.

PROOF. Note that if $2^s|k$ and $k/2^s \in \mathcal{O}(n-s)$, then $2^{i-1}(n - i + 1) \leqq k < 2^i(n - i)$ implies $2^{i-s-1}(n - i + 1) \leqq k/2^s < 2^{i-s}(n - i)$, i.e., $2^{(i-s)-1}((n-s) - (i-s) + 1) \leqq k/2^s < 2^{i-s}((n-s) - (i-s))$. With this observation, the proof is a straightforward induction using Lemma 7.3 and Corollary 7.6. Q.E.D.

Of course, this bound is not faithful, e.g., for k odd, $|F^{\cdot}/A(F,T)| = 1$ automoatically.

CHAPTER 8

THE SQUARE-CLASS GRAPH

T. Y. Lam has observed that a natural partial ordering may be imposed on $F^{\cdot}/T^{\cdot}$ for (F,T) a preordered field.

DEFINITION 8.1 (Lam). Let (F,T) be a preordered field. Define a relation $\prec$ on $F^{\cdot}/T^{\cdot}$ as follows: $aT^{\cdot} \prec bT^{\cdot}$ if and only if $a \in D_T\langle 1,b\rangle$. (For the notation $D_T\langle 1,a\rangle$, see Chapter 7.)

We show straightaway that $\prec$ is a partial ordering on $F^{\cdot}/T^{\cdot}$.

PROPOSITION 8.2 (Lam). $\prec$ is a partial ordering on $F^{\cdot}/T^{\cdot}$.

PROOF. $\prec$ is reflexive, for this amounts to the observation that $a \in D_T\langle 1,a\rangle$. For antisymmetry, suppose $a \in D_T\langle 1,b\rangle$ and $b \in D_T\langle 1,a\rangle$. Then for all $\sigma \in X/T$, $\sigma(b) = 1$ if and only if $\sigma(a) = 1$, i.e., $\sigma(a) = \sigma(b)$. Thus $aT^{\cdot} = bT^{\cdot}$ by Artin-Schreier theory. Finally. to obtain the transitive law, let $a \in D_T\langle 1,b\rangle$ and $b \in D_T\langle 1,c\rangle$. Then there exist $t_1,t_2,t_3,t_4 \in T$ with $a = t_1 + t_2b$ and $b = t_3 + t_4c$, so $a = (t_1 + t_2t_3) + (t_2t_4)c$, and $a \in D_T\langle 1,c\rangle$. (Alternately, use an argument as for antisymmetry, and modus ponens.) Q.E.D.

REMARK 8.3. The map $xT^{\cdot} \longmapsto -xT^{\cdot}$ is order-reversing, since $a \in D_T\langle 1,b\rangle \longrightarrow -b \in D_T\langle 1,-a\rangle$ ("contraposition").

We may think of the poset $(F^{\cdot}/T^{\cdot},\ )$ as a graph, called the square-class graph (terminology motivated by the case F pythagorean, $T = F^2$), with maximal element $-T^{\cdot}$ (since $D_T\langle 1,-1\rangle = F^{\cdot}/T^{\cdot}$) and minimal element $T^{\cdot}$ (by contraposition). For the remainder of the chapter, we assume $|F^{\cdot}/T^{\cdot}| < \infty$.

If we identify each coset $xT^{\cdot}$ with the corresponding sign-prescription (Chapter 7) then we may say

$$\{\varepsilon_{i,j} : 1 \leqq i \leqq q;\ 1 \leqq j \leqq m_i\} \prec \{\varepsilon'_{i,j} : 1 \leqq i \leqq q;\ 1 \leqq j < m_i\}$$

if and only if $\varepsilon'_{i,j} \leqq \varepsilon_{i,j}$ (where $\{\pm 1\}$ is ordered as a subset of $\mathbb{R}$).

EXAMPLE 8.4. Suppose X/T is SAP. Then all sign-prescriptions may be filled; so $F^{\cdot}/T^{\cdot}$ is in bijection with the power set of X/T, say by $xT^{\cdot} \longleftrightarrow \{P \in X/T : x \in P\}$. In this case, $(F^{\cdot}/T^{\cdot},\prec)$ is identified with the power set of X/T under $\supseteq$. Thus, for instance, if

$|F^{\cdot}/T^{\cdot}| = 8$ and $|X/T| = 3$, then the square-class graph looks like this:

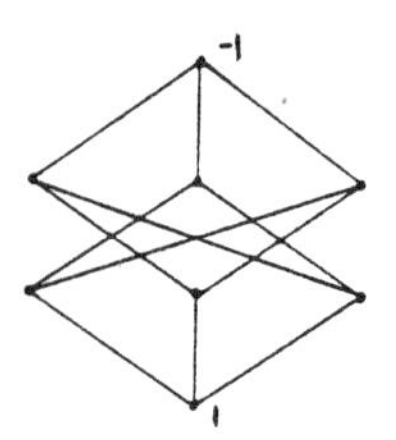

At the opposite extreme, if T is a fan, then $D_T\langle 1,a\rangle = T^{\cdot} \cup aT^{\cdot}$ unless $a \in -T$. So in this case, all elements of $F^{\cdot}/T^{\cdot}$ compare only with 1 and -1. For instance, if $|F^{\cdot}/T^{\cdot}| = 8$ and $|X/T| = 4$, then the square-class graph looks like:

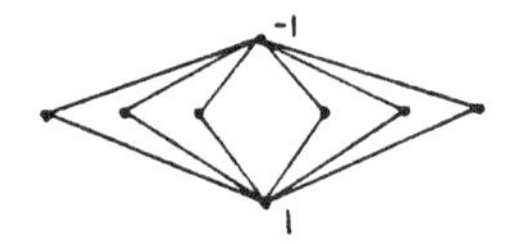

In order to understand the structure of the square-class graph in the general case, we must separate the usual cases.

PROPOSITION 8.5. Let (F,T) be a preordered field with $|F^{\cdot}/T^{\cdot}| < \infty$. Suppose X/T partitions into more than one Bröcker class. Denote by T_i, $1 \leqq i \leqq q$, the corresponding preorderings (i.e., T_i is the intersection of all orderings in the ith Bröcker class). Then $(F^{\cdot}/T^{\cdot},\prec)$ is isomorphic to

$$\prod_{i=1}^{q} (F^{\cdot}/T_i^{\cdot},\prec);$$

that is, the isomorphism of groups

$$F^{\cdot}/T^{\cdot} \longrightarrow \prod_{i=1}^{q} (F^{\cdot}/T_i^{\cdot})$$

of Proposition 3.2 (i) is order preserving, when the image is ordered as a direct product.

PROOF. From the sign-prescription viewpoint, the map above is, in the notation of Chapter 7,

$$\{\varepsilon_{i,j} : 1 \leqq i \leqq q,\ 1 \leqq j \leqq m_i\} \longmapsto (\{\varepsilon_{1,j} : 1 \leqq j \leqq m_1\}, \cdots).$$

That this is order preserving follows from the sign-prescription characterization of $\prec$ in Remark 8.3. Q.E.D.

OBSERVATION 8.6. The same type of decomposition can be effected if X/T consists of a single Bröcker class, provided that for v the finest valuation strongly compatible with T, $|\Gamma_v/v(T^\cdot)| = 1$; we then just replace F with $\overline{F}$, T with $\overline{T}$.

EXAMPLE 8.7. Let $|F^\cdot/T^\cdot| = 16$, $|X/T| = 5$. Since 5 is odd, if X/T is a single Bröcker class, then we are in the case of Observation 8.6. As the only decomposition of $5 \in \mathcal{O}(4)$ is $4 = 1 + 3$, $5 = 1 + 4$, the graph for $F^\cdot/T^\cdot$ is the direct product of a two-point graph with the graph constructed in Example 8.4 for $|F^\cdot/T^\cdot| = 8$, $|X/T| = 4$. The result is:

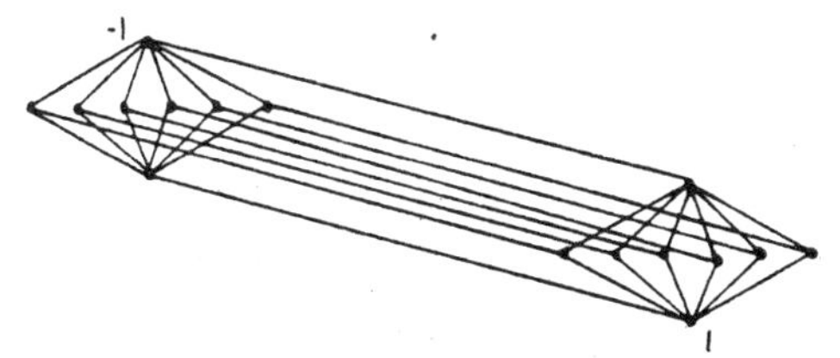

PROPOSITION 8.8. Let (F,T) be a preordered field with $|F^\cdot/T^\cdot| < \infty$. Suppose X/T consists of a single Bröcker class, and that $|\Gamma_v/v(T^\cdot)| = 2^s > 1$ for v the finest valuation strongly compatible with T. Then the graph for $F^\cdot/T^\cdot$ may be constructed from that for $\overline{F}^\cdot/\overline{T}^\cdot$ by adjoining to that graph $(2^s-1)|F^\cdot/T^\cdot|$ vertices whose only "connections" are to 1 and -1.

PROOF. With notations as in Chapter 7, the proofs of Propositions 7.4 and 7.5 show that $(\overline{x},\overline{\gamma})$ is rigid if $\overline{\gamma} \neq 0$, and that $\langle 1,(\overline{x},0)\rangle$ represents $(\overline{y},\overline{\gamma})$ if and only if $\overline{\gamma} = 0$ and $\overline{y} \in D_{\overline{T}}\langle 1,\overline{x}\rangle$. Thus, points of the form $(\overline{x},0)$ will give rise to a copy of the graph of $\overline{F}^\cdot/\overline{T}^\cdot$, and the remaining vertices connect only to 1 and -1. Q.E.D.

OBSERVATION 8.9. To say that $xT^\cdot$ "connects" only to 1 and -1 is to say that $\langle 1,x\rangle$ represents only 1 and x, and that $\langle 1,y\rangle$ represents x only if $y = -1$ or $y = x$, i.e., $\langle 1,-x\rangle$ represents only 1 and $-x$ by contraposition. Thus, such vertices correspond to nonelements of $A(F,T)$.

EXAMPLE 8.10. $|F^\cdot/T^\cdot| = 2^4$, $|X/T| = 6$. Then the square-class graph is:

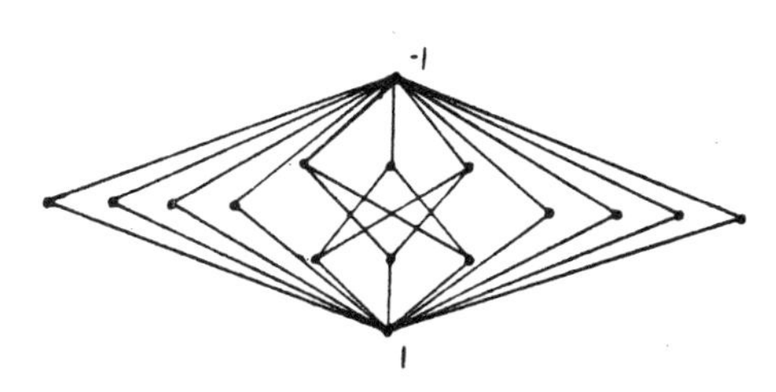

If G is a graph with t vertices and greatest and least elements, let us denote by $2G$ a graph with $2t$ vertices such that:

1. $2G$ contains a set of t vertices forming a copy of G.
2. The remaining t vertices are comparable only to the greatest and least elements of the copy of G (and sit between them).

COROLLARY 8.11. The set $\mathcal{G}$ of all isomorphism classes of square-class graphs for preordered fields (F,T) with T of finite index is described inductively as follows:

(i) $\mathcal{G}$ contains a 2-vertex graph with greatest and least element.

(ii) $G \in \mathcal{G}$ implies $2G \in \mathcal{G}$.

(iii) $G, H \in \mathcal{G}$ implies $G \times H \in \mathcal{G}$.

PROOF. That all square-class graphs are elements of $\mathcal{G}$ follows from 8.5, 8.6 and 8.8. For the converse, we claim that every graph in $\mathcal{G}$ is realizable by (F,F^2) where F is a pythagorean field of finite transcendence degree over $\mathbb{Q}$.

To start with, we can realize the two-point graph via a real closure of $\mathbb{Q}$. Given a realizable $G \in \mathcal{G}$, choose F as above realizing G. Then Proposition 8.8 shows that $2G$ is realized by a henselization of $F(x,x^{1/2},x^{1/4},x^{1/8},\ldots)$ with respect to the degree valuation. Finally, given realizable G and H, we may follow the construction of Theorem 4.1 to realize $G \times H$. Q.E.D.

COROLLARY 8.12. All square-class graphs of preordered fields (F,T) with $|F^{\cdot}/T^{\cdot}| < \infty$ are lattices. Further, the square-class graph for such (F,T) is distributive if and only if X/T is SAP.

PROOF. We recall that a graph is a lattice if any two elements have a least upper bound and a greatest lower bound; the lattice is called distributive if the distributive law holds for formation of lub's and glb's. It is straightforward to see that the operations $G \mapsto 2G$ and $(G,H) \mapsto G \times H$ take lattices to lattices.

For the second assertion, we have already seen in Example 8.4 that X/T SAP implies that the square-class graph is a distributive lattice. Suppose X/T is non-SAP. If X/T partitions into more than one Bröcker class, then at least one partitioning subspace will be non-SAP. Inductively, the corresponding square-class graph is non-distributive; but this graph is a direct factor of the square-class graph for (F,T). Similarly, we are done if X/T is a single Bröcker class with

$|\Gamma_v/v(T^\cdot)| = 1$, where v is the finest valuation strongly compatible with T. Finally, if X/T is a single Bröcker class with $|\Gamma_v/v(T^\cdot)| = 2^s > 1$, Proposition 8.8 applies, and we see that the square-class graph contains a sublattice of the form

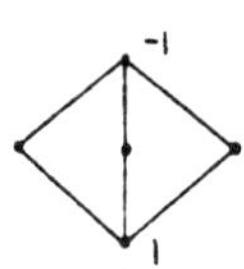

and so is not distributive. Q.E.D.

COROLLARY 8.13. Let (F,T) be a preordered field with X/T finite. Then for any $a, b \in F^\cdot$ there exists $c \in F^\cdot$ such that:

$$D_T\langle 1,a\rangle \cap D_T\langle 1,b\rangle = D_T\langle 1,c\rangle.$$

PROOF. c is just the "meet" of a and b. Q.E.D.

We wish to observe that the study of square-class graph is equivalent to the study of reduced Witt rings in the case $|F^\cdot/T^\cdot| < \infty$. For this it is essential to view a square-class graph as a group together with a partial ordering $\prec$. Rather than showing directly that the reduced Witt ring $W(F,T)$ is recoverable from the square-class graph, we show that X/T is recoverable. We can then reconstruct the reduced Witt ring via the theory of Marshall, as we have the abstract space of orderings associated to (F,T) intact.

PROPOSITION 8.14. Let (F,T) be a preordered field with $|F^\cdot/T^\cdot| < \infty$. Viewing an ordering in X/T as a subgroup of $F^\cdot/T^\cdot$, a subset P of $F^\cdot/T^\cdot$ is an ordering if and only if

(i) P is a subgroup of $F^\cdot/T^\cdot$.

(ii) $-1 \notin P$.

(iii) $aT^\cdot \prec bT^\cdot$ and $bT^\cdot \in P^\cdot$ implies $aT^\cdot \in P$.

(iv) P is maximal with respect to (i), (ii), and (iii).

PROOF. (i), (ii), (iii) and (iv) all hold for any ordering P. Let us show that P satisfies (i), (ii) and (iii) if and only if P is a preordering (containing T). (i), (ii), and (iii) are clear for preorderings. The only condition for preorderings not explicit among (i), (ii) and (iii) is closure under addition. But writing $a + b = a(1 + b/a)$, we see that closure under addition obtains by (i) and (iii). Since maximal preorderings are orderings, we are done. Q.E.D.

Note that the inductive construction of Corollary 8.11 may be modified to incorporate group structure; group theoretically, in operation (ii) we form the direct product of a given group with Z_2 (retaining the original ordering on elements of the form $(x,0)$); in operation (iii) we form a group theoretic (and order theoretic) direct product.

BIBLIOGRAPHY

1. Becker, E., and Bröcker, L., "On the description of the reduced Witt ring," Journal of Algebra 52 (1978), pp. 328-346.

2. Becker, E., and Köpping, E., "Reduzierte quadratische Formen und Semiordnungen reeller Korper," Abh. Math. Sem. Univ. Hamburg 46 (1977), pp. 143-177.

3. Berman, L., "The Kaplansky radical and values of binary quadratic forms over fields," Thesis (1978), University of California, Berkeley.

4. Bröcker, L., "Characterization of fans and hereditarily pythagorean fields," Mathematische Zeitschrift 151 (1976), pp. 149-163.

5. Bröcker, L., "Über die Anzahl der Anordnungen eines kommutativen Körpers," Archiv der Mathematik 29 (1977), pp. 458-464.

6. Craven, T., "The Boolean space of orderings of a field," Transactions of the American Mathematical Society 209 (1975), pp. 225-235.

7. Elman, R., and Lam, T. Y., "Quadratic forms over formally real fields and pythagorean fields," American Journal of Mathematics 94 (1972), pp. 1155-1194.

8. Elman, R., Lam, T. Y., and Prestel, A., "On some Hasse principles over formally real fields," Mathematische Zeitschrift 134 (1973), pp. 291-301.

9. Endler, O., Valuation Theory, Springer-Verlag, New York-Heidelberg-Berlin, 1972.

10. Knebusch, M., Rosenberg, A., and Ware, R., "Structure of Witt rings, quotients of abelian group rings, and orderings of fields," Bulletin of the American Mathematical Society 77 (1971), pp. 205-210.

11. Knebusch, M., Rosenberg, A., and Ware, R., "Signatures on semilocal rings," Journal of Algebra 26 (1973), pp. 208-250.

12. Lam, T. Y., The Algebraic Theory of Quadratic Forms, Benjamin, Reading, Massachusetts, 1973.

13. Lam, T. Y., "The theory of ordered fields," to appear in the Proceedings of the Algebra and Ring Theory Conference, ed. B. MacDonald, University of Oklahoma, March, 1979, Marcel Dekker, 1980.

14. Marshall, M., "Classification of finite spaces of orderings," Canadian Journal of Mathematics 31 (1979), pp. 320-330.

15. Marshall, M., "Quotients and inverse limits of spaces of orderings," to appear in Canadian Journal of Mathematics.

16. Marshall, M., "The Witt ring of a space of orderings," to appear in Transactions of the American Mathematical Society.

17. Marshall, M., "Spaces of orderings, IV," to appear in Canadian Journal of Mathematics.

18. Pfister, A., "Quadratische Formen in beliebigen Körpern," Inventiones Mathematicae 1 (1966), pp. 116-132.

19. Prestel, A., Lectures on Formally Real Fields, Monografias de Mathemática 22, Instituto de Matemática Pura e Aplicada, Rio de Janiero, 1975.

20. Reznick, B., Private Communication, Spring, 1979.

21. Ribenboim, P., Théorie des Valuations, Les Presses de l'Université de Montreal, 1965.

22. Stone, A. L., "Nonstandard Analysis in Topological Algebra," in Applications of Model Theory to Algebra, Analysis and Probability, ed. W. A. J. Luxemburg, Holt, 1969.

DEPARTMENT OF MATHEMATICS
HOLY NAMES COLLEGE
OAKLAND, CALIFORNIA 94619

Contemporary Mathematics
Volume 8, 1982

FINITELY GENERATED GROUPS OF DIVISIBILITY

Bruce Glastad[1]

and

Joe L. Mott

ABSTRACT. The group of divisibility $G(D)$ of an integral domain D is the group $K^*/U(D)$ partially ordered by $D^*/U(D)$ where K is the quotient field of D. Several classes of integral domains D are completely characterized by the order properties of $G(D)$; for example, D is a GCD-domain if and only if $G(D)$ is lattice ordered. However, under certain circumstances even when the order structure of $G(D)$ is disregarded quite a bit of information about D is deducible from the algebraic structure of $G(D)$ alone. For example, if $G(D)$ is a subgroup (just as an abelian group) of the group of additive rational numbers, then D is a valuation ring of Krull dimension one and the order on $G(D)$ must have been a total order. If $G(D)$ is finitely generated, then the integral closure $\bar{D}$ of D must be a Bezout domain with only finitely many maximal ideals. Moreover, $\bar{D}$ is a finite D-module and $U(\bar{D})/U(D)$ is a finite group. In particular, if $G(D)$ is a finitely generated free abelian group, then $G(D)$ and $G(\bar{D})$ are isomorphic as unordered abelian groups (but D need not be integrally closed). Domains with finitely generated groups of divisibility abound, for if D is any semi-local domain whose quotient field is a global field, then $G(D)$ is finitely generated. If D is a one-dimensional semi-local domain such that each residue field is finite, then $G(D)$ is finitely generated if and only if D is analytically unramified.

1. INTRODUCTION. If R is a commutative ring with identity

1980 Mathematics Subject Classification. 13A18, 13E05, 6F20

[1] The results of this paper formed part of Mr. Glastad's doctoral dissertation submitted to Florida State University and written under the direction of Professor Mott.

0271-4132/81/0000-0685/$05.25

and T is the total quotient ring of R, then the group of divisibility of R (also called the value group of R [11]) is the group $G(R) = U(T)/U(R)$ partially ordered by $R^*/U(R)$, where $U(R)$ denotes the group of units of R and R^* the regular elements of R. Now $G(R)$ is order isomorphic to the group of principal fractional ideals $P(R) = \{xR \mid x \in U(T)\}$ partially ordered by $\{xR \mid x \in R^*\}$.

Several classes of integral domains D are completely characterized by the <u>order</u> properties of $G(D)$, for example, a domain D is a GCD-domain if and only if $G(D)$ is lattice ordered. However, under certain circumstances, when the order structure of this group is disregarded, quite a bit of information about the domain D is deducible from the <u>algebraic structure</u> of the group $G(D)$ alone. For instance, if $G(D)$ is isomorphic (just as an abelian group) to a subgroup of the additive group of rational numbers - that is, $G(D)$ is torsion-free of rank one- then D must be a valuation ring of Krull dimension one and $G(D)$ must, in fact, be totally ordered. If $G(D)$ is finitely generated, then the integral closure of D must be a Bezout domain with finitely many maximal ideals. Results of this nature are what we are seeking to present in this paper.

2. PRELIMINARY RESULTS.

THEOREM 2.1. *If $G(D)$ has finite torsion free rank n, then D possesses at most n incomparable valuation overrings and each of these valuation overrings has dimension less than or equal to n. Therefore, $\bar{D}$, the integral closure of D, is the intersection of no more than n valuation rings and is a Bezout domain with finitely many prime ideals.*

PROOF: Let $\{V_i\}_{i=1}^m$ be a finite collection of non-trivial incomparable valuation overrings of D and set

$D_k = \cap_{i=1}^{k} V_i$, $k = 1,\ldots,m$. Each D_k is a Bezout domain with k maximal idelas [12, p. 38]. Moreover, D_{k-1} is a quotient ring of D_k and there is a canonical homomorphism from $G(D_k)$ onto $G(D_{k-1})$. Since D_{k-1} has fewer maximal ideals than D_k , the kernel of this homomorphism is non-trivial. Since D_m is an overring of D , $G(D_m)$ is a homomorphic image of $G(D)$. Hence $G(D_m)$ and each $G(D_k)$ has torsion free rank $\leq n$. In fact, rank $G(D_k) - 1 \geq$ rank $G(D_{k-1})$ and, therefore

$$n = \text{rank } G(D) \geq \text{rank } G(D_m) - (m-1) \geq \text{rank } G(D_1) .$$

Since D_1 is not a field, rank $G(D_1) \geq 1$. Hence $n \geq m$.

COROLLARY 2.2 _If_ $G(D)$ _is a torsion-free abelian group of rank one, then_ D _is a one-dimensional valuation ring._

PROOF: Theorem 2.1 implies that the integral closure $\bar{D}$ of D is a valuation ring of dimension one. Moreover, the canonical map of $G(D)$ onto $G(\bar{D})$ has trivial kernel since any non-trivial homomorphism of a subgroup of Q into a torsion-free abelian group is a monomorphism. This means that $U(D) = U(\bar{D})$ and, therefore, by the following lemma, that $D = \bar{D}$.

LEMMA 2.3. _Let_ S _be a quasi-local ring and_ R _be a subring. If_ $U(R) = U(S)$, _then_ $R = S$.

PROOF. An element x of S is either already a unit of S , and therefore it belongs to R , or x belongs to the maximal ideal of S , in which case $1 + x$ is a unit of S , whence $1 + x$, and so also x , belongs to R .

The following example demonstrates that, in the statement of Theorem 2.1, we cannot conclude that D is integrally closed in general.

EXAMPLE 2.4. Let F be a field and K a proper extension field of F . If $D = F + XK[[X]]$ is the ring of a formal power series over K with constant terms in F , then $G(D) = K^*/F^* \oplus Z$.

If K is a finite field, then $G(D)$ has torsion free rank one and a non-trivial torsion subgroup. Moreover, the integral closure of D is $K[[X]]$.

REMARK 2.5. In general, if a domain D is integrally closed, then D is the intersection of a family of valuation rings V_α and the group of divisibility of D can be embedded in the cartesian product of value groups $G(V_\alpha)$. Since each $G(V_\alpha)$ is totally ordered (and hence torsion free), $G(D)$ is torsion free.

3. FINITELY GENERATED GROUPS OF DIVISIBILITY: an application of a Theorem of Brandis. We turn our attention now to the case where $G(D)$ is finitely generated. This means, of course, that D is semi-quasi-local (Theorem 2.1) and therefore we consider the quasi-local case first. We shall have to separate the problem into two main cases determined by whether the residue field is finite or infinite. The most important tool in our investigation is the following Theorem of Brandis [3].

BRANDIS' THEOREM. *Let K be an infinite field and L an extension field. Moreover, let K^* and L^* denote their respective groups of units. If L^*/K^* is finitely generated, then $K = L$.*

Also, it can be observed that if K is a finite field and if L^*/K^* is finitely generated, then L is also a finite field.

Now let us record three useful, but simple results that we shall use repeatedly in the following work.

LEMMA 3.1. *Let S be a quasi-local ring and R a subring over which S is integral. If $U(S)/U(R)$ is finitely generated, then S is a finite R-module.*

PROOF. Suppose $x_1,\ldots,x_n$ are elements of $U(S)$ whose images generate $U(S)/U(R)$. Then clearly

$U(S) = U(R[x_1,\ldots,x_n,x_1^{-1},\ldots,x_n^{-1}])$; by Lemma 2.3, $S = R[x_1,\ldots,x_n,x_1^{-1}\ldots x_n^{-1}]$, and since S is integral over R the result follows.

LEMMA 3.2. *Let f be a homomorphism of the ring A onto the ring B whose kernel contains the Jacobson radical of A; let C be a subring of B and let E be the inverse image in A of C under f. Then $U(A)/U(E) \simeq U(B)/U(C)$.*

LEMMA 3.3 *Suppose that $D = \bigcap_{i=1}^{n} D_i$ where each D_i is an overring of D. Then $G(D)$ is finitely generated if and only if each $G(D_i)$ is finitely generated. In particular, a semi-quasi-local domain D has finitely generated group of divisibility if and only if $G(D_M)$ is finitely generated for each maximal ideal M of D.*

The proof is immediate from the observation that $G(D)$ is canonically embeddable in the direct product of the groups $G(D_i)$.

INFINITE RESIDUE FIELD CASE.

THEOREM 3.4. *Suppose that $G(D)$ is finitely generated and that the integral closure V of D is a valuation ring. Then D is quasi-local and if the residue field of D is infinite, $D = V$.*

PROOF. By hypothesis, V is integral over D and therefore D is quasi-local; denote by Q and M the maximal ideals of V and D respectively. For the same reason, MV is Q-primary, and we shall be done once we have shown that $MV = Q$. To see that this suffices, observe that $U(V)/U(D)$ finitely generated and V quasi-local imply (Lemma 3.1) that V is a finite D-module. Brandis' Theorem tells us that as $U(V/Q)/U(D/M) \simeq U(V)/U(D+Q)$ is finitely generated, and D/M is infinite, $V/Q = (D+Q)/Q$, that is, $V = D + Q$. Consequently, $M(V/D) = (Q+D)/D = V/D$, and then by Nakayama's Lemma, $D = V$.

Thus, assume $MV \neq Q$, and let $x \in Q - MV$ (whence, $xV \supset MV$). It is clear that $(1+xV)/(1+M)$ is a subgroup of $U(V)/U(D)$ and maps homomorphically onto $(1+xV)/(1+MV)$. One checks to see that the map taking $(1+xV)/(1+MV)$ onto $xV/(MV+x^2V)$ by sending $(1+xa)(1+MV)$ to $xa + (MV+x^2V)$, $a \in V$, is a well defined homomorphism from the multiplicative group on the left-hand side onto the additive group on the right-hand side; denoting this latter group by N, we deduce that N is finitely generated as a group. But N is a module over V, annihilated by xV, so it may be regarded as a module over V/xV, and thus in turn as a module over $(D+xC)/xV \simeq D/M$, the last isomorphism resulting from the fact that $xV \cap D = M$. Hence, the additive group of N is isomorphic to a direct sum of copies of the additive group of D/M, an infinite field, and being also finitely generated, $N = 0$. Thus, $xV = x^2V + MV$ for each $x \in Q - MV$. From $V = D + Q$ we see that V/D and Q/M are isomorphic as D-modules, whence Q/M is a finitely generated D-module which maps onto Q/MV. Now M annihilates Q/MV, and so Q/MV may be regarded as a finite dimensional vector space over $D-M$: xV/MV is a finite dimensional subspace whose structure as a D-module is essentially the same as its structure as a D/M-module; in particular, xV/MV is a finitely generated D-module and *a fortiori*, a finitely generated V-module. Taking this together with what we have already proved, we have $xV(xV/MV) = (x^2V+MV)/MV = xV/MV$ for each $x \in Q - MV$. A final application of Nakayama's Lemma then leads to the conclusion that $xV = MV$, contradicting the choice of x. Hence, $MV = Q$, and all is proved.

We are now ready to prove a more general result.

THEOREM 3.5. *Suppose that* D *is quasi-local with maximal ideal* M *and* $G(D)$ *is finited generated.* *If* D/M *is infinite,*

<u>then</u> D <u>is a valuation ring</u>. <u>Therefore</u>, $G(D)$ <u>is totally ordered in this case</u>.

<u>PROOF</u>: Let $Q_1,\ldots,Q_v$ be the maximal ideals of $\bar{D}$. For each $i = 1,\ldots,v$, $U(\bar{D}/Q_i)/U(D/M) \simeq U(\bar{D}_{Q_i})/U(D+Q_i\bar{D}_{Q_i})$ from the isomorphism between $\bar{D}_{Q_i}/Q_i\bar{D}_{Q_i}$ and $\bar{D}/Q_i$. Thus $U(\bar{D})/Q_i)/U(D/M)$ is finitely generated and therefore by Brandis' Theorem, $\bar{D}/Q_i = D/M$ for each i. Denote by h the diagonal map sending $\bar{D}$ onto $\oplus \sum_{i=1}^{v} \bar{D}/Q_i$ and by S the inverse image under h of $\Delta = \{(a+Q_1,a+Q_2,\ldots,a+Q_v) : a \in D\}$. Then $S = D + \cap_{i=1}^{v} Q_i$, and $U(\bar{D})/U(S) \simeq U(\oplus \sum_{i=1}^{v} D/Q_i)/U(\Delta)$ is finitely generated. Since the residue fields are infinite, we are confronted with the following situation: F is in infinite field $(\simeq D/M)$, $K_i (=D/Q_i)$, $i=1,\ldots,v$, are fields isomorphic to F, say $\sigma_i(F) = K_i$, and $U(K_1 \oplus\ldots\oplus K_v)/\{(\sigma_1(f),\ldots,\sigma_v(f) : f \in F^*\}$ is finitely generated. A quick calculation shows that the map sending the element $(\alpha_1,\ldots,\alpha_v) \in K_1^* \oplus\ldots\oplus K_v^*$ onto the element $(\sigma_2(\sigma_1^{-1}(\alpha_1^{-1}))\alpha_2,\ldots,\sigma_v(\sigma_1^{-1}(\alpha_1^{-1}))\alpha_v)$ is a homomorphism from $U(K_1\oplus\ldots\oplus K_v)$ onto $U(K_2)\oplus\ldots\oplus U(K_v)$ with kernel $\{(\sigma_1(f),\ldots,\sigma_v(f) : f \in F^*\}$. Hence, if $v > 1$, we must have $K_2^* \oplus\ldots\oplus K_v^*$ finitely generated, whence $K_2^* \simeq U(D/M)$ is finitely generated, and this occurs only if D/M is finite. (This is obvious enough if char D/M is 0, since Q^* is hardly finitely generated; if char D/M is p, and W is the prime field of D/M, then if $x \in D/M$ is transcendental over W it is clear that $W(x)^*$ is not finitely generated because of the existence of infinitely many irreducible elements in $W[x]$; if D/M is algebraic over W and $U(D/M)$ is finitely generated, then D/M is finite.) This contradiction forces $v = 1$, that is, $\bar{D}$ is quasi-local and thus a valuation ring by Theorem 2.1. Theorem 3.4 now applies and we conclude that D is a valuation ring.

We apply this theorem to obtain the following golbal result:

COROLLARY 3.6. Suppose that $G(D)$ is finitely generated. If P is a prime ideal of D, then D_P is a valuation ring if either P is not maximal or P is maximal and D/P is infinite. If D/M is infinite for each maximal ideal of D, then D is a semi-quasi-local Bezout domain and $G(D)$ is lattice-ordered.

FINITE RESIDUE FIELD CASE.

Maintaining the same notation as above, we now prove two results analogous to Theorems 3.4 and 3.5 except that in this case the residue field is finite.

THEOREM 3.7. Suppose that $G(D)$ is finitely generated and that the integral closure V of D is a valuation ring. Then D is quasi-local and if the residue field of D is finite, then $U(V)/U(D)$ is a finite group.

PROOF: As in the proof of the Theorem 3.4, we conclude that V is a finitely generated D-module, and hence, the conductor, f, of D in V is non-zero. If P is a non-maximal prime ideal of D, then $G(D_P)$ is finitely generated, D_P has infinite residue field, and the integral closure of D_P is V_{D-P}, a valuation ring. Hence, by Theorem 3.4, D_P is a valuation ring, and thus integrally closed. By [15, p.269], $f \not\subseteq P$, P a non-maxiaml prime; it follows that f is M-(equivalently, Q-) primary. From the isomorphism between $(U(V)/U(D))/(U(D+Q)/U(D))$ and $U(V)/U(D+Q)$, we see that necessary and sufficient conditions for $U(V)/U(D)$ to be finite are that UD+Q)/U(D) and $U(V)/U(D+Q)$ be finite. Since $U(V)/U(D+Q)$ is isomorphic to $U(V/Q)/U(D/M)$, and D/M is finite and V is a finite D-module, we see that $U(V)/U(D+Q)$ is finite. We conclude the proof by showing that $U(D+Q)/U(D)$ is finite.

Let $d+q$ be an element of $U(D+Q)$. Then d is an element

of $U(D)$, and $(d+q)U(D) = (1+q')U(D)$, for some $q' \in Q$. Let p be the characteristic of D/M and let $1+q \in 1+Q$. Since f is both M- and Q-primary, and $p \in M$, $q \in Q$, there exist positive integers r,s such that p^r and q^s belong to f. Now for all positive integers t, $(1+q)^{p^t} =$ $1+p^t q+\binom{p^t}{2}q^2+\ldots+\binom{p^t}{s-1}q^{s-1}+q^s v$, for some $v \in v$. By choosing t large enough, we can guarantee that p^r divides $\binom{p^t}{i}$, $i=1,\ldots,s-1$, and thus $(1+q)^{p^t} \in 1+f \subseteq U(D)$, which shows that $U(D+Q)/U(D)$ is torsion and being a subgroup of $G(D)$ it is already finitely generated, and hence is finite.

THEOREM 3.8. <u>Suppose that</u> D <u>is quasi-local with maximal ideal</u> M, <u>that</u> $G(D)$ <u>is finitely generated, and, moreover, suppose that</u> D/M <u>is finite</u>. <u>Then</u> $U(\bar{D})/U(D)$ <u>is a finite group and</u> $\bar{D}$ <u>is a finite</u> D-<u>module</u>, <u>where</u> $\bar{D}$ <u>is the integral closure of</u> D.

<u>PROOF</u>: Use the same notation and set up as in the proof of Theorem 3.5. Then because $U(\bar{D}/Q^i)/U(D/M)$ is finitely generated and $\bar{D}/Q_i$ is integral over D/M, a finite field, it follows that $\bar{D}/Q_i$ is finite for each i. Consequently, $U(\bar{D})/U(S)$ is finite and $\bar{D}$ is a finite S-module.

To show that $U(\bar{D})/(D)$ is finite, we need now merely show that $U(S)/U(D)$ is finite. Since $S = D + \bigcap_{i=1}^{v} Q_i$, S is quasi-local with maximal ideal $\bigcap_{i=1}^{v} Q_i$, S is integral over D, and $U(S)/U(D)$ is finitely generated. By Lemma 3.1 S is a finite D-module, and hence, the conductor, f, of D in S is non-zero. Furthermore, since $\bar{D}$ is a finitely generated S-module, and S is a finitely generated D-module, $\bar{D}$ is a finitely generated D-module. The lemma cited [15, p. 269] in the proof of Theorem 3.7 is valid for S, even though S is not the integral closure of D ; its proof only requires that S be a finite D-module. Also, if P is a non-maximal prime ideal of D , then Corollary 3.6 quarantees that D_P is a valuation ring and thus is integrally closed. It follows that f is M-(equivalently, $\overset{v}{\underset{i=1}{\cap}} Q_i$-) primary, and the rest of the proof is now accomplished by the last paragraph of the proof of Theorem 3.7.

COMBINED RESULTS

We combine Theorems 3.4 through 3.8 to obtain a global result.

THEOREM 3.9. _Let_ D _be a domain such that_ $G(D)$ _is finitely generated_. _Then_ $\bar{D}$ _is a finite_ D-_module and_ $U(\bar{D})/U(D)$ _is a finite group_.

PROOF: The first assertion follows from the more general result that if R is a semi-quasi-local ring and M is an R-module, then M is a finitely generated R-module if and only if M_P is a finitely generated R_P-module for each maximal ideal P of R ; see [4, p.54] . By Theorems 3.4 and 3.8 $\bar{D}_{D-P}$, the integral closure of D_P , is a finitely generated D_P-module for each maximal ideal P of D .

The second assertion also follows from a more general result: let G be an abelian group, $G_1,\ldots G_n$ subgroups of G , and $H_1,\ldots,H_n$ subgroups of the G_i ; if G_i/H_i is finite for each i , then $(\bigcap_1^n G_i)/(\bigcap_1^n H_i)$is finite. By induction, it suffices to consider the case $n = 2$. Thus, suppose that G_1/H_1 and G_2/H_2 are finite. Then $G_1 \cap G_2/G_1 \cap H_2$ is isomorphic to a subgroup of G_2/H_2 and $G_1 \cap G_2/H_1 \cap G_2$ is isomorphic to a subgroup of G_1/H_1 ; by Poincare's Theorem [8, p.62], $G_1 \cap G_2/G_1 \cap H_2 \cap H_1 \cap G_2 = G_1 \cap G_2/H_1 \cap H_2$ is finite. If $P_1,\ldots,P_k$ are the maximal ideals of D, then again by Theorems 3.4 and 3.8, $U(\bar{D}_{D-P_i})/U(D_{P_i}) \simeq G_i/H_i$ is finite for each i. Letting K^* play the role of G in the foregoing discussion, we conclude that $\bigcap_{i=1}^{k} U(\bar{D}_{D-P_i})/\bigcap_{i=1}^{k} U(D_{P_i}) = U(\bar{D})/U(D)$ is finite, this last equality resulting from

$$\bigcap_{i=1}^{k} \bar{D}_{D-P_i} = \bar{D} \quad \text{and} \quad \bigcap_{i=1}^{k} D_{P_i} = D .$$

From the exactness of the sequence $\{1\} \to U(\bar{D})/U(D) \to G(D) \to G(\bar{D}) \to \{1\}$ we obtain the following:

COROLLARY 3.10. *Let* $G(D)$ *be finitely generated. Then the torsion-free rank of* $G(D)$ *is the same as the torsion-free rank of* $G(\bar{D})$.

COROLLARY 3.11. *If* $G(D)$ *is finitely generated free, then* $U(D) = U(\bar{D})$; *consequently,* $G(D)$ *and* $G(\bar{D})$ *are isomorphic as unordered abelian groups.*

It seems appropriate at this point to raise the question: if $G(D)$ is finitely generated and $G(D) = G(\bar{D})$, is $D = \bar{D}$? The following example provides a negative response, but this example has another implication, and to understand its significance, we discuss the answer to another question: what domains have group of divisibility $G = Z \oplus Z$? Three examples readily come to mind: First, embed G into the group of additive real numbers and give G the induced total order T_1, then let D be a valuation ring of dimension one with G ordered by T_1 as its group of divisibility. Second, give G the lexicographic ordering T_2 and let D be a valuation ring of Krull dimension two with group of divisibility G, ordered by T_2. Third, let D be the intersection of two rank one discrete valuation rings (DVR's). Then D is a Bezout domain with two maximal ideals and the group of divisibility is G with the cardinal ordering, that is, $(a,b) \le (c,d)$ if and only if $a \le c$ and $b \le d$. Then we might conjecture that these three examples are the only domains with group of divisibility $Z \oplus Z$ as an unordered group. But the following example also lays this conjecture to rest.

EXAMPLE 3.12. Let $k = GF(2)$, and let V_1 and V_2 be distinct DVR's of the form $k + M_1$, $k + M_2$, where M_1 and M_2 are the respective maximal ideals of V_1 and V_2. For example, let $V_1 = k[x]_{(x)}, V_2 = k[x]_{x+1)}$.) Set $D = k + (M_1 \cap M_2)$. Then $\bar{D} = V_1 \cap V_2$ is integrally closed and is integral over D because they share the common ideal $M_1 \cap M_2$ and the image of $V_1 \cap V_2$ modulo this ideal is isomorphic to the finite ring $k \oplus k$ which is certainly integral over the image of D modulo $M_1 \cap M_2$ (see [6, p. 91]). Indeed, if T is the canonical map from $\bar{D}$ onto $D/M_1 \cap M_2 = k \oplus k$ and if k' denotes k immersed as the diagonal of $k \oplus k$, then $D = \sigma^{-1}(k')$. Hence, $U(\bar{D})/U(D) \simeq U(\bar{D}/M_1 \cap M_2)/U(D/M_1 \cap M_2) = \{1\}$ since the units of $k \oplus k$ coincide with the units of the diagonal of $k \oplus k$. Therefore, it follows that $G(D)$ is isomorphic (as an unordered group) to the direct sum of two copies of Z. Note too, that since the diagonal of $k \oplus k$ is a proper subring of $k \oplus k$, $D \neq \bar{D}$. Before leaving this example, let us observe that $GF(2)$ is intrinsic to its construction. Thus, suppose that (D,M) is quasi-local, $D \neq \bar{D}$, $G(D)$ is finitely generated, and $U(D) = U(\bar{D})$. Theorem 3.5 guarantees that D/M is finite and Lemma 2.3 that $\bar{D}$ is not quasi-local. Let $M_1, \ldots, M_n$ be the maximal ideals of $\bar{D}$, $n \geq 2$, and $N = \bigcap_{i=1}^{n} M_i$; by hypothesis $U(D) = U(\bar{D})$, and therefore $\{1\} = U(\bar{D})/U(D+N) \simeq U(\bar{D}/N)/U(D+N/N) \simeq (U(\bar{D}/M_1) \oplus \ldots \oplus U(D/M_n))/\{(\sigma_1(\alpha), \sigma_2(\alpha), \ldots, \sigma_n(\alpha)) : \alpha \in (D/M)^*\}$, where σ_i is the canonical embedding of D/M in $\bar{D}/M_i$. Suppose that $D/M \neq GF(2)$, and let $\gamma \in (D/M)^* - \{1\}$. By the preceding, there exists $\alpha \in (D/M)^*$ such that $(\sigma_1(\gamma), 1, 1, \ldots, 1) = (\sigma_1(\alpha), \sigma_2(\alpha), \ldots, \sigma_n(\alpha))$; from $\sigma_2(\alpha) = 1$, we conclude that

$\alpha = 1$ and hence that $\sigma_1(\alpha) = \sigma_1(\gamma) = 1$, a contradiction.

Now let us consider this question: If $G(D)$ is finitely generated, and L is finite algebraic extension field of K, and J is a ring containing D and contained in L, when can we say that $G(J)$ is finitely generated? The most general result known to us is our next theorem.

THEOREM 3.13. *Let the notation be as in the preceding paragraph. If J is the integral closure of D in L, and $G(D)$ is finitely generated, then $G(J)$ is finitely generated.*

PROOF: We know that J (which is also the integral closure of $\bar{D}$ in L) is Prüfer and has only finitely many prime ideals; see [6, p. 277] and [6, p. 121]. Hence, J is the intersection of a finite number of valuation rings on L. By Lemma 3.3 our problem is now reduced to showing that each valuation overring of J has finitely generated value group. Let V be such a valuation ring, and let $W = K \cap V$. Then $G(W)$ is a subgroup of $G(V)$ and has index $\leq [L:K]$ by [16, p. 52]; since W is a valuation overring of D, $G(W)$ is finitely generated, and it follows that $G(V)$ is also.

So far, it has been of considerable importance to us that $\bar{D}$ is a finite D-module if $G(D)$ is finitely generated; in the terminology of Matsumura [10, p.231], D is N-1. In [10, Chapter 12], [5, pp. 110-119], and [12, p.112], a large number of results are given for a domain satisfying the stricter condition (N-2) that its integral closure in any finite algebraic extension field of its quotient field is finitely generated as a module. In line with the last theorem, one might hope that if $G(D)$ is finitely generated, then D satisfies N-2 as well. Unfortunately, such is not the case. Thus, D may be a rank one discrete valuation ring, whose integral closure in a quadratic extension of K is also a valuation ring, but is not a finitely generated

D-module; see [7, p. 70].

We conclude this section with a rather restricted result which will be of use later.

THEOREM 3.14. *Let* D *be a one-dimensional semi-local domain such that the residue field of* D *with respect to each maximal ideal is finite. If* J *, the integral closure of* D *in* L *, a finite algebraic extension field of* K *, is a finite* D-*module and* G(J) *is finitely generated, then* G(D) *if finitely generated.*

PROOF: $J \cap K = \bar{D}$ implies $U(J) \cap U(K) = U(\bar{D})$, and hence, $K^*/U(\bar{D})$ may be considered as a subgroup of $L^*/U(J)$, whence $G(\bar{D})$ is finitely generated. Since J was assumed to be a finite D-module, and $\bar{D}$ is a D-submodule, D Noetherian implies $\bar{D}$ is a finite D-module. It follows that there is no loss of generality in assuming $L = K$ and $J = \bar{D}$.

Let M be a maximal ideal of D. Then $\bar{D}_{D-M}$ is a finitely generated D_M-module, has finitely generated group of divisibility, and the conductor, f, of D_M in $\bar{D}_{D-M}$ is non-zero. Hence, either $f = D_M$, or $U(\bar{D}_{D-M})/U(D_M) \simeq U(\bar{D}_{D-M}/f)/U(D_M/f)$ is finite [9, p. 561]; in either case, $G(D_M)$ is finitely generated. By Lemma 3.3, G(D) is finitely generated.

So far, except through the construction of special examples, we have not had much to say about the occurrence of domains with finitely generated groups of divisibility. The next corollary shows that such domains abound. Recall that a global field is either a finite algebraic extension of Q, or an algebraic function field in one variable over a finite field.

COROLLARY 3.15. *Let* D *be a semi-quasi-local domain whose quotient field* K *is a global field. Then* G(D) *is finitely generated.*

PROOF: The well-known generalization of the Krull-Akizuki Theorem guarantees that D is one-dimensional and Noetherian. Moreover, D contains either Z or $F[y]$, where F is a finite field and y is transcendental over F, and therefore all its residue fields are finite; see [9,p. 561]. In the case that char $D = 0$, we have $D \supseteq Z$, and $\bar{D} \supseteq \bar{Z}$, $\bar{Z}$ the integral closure of Z in K, and we know that $\bar{Z}$ is a finite Z-module since K/Q is separable; see [15, p. 265]. In the case that $D \supseteq F[y]$, we again obtain the integral closure of $F[y]$ in K as a finite $F[y]$-module as a consequence of the Noether Normalization Theorem [15, p. 268]. Hence, if we let D_0 denote either Z or $F[y]$, and $\bar{D}_0$ its integral closure in K, we see that $\bar{D}_0 = D_0x_1+\ldots+D_0x_n$, where $x_i \in \bar{D}_0$. Since D_0 is Dedekind, so is $\bar{D}_0$, and consequently every overring of $\bar{D}_0$ is integrally closed. It follows that $D[x_1,\ldots,x_n]$ is an integrally closed domain integral over D with quotient field K, and therefore $\bar{D} = D[x_1,\ldots,x_n]$ is a finite D-module. Moreover, D is a semi-local Dedkind domain so that $G(\bar{D})$ is finitely generated; thus the conclusion is immediate form Theorem 3.14.

4. APPLICATION TO ONE-DIMENSIONAL DOMAINS. We saw in section 3 that if $G(D)$ is finitely generated, then the integral closure $\bar{D}$ of D is a finite D-module (that is, D satisfies N-1). Nagata observes in an exercise [12, p. 122] that a one-dimensional semi-local domain D is N-1 if and only if the completion of D under the Jacobson radical topology is reduced, or in other words, if and only if D is analytically unramified. In fact, the proof that D is analytically unramified follows easily from the observation that under the circumstances the completion of D is a subring of the completion of $\bar{D}$ [16, p.275-7] and

the completion of $\bar{D}$ is a direct sum of valuation rings and so certainly is reduced. Nagata has shown [12, p. 114] that an analytically unramified semi-local domain satisfies N-1 .

Thus, if D is a one-dimensional semi-local domain and if $G(D)$ is finitely generated, then D is analytically unramified. Moreover, the converse holds in the case that the residue fields are all finite. For in this case, by the exercise of Nagata quoted above, D is N-1 and since $\bar{D}$ is a semi-local Dedekind domain, $G(\bar{D})$ is finitely generated. To conclude $G(D)$ is finitely generated, we cite Theorem 3.14.

In summary, we have the following result:

THEOREM 4.1. *Suppose that* D *is a one-dimensional semi-local domain such that each residue field is finite. Then the following are equivalent:*

(i) $G(D)$ *is finitely generated.*

(ii) D *is analytically unramified.*

(iii) $\bar{D}$ *is a finite* D-*module.*

(iv) *The conductor of* D *in* $\bar{D}$ *is non-zero.*

As we have just seen, for certain rings R , the completion of R and the group of divisibility of R are interrelated. That this should be so is not suprising in view of the following:

PROPOSITION 4.2. *If* (R,M) *is a one-dimensional local ring, and if* (R',M') *is its completion, then* $G(R)$ *and* $G(R')$ *are order isomorphic as partially ordered abelian groups.*

PROOF. Let x be a regular element of R' ; the one-dimensionality of R' forces xR' to be an open ideal, hence an extended ideal of R — say $xR' = AR'$ where A is an ideal of R . That A can be taken to be principal is assured by [1, p.73]. Then $xR' = x_0R'$ where $x_0 \in R$. It follows that if x,y are regular elements of R' , then

$f((x_0/y_0U(R)) = (x/y)U(R')$ defines an order isomorphism from $G(R)$ onto $G(R')$.

REFERENCES

1. N. Bourbaki, Éléments De Mathématique, Algebre Commutative, Chapt. III, IV, Hermann, Pairs, 1967.

2. ____________, Éléments De Mathématique, Algebre Commutative, Chapt. V, VI, Hermann, Paris, 1964.

3. A. Brandis, Über die multiplikiative Struktur von Körpererweiterungen, Math. Z., 87(1965), 71 - 73.

4. E. D. Davis, Overrings of commutative rings I, Noetherian overrings, Trans. Amer. Math. Soc., 104(1962), 52 - 61.

5. J. Dieudonné, Topics in Local Algebra, Notre Dame Mathematical Lectures Number 10, University of Notre Dame Press, 1967.

6. R. Gilmer, Multiplicative Ideal Theory, Marcel Dekker, New York, 1972.

7. I. Kaplansky, Commutative Rings, Allyn and Bacon, Boston, Mass., 1969.

8. A. G Kurosh, The Theory of Groups, Vol. 1, Chelsea Publishing Company, New York, 1960.

9. K. B. Levitz and J. L. Mott, Rings with finite norm property, Can. J. Math., Vol. 24, 1972, 557 - 565.

10. H. Matsumura, Commutative Algebra, W. A. Benjamin, Inc., New York, 1970.

11. J. L. Mott and M. Schexnayder, Exact sequences of semi-value groups, J. Reine Angew. Math., 283/284(1976), 388-401.

12. M. Nagata, Local Rings, Wiley (Interscience), New York, 1962.

13. D. G. Northcott, Lessons on Rings, Modules and Multiplicities, Cambridge University Press, New York, 1968.

14. P. Samuel, Algebre Locale, Gaitier-Villar, Paris, 1953.

15. O. Zariski and P. Samuel, Commutative Algebra, Vol. I, Van Nostrand, Princeton, New Jersey, 1958.

16. ______________________, Commutative Algebra, Vol. 2, Van Nostrand, Princeton, New Jersey, 1960.

UNIVERSITY OF MISSISSIPPI
OXFORD, MISSISSIPPI 38655

FLORIDA STATE UNIVERSITY
TALLAHASSEE, FLORIDA 32306

Contemporary Mathematics
Volume 8, 1982

ON QUADRATIC FORMS AND ABELIAN VARIETIES OVER FUNCTION FIELDS

Albrecht Pfister

1. INTRODUCTION.

This paper is motivated by a recent article of A. Brumer [B]. Our main concern is the u-invariant $u(F)$ of a field F of transcendence degree 2 over a real closed field R. It is known that $u \leq 6$ and conjectured that $u \leq 4$. The statement "$u \leq 4$" is equivalent to a kind of "splitting property" or "period = index equality" in the Brauer group $B(F)$. Therefore a detailed study of $B(F)$ seems to be necessary. This is intimately related with the theory of abelian varieties over K where $K \subset F$ is an intermediate field of transcendence degree 1 over R. In a typical case there is an exact sequence of Scharlau [Sch] which shows that $B(F)$ contains $H^1(\overline{C})$ with $\overline{C}$ the divisor class group of $\overline{K}F/\overline{K}$. $H^1(\overline{C})$ may perhaps be attacked by the method of Ogg [O] and Šafarevič [S] though their results are not immediately applicable. In any case a lot more has to be done in order to decide the problem on the u-invariant. I hope to come back to this topic in a further paper.

2. TERMINOLOGY AND PREREQUISITES.

R denotes a real closed field, $C = \overline{R} = R(i)$ its algebraic closure. For a field F containing R the transcendence degree of F over R is denoted by $tr(F/R)$. If F is (formally) real the real closures of F are denoted by F_α where α runs through the orderings of F. If F is nonreal $s(F)$ is the level of F, i.e., the minimal number s such that -1 is a sum of s squares in F. With these notations the following results and conjectures are well known.

THEOREM OF TSEN [T] - Lang [L1]: Let $C \subset F$, $tr(F/C) = m$. Then F is a C_m-field. In particular any form f of degree d over F has a non-trivial zero in F provided the number of variables of f is bigger than d^m.

1980 Mathematics Subject Classification. 10C04, 14G30, 14K05.

0271-4132/81/0000-0802/$05.00

COROLLARY: Let f be a quadratic form in n variables over F. If $n > 2^m$, f is isotrophic in F. If $n = 2^m$, f is universal in F.

CONJECTURE OF S. LANG [L2]: Let F be nonreal, $tr(F/R) = m$. Then F is a C_m-field.

This conjecture is still open even for $m = 1$. If we restrict ourselves to quadratic forms we get

CONJECTURE 1: Let F be nonreal, $tr(F/R) = m$. Then every quadratic form over F in at least 2^m variables is universal in F.

This conjecture and the slightly more general Conjecture 2 below are true for $m = 1$, but still open for $m \geq 2$.

Let $W = W(F)$ be the Witt group of F. It consists of the similarity classes $\tilde{q}$ of regular quadratic forms q over F. Let $W_t = W_t(F)$ be the torsion subgroup of W. One knows that

$$W_t = \ker(W(F) \longrightarrow \prod_{\alpha} W(F\alpha)).$$

Finally let

$$u = u(F) = \max \{\dim q : q \text{ anisotropic quadratic form over } F, \tilde{q} \in W_t(F)\}$$

be the u-invariant of F in the sense of Elman-Lam.

CONJECTURE 2: Let $tr(F/R) = m$. Then $u(F) \leq 2^m$.

We are particularly interested in the case $m = 2$ of this conjecture since this is probably the "easiest" yet unsolved case of all the conjectures.

3. FIRST RESULTS ON $u(F)$ AND $W(F)$.

In this part we collect some results on u and W which are mainly known, maybe not all of them in the present form.

PROPOSITION 1 [EL4, Th.6.1]: If $tr(F/R) = m$, then $I^{m+1}(F)$ is torsion free. Here $I(F) \subset W(F)$ denotes the maximal ideal of all $\tilde{q}$ with $\dim q$ even. For $m = 1$ this proposition implies that a two-dimensional form $q \in W_t(F)$ is universal. Therefore Conjecture 2 is true for $m = 1$.

COROLLARY: Let $T = T(F)$ be the set (subgroup of $\dot{F}$) of totally positive elements of F. For $t \in T$ the form $<1, -t>$ is torsion, hence $2^m \times <1, -t> \sim 0$, hence t is a sum of 2^m squares in F.

PROPOSITION 2 [EL3, Th.3]: Let F be any field with $\operatorname{char} F \neq 2$.

The following statements are equivalent:

(i) $I^3(F)$ is torsion free.

(ii) Quadratic forms over F are classified (up to equivalence) by dimension, discriminant, Clifford invariant and total signature.

(iii) Quaternion forms $\langle 1,a,b,ab\rangle$ over F represent every $t \in T(F)$.

COROLLARY: $I^3(F)$ is torsion free iff torsion forms over F are classified by dimension, discriminant and Clifford invariant.

It is clear that Proposition 2 and its corollary apply for a field F with $tr(F/R) = 2$. Nevertheless this does not lead to an immediate proof of Conjecture 2 for $m = 2$. The reason is that the Clifford invariant is a fairly difficult object under these circumstances.

For a better investigation of the u-invariant it is helpful to make the following definition (after Elman-Lam).

DEFINITION: $u_i = u_i(F) = \max \{\dim q : q$ anisotropic quadratic form over F, $2^i x \tilde{q} = 0$ in $W(F)\}$.

Then $0 = u_0 \leq u_1 \leq \ldots$, and since W has only 2-power-torsion, $u = \sup u_i$.

PROPOSITION 3 [EL1, Th.4.1]: Suppose $0 < u_1 < \infty$. Then

$$u_i = (1 + \frac{1}{2} + \cdots + \frac{1}{2^i})\, u_1 < 2u_1 \quad \text{for} \quad i \geq 1.$$

In particular $u_i \leq 2u_1 - 1$ for all i and $u \leq 2u_1 - 1$.

PROPOSITION 4 [EL1, Prop.1.4 and Remark 4.2]: If $I^3(F)$ is torsion free and if $1 < u(F) < \infty$ then $u(F) = u_1(F)$ and $u(F)$ is even.

PROPOSITION 5 [EL1, Th.4.11]: If $tr(F/R) = 2$ then $u(F) \in \{0,1,2,4,6\}$.

REMARK: $u(F) = 0$ resp. 1 iff F is pythagorean and real resp. nonreal.

Let $T_i = T_i(F)$ denote the set of elements $t \in F$ which are sums of i squares in F. For $tr(F/R) = 2$ we have the following sequence of subgroups of $\dot{F}$:

$$\dot{F}^2 = T_1(F) < T_2(F) < T_4(F) = T(F) < \dot{F}.$$

PROPOSITION 6: Suppose $tr(F/R) = 2$. The following statements are equivalent:

(i) $u(F) \leq 4$.

(ii) Any form $<1, -t_1> \oplus a <1, -t_2>$ with $t_1, t_2 \in T_2F$, $a \in \dot{F}$ is universal in F.

PROOF: 1) Suppose $u(F) \leq 4$. Then

$$\phi := <1, -t_1> \oplus a < 1, -t_2> \oplus b <1, -t_1t_2>$$

with t_1, t_2, a as above and $b \in \dot{F}$, is isotropic in F, since $2 \times \phi \sim 0$. Therefore $\phi \cong <1, -1> \oplus \psi = bt_1t_2 <1, -1> \oplus \psi$ with $\dim \psi = 4$. We also have $d(\phi) = 1$, hence $\tilde{\phi} \in I^2(f) \cap W_t(F)$, hence $\phi \otimes <1, -c> \in I^3(F) \cap W_t(F) = 0$ for each $c \in \dot{F}$. This gives $\phi \cong c\phi$. Witt's cancellation theorem implies $\psi \cong c\psi$. In particular, ψ is universal in F. But then

$$<1, -t_1> \oplus a <1, -t_2> \oplus <b> \cong <bt_1t_2> \oplus \psi$$

is isotropic. This shows that $<1, -t_1> \oplus a <1, -t_2>$ is universal.

2) For the converse we may assume that $i = \sqrt{-1} \notin F$. Let ϕ be a quadratic form over F (not necessarily anisotropic) with $2 \times \phi \sim 0$. Then ϕ has even dimension $2n$ and it is easily shown that

$$\phi \cong \bigoplus_{i=1}^{n} a_i <1, -t_i>$$

with $a_i \in \dot{F}$, $t_i \in T_2(F)$.

Without loss of generality we can take $a_1 = 1$. Now statement (ii) implies that ϕ is isotropic if $n \geq 3$. Therefore $u_1(F) \leq 4$ by the definition of u_1, hence $u(F) \leq 4$ by Propositions 3 and 4.

4. THE BRAUER GROUP $B(F)$ OF F FOR $tr(F/R) = 2$.

We use the following notations:

$\bar{I}^m = I^m/I^{m+1}$ for $m \geq 1$;

$I_t^m = I^m \cap W_t = \ker (I^mF \longrightarrow \prod_\alpha I^m(F_\alpha))$;

$\bar{I}_t^m$ = image cf I_t^m under the canonical map $I^m \longrightarrow \bar{I}^m$;

$B = B(F)$ = Brauer group of F;

$B_2 = B_2(F)$ = subgroup of elements of order ≤ 2;

$Q = Q(F)$ = subgroup of B_2 which is generated by the classes of quaternion algebras (a,b) with $a, b \in \dot{F}$.

A suffix t denotes the elements of the relevant group vanishing in all real closures F_α of F, e.g.,

$$B_{2t}(F) = \ker (B_2(F) \longrightarrow \prod_\alpha B_2(F_\alpha)).$$

REMARK: Since B and its subgroups are torsion groups anyway the suffix t should not be interpreted as an abbreviation of "torsion elements" here!

$$\begin{aligned} B_s(F) = B_{2s}(F) &= \{c \in B(F) : c \text{ has a quadratic splitting field}\} \\ &= \{(a,b) \in B(F) : a,b \in \dot{F}\} \\ B_{st}(F) = B_s \cap B_{2t} &= \{(a,b) \in B(F) : \langle 1, -a, -b, ab\rangle \in W_t(F)\} \\ &= \{(a,b) \in B(F) : b \text{ represented by } 2 \times \langle 1, -a\rangle\} \\ &\quad \text{for } \operatorname{tr}(F/R) = 2. \end{aligned}$$

The last equality follows from Proposition 1.

REMARK: B_s resp. B_{st} are not subgroups of B_2 resp. B_{2t} unless they are equal to these groups.

PROPOSITION 7: If $\operatorname{tr}(F/R) = 2$ then

$$Q(F) = \{(-1,a) \cdot (b,c) \in B : a,b,c \in \dot{F}\}$$

PROOF. Let $a,b,c,d \in \dot{F}$ and let $(a,b) \cdot (c,d)$ be a product of 2 quaternion algebras in $B(F)$. Then $(a,b) \cdot (c,d) = c(\phi)$ where $\phi = \langle -a, -b, ab, c, d, -cd\rangle$ and $c : W(F) \longrightarrow B(F)$ is the Clifford invariant. Since $d(\phi) = 1$ we have $c(\phi) = c(-a\phi)$. Put $-a\phi = \langle 1\rangle \oplus \psi$, $\dim \psi = 5$. ψ is isotropic over $K = F(i)$. This implies $\psi = \langle -b', -b', c', d', -c'd'\rangle$ over F since $d(\psi) = -1$. (Or ψ is isotropic over F.) Therefore $c(\phi) = c(\langle 1\rangle \oplus \psi) = (-1, b') \cdot (c',d')$. As any element of $Q(F)$ is a product of quaternion algebras an immediate induction on the number of factors gives the result.

REMARK: Without further assumptions this result cannot be improved as is seen by the following example: Let $F = R(x,y)$ be the rational function field in 2 variables over R. Let $\phi = \langle 1, 1, 1, x, y, -xy\rangle$, $c(\phi) = (1, -1) \cdot (x, y)$. By an easy degree argument in the extension field $R((x))((y))$ containing F one sees that ϕ is anisotropic. Therefore $c(\phi)$ is not a quaternion algebra and cannot split in a quadratic extension of F. ϕ is a totally indefinite form over F, but not a torsion form. Similarly one gets anisotropic totally indefinite forms over F of arbitrary dimension >1.

PROPOSITION 8: If $tr(F/R) = 2$ then $\overline{I}^2(F) \cong Q(F) = B_2(F)$.

PROOF: We use the results of Arason [A], especially his Satz 3.5, Cor. 3.8, Cor. 4.6 and Satz 4.19. We have to show that for any quadratic extension $E = F(\sqrt{d})$ the sequence

$$\overline{I}^1(F) \xrightarrow{\mu} \overline{I}^2(F) \xrightarrow{i} \overline{I}^2(E) \xrightarrow{s^*} \overline{I}^2(F)$$

is exact. Here μ denotes multiplication by $<1, -d>$ and s^* denotes the Scharlau transfer. Since $I^3(F(\sqrt{-1})) = 0$ the exactness at the place $\overline{I}^2(F)$ and the isomorphism $\overline{I}^2(F) \cong Q(F)$ follow from the results cited above. Take now $\overline{\phi} \in \overline{I}^2(E)$ with $s^*(\overline{\phi}) = 0$. By Proposition 8 $c(\overline{\phi}) = (-1,\alpha)(\beta,\gamma)$ with suitable $\alpha, \beta, \gamma \in \dot{E}$. Since $\overline{I}^2(E) = Q(E)$ we may take

$$\phi = <1, -\alpha, -a, \beta, \gamma, -\beta\gamma> = <1> \oplus \psi.$$

Then $s^*(\psi) = s^*(\phi) \in I^3(F)$. As $\dim (s^*\psi) = 10$, $s^*(\psi)$ must be isotropic which implies $\psi = <a> \oplus \chi$ with $a \in \dot{F}$, $\dim \chi = 4$. From $s^*(\chi) \in I^3(F)$ we conclude by [A, Zusatz zu 2.4] that $s^*(\phi) = s^*(\chi) = s^*(\omega)$ with $\omega \in I^3(E)$. Hence ϕ and ω differ by a form ϕ_0 coming from $F: \phi \sim \phi_0 \oplus \omega$, $\phi_0 \in W(F)$. Comparing dimensions and discriminants shows that automatically $\phi_0 \in I^2(F)$. This gives $\overline{\phi} = i(\overline{\phi}_0)$ with $\overline{\phi}_0 \in \overline{I}^2(F)$.

COROLLARY: If $tr(F/R) = 2$ then $I_t^2(F) \cong \overline{I}_t^2(F) = B_{2t}(F)$.

This is clear since $I_t^3(F) = 0$.

REMARK: Compare [EL 2].

PROPOSITION 9: Let $tr(F/R) = 2$. Then $u(F) \leq 4$ iff $B_{2t}(F) = B_{st}(F)$.

PROOF: 1) Any element of $B_{2t}(F)$ has the form $(-1, a)(b,c) = c(\phi)$ with $\phi = <1, -a, -a, b, c, -bc> \in W_t(F)$. If $u(F) \leq 4$ then ϕ is isotropic which gives $c(\phi) = (a', b') \in B_{st}(F)$.

2) Assume now that $u(F) = u_1(F) = 6$. By Proposition 7 there must exist an anisotropic 5-dimensional form $\psi = <1, -t_1> \oplus a <1, -t_2> \oplus <b>$ with $t_1, t_2 \in T_2(F)$. The proof of Proposition 7 shows that

$$\phi = <1, -t_1> \oplus a <1, -t_2> \oplus b <1, -t_1 t_2>$$

is anisotropic, too. Satz 14 of [P] implies that $c(\phi)$ is not a quaternion algebra, hence $c(\phi) \notin B_{st}(F)$. On the other hand $\tilde{\phi} \in I_t^2(F)$, therefore $c(\phi) \in b_{2t}(F)$.

REMARK: Compare [EL 2, Prop. 6.4] and [ELTW, Cor. 3.15].

5. COHOMOLOGY

Here we repeat some general results of the paper [Sch] concerning $B(F)$ as well as the necessary notation. Later we specialize to $tr(F/R) = 2$. Let K be a field of characteristic 0, F/K an algebraic function field in one variable over K (with K algebraically closed in F). Denote resp. by

$$H, D, C, C_0, J, U, CJ, CU$$

the groups of principal divisors, divisors, divisor classes, divisor classes of degree 0, ideles, unit ideles, idele classes and unit idele classes of F/K. Then one has the following commutative diagram with exact rows and columns:

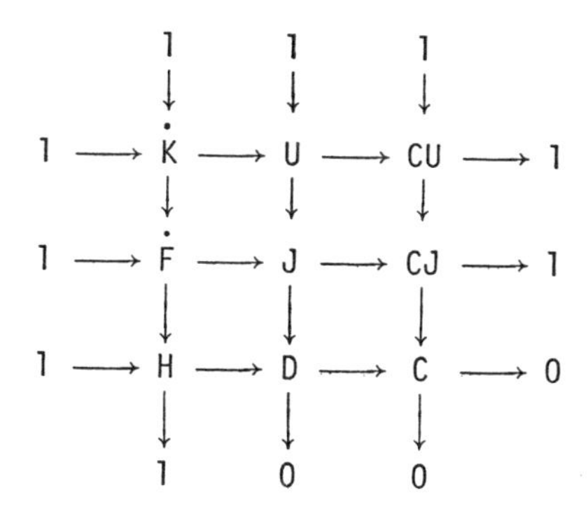

Let $\overline{K}$ be the algebraic closure of K, $\overline{K}F = \overline{F}$ the compositum of $\overline{K}$ and F, G the Galois group of $\overline{K}/K$ or $\overline{F}/F$. The objects belonging to the function field $\overline{F}/\overline{K}$ are denoted by $\overline{H}, \overline{D}, \cdots$ The corresponding diagram for $\overline{F}/\overline{K}$ is a commutative diagram of G-modules and induces long exact cohomology sequences from each row or column.

For any discrete valuation ν_ζ of F/K we let F_ζ denote the completion of F with respect to the prime divisor ζ, K_ζ the residue class field, G_ζ the Galois group of $\overline{K}/K_\zeta$. Finally, g is the genus of F/K (or $\overline{F}/\overline{K}$), d is the smallest positive degree of a divisor in D, $\chi(G) = \mathrm{Hom}\,(G, \mathbb{Q}/\mathbb{Z})$ is the character group of G and $H^m(\overline{A}) = H^m(G, \overline{A})$ for any G-module A.

By Hilbert 90 we have $H^1(\dot{\overline{K}}) = H^1(\dot{\overline{F}}) = 0$, by the Theorem of Tsen we have $B(\overline{F}) = 0$, hence $B(F) \cong H^2(\dot{\overline{F}})$.

We get the following exact sequences:

$$(1) \quad 0 \longrightarrow H^1(\overline{H}) \longrightarrow B(K) \xrightarrow{\rho} B(F) \xrightarrow{\alpha} H^2(\overline{H}) \longrightarrow H^3(\dot{\overline{K}});$$

$$(2) \quad 0 \longrightarrow H^1(\overline{D}) \longrightarrow H^1(\overline{C}) \xrightarrow{\delta} H^2(\overline{H}) \xrightarrow{\beta} H^2(\overline{D}) \xrightarrow{\sigma} H^2(\overline{C}) \longrightarrow H^3(\overline{H})$$

By Shapiro's Lemma $H^m(\bar{D}) = \coprod_\zeta H^m(G_\zeta, \mathbb{Z})$, in particular $H^1(\bar{D}) = 0$ (since G_ζ is profinite), $H^2(G_\zeta, \mathbb{Z}) \cong H^1(G_\zeta, \mathbb{Q}/\mathbb{Z}) \cong \chi(G_\zeta)$ from the exact G_ζ-module sequence $0 \longrightarrow \mathbb{Z} \longrightarrow \mathbb{Q} \longrightarrow \mathbb{Q}/\mathbb{Z} \longrightarrow 0$.

$\chi(G_\zeta)$ is related to the Brauer group $B(F_\zeta)$ by the exact sequence of Witt:

$$(3) \qquad 0 \longrightarrow B(K_\zeta) \longrightarrow B(F_\zeta) \xrightarrow{\nu_\zeta^*} \chi(G_\zeta) \longrightarrow 0.$$

If $i_\zeta : B(F) \longrightarrow B(F_\zeta)$ is the natural map, then

$$\beta \circ \alpha = \coprod_\zeta \nu_\zeta^* \circ i_\zeta : B(F) \longrightarrow H^2(\bar{D}) = \coprod_\zeta \chi(G_\zeta).$$

We now consider various special cases.

A) $K = R$ real closed. Then $G \cong \mathbb{Z}/2\mathbb{Z}$, $B(K) \cong \mathbb{Z}/2\mathbb{Z}$, $B(F(i)) = 0$, $B(F) = \{(-1, a) : a \in \dot{F}\} \cong \dot{F}/T_2(F) = \dot{F}/T(F)$, $H^3(\dot{K}) = H^1(\dot{K}) = 0$, $H^3(\bar{H}) = H^1(\bar{H})$.

One has to distinguish two cases:

a) F is nonreal. Then $T_2F = \dot{F}$ which implies $B(F) = H^2(\bar{H}) = H^1(\bar{C}) = 0$ and $H^1(\bar{H}) = B(K) = \mathbb{Z}/2\mathbb{Z}$. Further $d = 2$ since $K_\zeta = \bar{K} = R(i)$ for all ζ, G_ζ trivial, $\chi(G_\zeta) = 0$ for all ζ, $H^2(\bar{D}) = H^2(\bar{C}) = 0$.

b) F is real. Then ρ is injective, hence $H^1(\bar{H}) = 0$, $H^2(\bar{H}) \cong \dot{F}/\pm T(F)$.

Sequence (2) reduces to

$$0 \longrightarrow H^1(\bar{C}) \longrightarrow \dot{F}/\pm T(F) \longrightarrow \coprod_{\zeta \text{ real}} \mathbb{Z}/2\mathbb{Z} \longrightarrow H^2(\bar{C}) \longrightarrow 0.$$

$H^1(\bar{C})$ has order 2^{r-1} where r is the number of real connected components of the Riemann surface of F/R. See e.g., [G].

In both cases a) and b) the following results holds: $H^1(\overline{CJ}) = 0$. See e.g., [R].

B) $\operatorname{tr}(K/R) = 1$. Local results.

The Propositions 7, 8 and 9 of Part 4 hold for F_ζ instead of F as well since the proofs used only the fact that $F(i)$ is a C_2-field. In addition the theory of Springer [Sp] for quadratic forms over fields which are complete under a discrete valuation implies the following results:

Any anisotropic quadratic form q over F_ζ induces two ansiotropic residue class forms q_1, q_2 over K_ζ. q is a torsion form (i.e., $\tilde{q} \in W_t(F_\zeta)$) iff q_1 and q_2 are torsion forms. $u(F_\zeta) = 2u(K_\zeta) \leqq 4$.

$$B_{2t}(F_\zeta) = B_{st}(F_\zeta) = \{(a, b) : a, b \in \dot{F}_\zeta,\ b \text{ represented by } 2 \times \langle 1, -a\rangle\}.$$

Since $(a,b) = c(\phi)$ for $\phi = \langle 1, -a, -b, ab\rangle \in I_t^2(F_\zeta)$ we see that $c(\phi) = 1$ if a, b are both units in F_ζ. Otherwise w.l.o.g. a is a unit, b a prime element of F_ζ. Then the residue class $\bar{a}$ of a lies in $T(K_\zeta) = T_2(K_\zeta)$ and $\langle 1, -\bar{a}\rangle$ is universal in K_ζ. ϕ is independent of the prime element b. a is in $T_2(F_\zeta)$ but (a,b) depends only on the square class of $\bar{a}$ in $\dot{K}_\zeta/\dot{K}_\zeta^2$. Since $B(K_\zeta)$ is known from A) we have proved:

PROPOSITION 10: Suppose $tr(K/R) = 1$, F/K algebraic function field, F_ζ completion of F with respect to a prime divisor ζ on F/K, K_ζ residue class field of F_ζ. Then

$$B_t(K_\zeta) = 0, \quad B(K_\zeta) = B_2(K_\zeta) = \dot{K}_\zeta/T(K_\zeta), \quad B_{2t}(F_\zeta) = B_{st}(F_\zeta) \cong T(K_\zeta)/\dot{K}_\zeta^2.$$

In the exact sequence (3) the restriction of ν_ζ^* to $B_{2t}(F_\zeta)$ is an isomorphism.

C) $tr(K/R) = 1$. Global results.

The field $K(i)$ is a C_1-field, hence of cohomological dimension 1. If G' denotes the subgroup of G of index ≤ 2 which corresponds to $K(i)$, any cohomology group $H^m(G', \bar{A})$ with $m \leq 3$ and an arbitrary G'-module $\bar{A}$ vanishes. The restriction-inflation sequence then leads to to the following results:

$$H^3(\dot{\bar{K}}) = H^1(\dot{\bar{K}}) = 0, \quad H^3(\bar{H}) = H^1(\bar{H}).$$

Together with $B(K) = \{(-1, a) : a \in \dot{K}\} \cong \dot{K}/T(K)$ (see Part A) and $H^1(\bar{H}) = \ker \rho$ we find $H^1(\bar{H}) = \dot{K} \cap T_2(F)/T(K)$.

The other terms in the exact sequences (1), (2) are more difficult. We distinguish two cases:

a) K is nonreal.

Here $T(K) = \dot{K}$, hence $B(K) = H^1(\bar{H}) = 0$.

(1) takes the form $B(F) \xrightarrow[\cong]{\alpha} H^2(\bar{H})$;

(2) takes the form

$$0 \longrightarrow H^1(\bar{C}) \longrightarrow H^2(\bar{H}) \longrightarrow \coprod_\zeta \chi(G_\zeta) \longrightarrow H^2(\bar{C}) \longrightarrow 0$$

exact.

$H^2(\bar{C})$ can be determined from the exact sequence $0 \longrightarrow \bar{C}_0 \longrightarrow \bar{C} \longrightarrow \mathbb{Z} \longrightarrow 0$ which implies:

$$\bar{C}^G \longrightarrow Z \longrightarrow H^1(\bar{C}_0) \longrightarrow H^1(\bar{C}) \longrightarrow 0 \longrightarrow H^2(\bar{C}_0) \longrightarrow H^2(\bar{C}) \longrightarrow H^2(\mathbb{Z})$$

is exact. Here $H^2(\bar{C}_0) = 0$ by a theorem of Tate, $H^2(Z) = \chi(G)$.

We arrive at the exact sequence

$$(4) \quad 0 \longrightarrow H^1(\bar{C}) \longrightarrow B(F) \longrightarrow \coprod_{\zeta} (G_\zeta) \xrightarrow{\sigma} \chi(G) \longrightarrow 0$$

which also appears in [Sch, Satz 3.1]. Note that $\chi(G_\zeta) \cong B(F_\zeta)$ since $B(K_\zeta) = 0$ and that σ is the sum of the natural maps $\chi(G_\zeta) \longrightarrow \chi(G)$ induced from the transfer maps $G^{ab} \longrightarrow G_\zeta^{ab}$.

We are mainly interested in $B_2(F) = B_{2t}(F)$ (since F is nonreal). Let us introduce the group

$$N = \ker\, (B_2(F) \longrightarrow \coprod_{\zeta} B_2(F_\zeta)).$$

From (4) we deduce: $N = H^1(\bar{C}) \cap B_2(F) = H^1(\bar{C})_2$.

Here we view $H^1(\bar{C})$ as a subgroup of $B(F)$ and denote by $H^1(\bar{C})_2$ the elements of order ≤ 2 in $H^1(\bar{C})$.

The exact sequence $\bar{C}^G \longrightarrow \mathbb{Z} \longrightarrow H^1(\bar{C}_0) \longrightarrow H^1(\bar{C}) \longrightarrow 0$ together with $C \subset \bar{C}^G$ and $\mathrm{im}\,(C) = d\mathbb{Z} \subset \mathbb{Z}$ implies that $H^1(\bar{C})$ is a factor group of $H^1(\bar{C}_0)$, of finite index dividing d. C_0 is the Jacobian variety of the function field F/K, hence an abelian variety of dimension g defined over K. $H^1(\bar{C}_0)$ is the Weil-Châtelet group of principal homogeneous spaces over C_0. If K contains $i = \sqrt{-1}$ then $H^1(\bar{C}_0)$ is known in principle by the work of Ogg [O] and Šafarevič [S]. Furthermore Proposition 9 implies that every element of $H^1(\bar{C})_2 \subset B_2(F)$ is split by a quadratic extension of F, i.e., order and index agree in $H^1(\bar{C})_2$. For $i \notin K$ which is the case of our main interest I do not know any details about $H^1(\bar{C}_0)$.

REMARK: If $s(K)$ denotes the level of K then $s(K) \leq 2$ for K nonreal and $s(F) = s(K)$ since K is algebraically closed in F. Conversely, if $s(F) = 2$ we have $-1 = a^2 + b^2$ for some $a,b \in \dot{F}$.

Put $\frac{b}{a} = x \in \dot{F}$. Then $-1 = a^2(1 + x^2)$ and F contains $\frac{1}{a} = \sqrt{-(1 + x^2)}$.

x is transcendental over R.

As K can be any subfield of F with $\mathrm{tr}(K/R) = 1$ we may assume that K is the algebraic closure of $R(x, \sqrt{-(1 + x^2)})$ in F.

b) K is real.

This case occurs necessarily if F is real or if F is nonreal with $s(F) = 4$. An example is given by $F = R(x, y, \sqrt{-f})$ where f is the Motzk in polynomial $1 - 3x^2y^2 + x^4y^2 + x^2y^4$. By [CEP] f is not a sum of 3 squares in $R(x,y)$. This immediately implies $s(F) = 4$.

Similarly to Case a) we introduce the group

$$N = \ker (B_{2t}(F) \longrightarrow \coprod_{\zeta} B_{2t}(F_{\zeta})).$$

By the corollary of Proposition 8 we have

$$N \cong \ker (I_t^2(F) \longrightarrow \coprod_{\zeta} I_t^2(F_{\zeta})),$$

by Proposition 10 we have

$$N = \ker (\beta \circ \alpha \mid B_{2t}(F)) \quad \text{since} \quad \beta \circ \alpha = \coprod_{\zeta} \nu_{\zeta}^{*} \circ i_{\zeta}.$$

If F is real with a fixed ordering α and $q \in W(F)$ is a quadratic form with $\operatorname{sign}_{\alpha}(q) = s_{\alpha} \in \mathbb{Z}$ then a result of Prestel [Pr, Cor.1] shows that F/K admits a real completion F_{ζ} with an ordering α' such that $\operatorname{sign}_{\alpha'}(q_{F_{\zeta}}) = s_{\alpha}$. Therefore an element of $B_2(F)$ which is mapped into $\coprod_{\zeta} B_{2t}(F_{\zeta})$ under $\coprod_{\zeta} i_{\zeta}$ automatically lies in $B_{2t}(F)$. The exact sequences (1), (2) also show that

$$\alpha(N) = \alpha(B_2) \cap H^1(\overline{C}) = H^1(\overline{C})_2' \subset H^1(\overline{C})_2.$$

This gives an exact sequence

$$(5) \qquad 0 \longrightarrow B(K) \cap B_{2t}(F) \longrightarrow N \xrightarrow{\alpha} H^1(\overline{C})_2' \longrightarrow 0.$$

From $B(K) \cong \dot{K}/T(K)$ we deduce $B(K) \cap B_{2t}(F) \cong \dot{K} \cap T(F)/T(K)$.

In the case where K is real but F is nonreal this is just $B(K)$. N is slightly more complicated than in Case a) here but again the essential and presumably big "part" of N is given by the subgroup $H^1(\overline{C})_2'$ of $H^1(\overline{C})_2$. I do not know whether $H^1(\overline{C})_2'$ equals $H^1(\overline{C})_2$ in Case b) but there seems to be no particular reason that it does.

Needless to say that better information on the cohomology groups in sequence (2), especially on $H^1(\overline{C})$ would be very welcome. Possibly some other long exact sequences derived from the commutative diagram at the beginning come into play. Special interest may be in the cohomology groups of the idele class group $\overline{CJ}$ because they determine the "class field theory" of F/K.

We now come back to our main problem whether $u(F) \leq 4$ or not. By Proposition 9 one has to investigate whether an element $c \in B_{2t}(F)$ splits in a suitable quadratic extension of F. Locally in F_ζ we have $c = (a_\zeta, b_\zeta)$ with a unit a_ζ and a prime element b_ζ and $c = 1$ in F_ζ for almost all ζ, say for $\zeta \notin S$ where S is a finite set of prime spots on F. Let $b \in \dot{F}$ be an element which is prime at all $\zeta \in S$. Then $c = (a_\zeta, b)$ for $\zeta \in S$, $c = 1$ for $\zeta \notin S$.

By the weak approximation theorem there exists $a \in \dot{F}$ such that $c = (a,b)$ for $\zeta \in S$, $c = 1$ for $\zeta \notin S$. The trouble is that in general $c \neq (a,b)$ for finitely many $\zeta \notin S$. Put $E = F(\sqrt{a})$ or $E = F(\sqrt{b})$. We have $c = 1$ in all completions of E which gives $c_E \in \ker (B_{2t}(E) \longrightarrow \coprod_R B_{2t}(E_R))$.

Hence we can always find a lot of quadratic extensions E of F such that the given element $c \in B_{2t}(F)$ belongs to N_E over E but we need an extension E with $c = 1$ over E. The groups N_E where E runs through all quadratic extensions of F are therefore a kind of obstruction to our problem $u(F) \leq 4$. It seems that N_E is usually far from being zero. Another problem which turns up together with the consideration of $c \in B_{2t}(F)$ over the completions F_ζ is the following:

Is it possible to find $a, b \in \dot{F}$ such that $c = (a,b)$ over F_ζ for all ζ with at most one exception? Of a similar nature is the following.

OPEN PROBLEM: Let K be any field, let $a, b \in K[x]$ be relatively prime. Is there a prime polynomial $p \in [x]$ with $p \equiv a \bmod b$? A necessary condition on K for this to hold is that K be neither real nor algebraically closed. Is this condition sufficient?

6. TWO EXAMPLES.

In this last part I want to illustrate the theory by two examples.

1) The example of Brumer [B].

Here K is nonreal, $\mathrm{tr}(K/R) = 1$, which implies that K is of cohomological dimension 1. q_1, q_2 are quadratic forms over K in 5 common variables. One wants to show that the equation $q_1 = q_2 = 0$ has a nontrivial solution in K. This is equivalent to showing that the single quadratic form $q = q_1 + xq_2$ over the rational function field $F = K(x)$ is isotropic. The pencil determined by q_1, q_2 contains degenerate forms. W.l.o.g. one can suppose that q_2 is degenerate. Then the discriminant of q is a square free polynomial $b = b(x) \in K[x]$ of degree ≤ 4. It is easily seen that the Clifford invariant $c(q)$ of q satisfies the following local conditions over the completions F_ζ of F/K:

$$c = 1 \text{ over } F_\zeta \text{ if } \zeta \nmid b(x),\ \zeta \neq \tfrac{1}{x},\quad c = (a,b) \text{ over } F_\zeta \text{ if } \zeta | b(x)$$

for suitable $a \in \dot{F}$. Since a matters only mod b, we can take a polynomial in x of degree ≤ 3 for a. Since a can be multiplied by a-square and $u(K) \leq 2$, a can be reduced to a polynomial of degree ≤ 2. Then $c(q)$ vanishes over all completions of the field $E = F(\sqrt{a})$. But E is of genus 0 and together with $u(K) \leq 2$ it follows that E is a rational function field over K. Hence $\overline{C(E)} = \mathbb{Z}$, $H^1(\overline{C(E)}) = 0$ and $c(q)$ vanishes in $B(E)$. This implies that $c(q)$ is a quaternion algebra over F and that q is isotropic over F.

2) The example of Cassels-Ellison-Pfister [CEP].

Here the original question is whether the Motzkin polynomial

$$f(x,y) = 1 - 3x^2y^2 + x^2y^4 + y^2x^4 \in R[x,y]$$

is a sum of 3 squares in $R(x,y)$. Put $K = R(x)$, $K' = C(x) = K(i)$. Let $\mathcal{D}$ be the abelian curve defined over K by

$$\eta_1^2 = \xi_1(\xi_1^2 + 4x^2(x^2 - 3)\xi_1 + 16x^2)$$

and let $F = K(\xi_1, \eta_1)$ be the function field of $\mathcal{D}$, $F' = F(i)$.

The following results have been proved in [CEP]:

a) The group $\mathcal{D}_{K'}$ of K'-rational points on $\mathcal{D}$ is finitely generated (by the Mordell-Weil Theorem) of rank 1 and $\mathcal{D}_K = \mathcal{D}_{K'} \cong \mathbb{Z}/2\mathbb{Z} \times \mathbb{Z}$ with generators $(\xi_1, \eta_1) = (0,0)$ of order 2, $(\xi_1, \eta_1) = (4, 8(x^2 - 1))$ of infinite order.

b) The elliptic curve E defined by

$$\eta^2 = x(x - 1)\xi^4 - 2x^2(x^2 - 3)\xi^2 + x(x - 1)(x + 1)^2(x^2 - 4)$$

is a homogeneous space of $\mathcal{D}$ [not of the curve C occurring in [CEP], this is an error!]. It has no K'-rational points and belongs to an element of order 2 in the Tate-Šafarevič-group $\text{Ш}(K', \mathcal{D})$.

c) If f were a sum of 3 squares in F then E would have a K'-rational point.

The following further results can be proved:

d) F is real since it is a cubic extension of the real field $K(\eta_1)$ which is purely transcendental over K. F/K has prime divisors of degree 1. They correspond bijectively to the K-rational points on $\mathcal{D}$.

e) As before let $\overline{K}$ be the algebraic closure of K, let G be the Galois group of $\overline{K}/K$, G' the Galois group of $\overline{K}/K'$.

Let C, C', $\overline{C}$ be the group of divisor classes of F/K, F'/K', $\overline{F}/\overline{K}$; $d = 1$ and the exact sequence $0 \longrightarrow \overline{C}_0^G \longrightarrow \overline{C}^G \longrightarrow \mathbb{Z} \longrightarrow H^1(\overline{C}_0) \longrightarrow H^1(\overline{C}) \longrightarrow 0$ imply $\overline{C}^G = C$, $H^1(G,\overline{C}) = H^1(G,\overline{C}_0) = H^1(K,\mathcal{D})$ where the latter is the Weil-Châtelet group of $\mathcal{D}$ over K. Similarly $H^1(G',\overline{C}) = H^1(K',\mathcal{D})$.

f) The theory of Ogg-Šafarevič gives an exact sequence

$$0 \longrightarrow ш(K',\mathcal{D}) \longrightarrow H^1(K',\mathcal{D}) \longrightarrow \coprod_p H^1(K'_p,\mathcal{D}) \longrightarrow 0.$$

Here we denote by K'_p the completions of the function field K'/C. They are given by the normed prime polynomials $p = p(x) \in C[x]$ together with $p = \frac{1}{x}$ for the place at infinity. (These completions are not to be confused with the completions F'_ζ of F'/K' where the valuation ν_ζ is trivial on K'!) For our curve $\mathcal{D}$ from (*) we get explicitly:

$H^1(K'_p,\mathcal{D}) \cong (\mathbb{Q}/\mathbb{Z})^2$ if $\mathcal{D}$ has good reduction mod p which is the case for $p \neq x - 0$, $x \pm 1$, $x \pm 2$, $\frac{1}{x}$,

$H^1(K'_p,\mathcal{D}) \cong (\mathbb{Q}/\mathbb{Z})^1$ for $p = x \pm 1$, $x \pm 2$, $\frac{1}{x}$,

$H^1(K'_p,\mathcal{D}) = 0$ $p = x$.

$ш(K',\mathcal{D}) = (\mathbb{Q}/\mathbb{Z})^2$ from the main formula of [O] or [S].

g) $H^1(K,\mathcal{D})$ and $H^1(K',\mathcal{D})$ are related by the restriction-inflation sequence

$$0 \longrightarrow H^1(G/G',\mathcal{D}_{K'}) \longrightarrow H^1(K,\mathcal{D}) \longrightarrow H^1(K',\mathcal{D}).$$

From $\mathcal{D}_{K'} = \mathcal{D}_K$ it follows that G/G' operates trivially on $\mathcal{D}_{K'}$ and that $H^1(G/G',\mathcal{D}_{K'}) = H^1(G/G', \mathbb{Z}/2) \times H^1(G/G',\mathbb{Z}) \cong \mathbb{Z}/2$.

Here the non-trivial element of G/G' is just complex conjugation, the non-trivial element of $H^1(G/G',\mathcal{D}_{K'})$ is given by the point $(0,0)$ of order 2 on $\mathcal{D}$. Its image in $H^2(G,\overline{H})$ is determined by the principal divisor (ξ_1) and the pre-images of this element under α are the quaternion algebras $(-1, a\cdot\xi_1)$ in $B(F)$ with $a \in \dot{K}$. They are not in $B_{2t}(F)$ however. Also one sees that $B(K) \cap B_t(F) = 0$; therefore,

$$N = \ker (B_{2t}(F) \longrightarrow \coprod_\zeta B_{2t}(F_\zeta)) \cong H^1(\overline{C})'_2.$$

Under the map $H^1(G,\overline{C}) \longrightarrow H^1(G',\overline{C})$, $H^1(\overline{C})_2'$ is injected into $H^1(G',\overline{C})$.

h) The homogeneous space $E \in H^1(G',\overline{C})$ goes into $(x(x-1), \xi_1)$ under the imbedding $H^1(G',\overline{C}) \longrightarrow B(F')$. All elements of $B_2(F')$ are of course quaternion algebras over F'. It seems likely that all elements of $B_{2t}(F)$ are again quaternion algebras over F but I have not been able to decide this.

In conclusion let me remark that many of the questions treated here essentially belong to the diophantine theory of surfaces over $R(x)$. Over number fields instead of $R(x)$ attempts in a similar direction have been taken up in the work of Colliot-Thélène, Coray and Sansuc (see [CCS]). In particular they define a so-called Manin obstruction to the validity of a local-global principle.

BIBLIOGRAPHY

[A] J. K. Arason: Cohomologische Invarianten quadratischer Formen. J. of Alg. 36, 448-491 (1975).

[B] A. Brumer: Remarques sur les couples de formes quadratiques. C. R. Acad. Sci., Paris 286, sér A, 679-681 (1978).

[CEP] J. W. S. Cassels, W. J. Ellison, and A. Pfister: On sums of squares and on elliptic curves over function fields. J. Number Th. 3, 125-149 (1971).

[CCS] J. L. Colliot-Thélène, D. Coray, J. J. Sansuc: Descente et principe de Hasse pour certaines variétés rationnelles. To appear.

[EL 1] R. Elman and T. Y. Lam: Quadratic forms and the u-invariant I. Math. Zeitschr. 131, 283-304 (1973).

[EL 2] ____________________: Pfister forms and their applications, J. Number Th. 5, 367-378 (1973).

[EL 3] ____________________: Classification theorems for quadratic forms over fields. Comm. Math. Helv. 49, 373-381 (1974).

[EL 4] ____________________: Quadratic forms under algebraic extensions. Math. Ann. 219, 21-42 (1976).

[ELTW] R. Elman, T. Y. Lam, J. P. Tignol, and A. Wadsworth: Witt rings and Brauer groups under multiquadratic extensions I. To appear.

[G] W. D. Geyer: Ein algebraischer Beweis des Satzes von Weichold über reelle algebraische Funktionenkörper. In "Algebraische Zahlentheorie" (Ed. H. Hasse and P. Roquette), Bibl. Institut Mannheim 1966.

[L 1] S. Lang: On quasi-algebraic closure. Ann. of Math. 55, 373-390 (1952).

[L 2] ______: The theory of real places. Ann. of Math. 57, 378-391 (1953).

[O] A. P. Ogg: Cohomology of abelian varieties over function fields. Ann. of Math. 76, 185-212 (1962).

[P] A. Pfister: Quadratische Formen in beliebigen Körpern. Invent. Math. 1, 116-132 (1966).

[Pr] A. Prestel: Local-global principles for quadratic forms over function fields. Proc. Conf. Quadr. Forms, Kingston 1976, Queen's Papers 46, 595-612 (1977).

[R] P. Roquette: Splitting of algebras by function fields of one variable. Nagoya Math. J. 27, 625-642 (1966).

[S] I. R. Šafarevič: Principal homogeneous spaces defined over a function field. Amer. Math. Soc. Transl. 37, 85-114 (1964).

[Sch] W. Scharlau: Über die Brauer-Gruppe eines algebraischen Funktionenkörpers in einer Variablen, J. r. a. Math. 239/240, 1-6 (1969).

[Sp] T. A. Springer: Quadratic forms over fields of discrete valuation. Indag. Math. 17, 352-362 (1955).

[T] C. Tsen: Quasi-algebraisch-abgeschlossene Funktionenkörper. J. Chinese Math. Soc. 1, 81-92 (1936).

FACHBEREICH MATHEMATIK DER UNIVERSITAT MAINZ
6500 MAINZ
WEST GERMANY

Contemporary Mathematics
Volume 8, 1982

ORDER EXTENSIONS AND REAL ALGEBRAIC GEOMETRY

D. W. Dubois and T. Récio

1. A Problem On Orderings Under Field Extensions.

Since the papers on Artin and Artin-Schreier first appeared, a central theme in the theory of ordered fields has been the problem of extension of an order from a field to an extension field; or, slightly more generally, given a field embedding $\rho : T \to F$, one studies the map ρ^* from the order space $\Omega(F)$ to the order space $\Omega(T)$, defined by the correspondence $\alpha \to \rho^{-1}(\alpha \cap \rho(T))$, where α is any order of F, i.e., $\alpha \varepsilon \Omega(F)$. In the usual topology of Harrison ρ^* is continuous and, as recently shown by Elman, Lam and Wadsworth (E-L-W), in case $F/\rho(T)$ is finitely generated, ρ^* is an open mapping. This open mapping theorem is a beautiful generalization of the classical theorems of Sturm and Sylvester concerning real roots of a real polynomial, since their alogrithms, in their essential finiteness, yield "open" conditions. In fact, this is implicitly used in the [E-L-W] proof, once in the formulation of Olga Taussky [T] and again the the form of Tarski's principle (cf. Prestel's contribution in Section 4 bis of [E-L-W]). Since $\text{Im}\,\rho^*$, being the image of the compact space $\Omega(T)$, is itself compact, hence closed, it is in fact a finite union of basic open sets $\Omega_{E_i} = \Omega_{E_i}(T) = \{\alpha \varepsilon \Omega(T) ; \alpha \supset E_i\}$, where each E_i is a finite subset of $T \setminus \{0\}$; or, what amounts to the same thing, $\text{Im}\,\rho^*$ is a clopen. Conversely in [E-L-W] it is shown that given any clopen H in $\Omega(T)$ there exists an F, and an embedding $\rho : T \to F$, with $F/\rho T$ finitely generated and $H = \text{Im}\,\rho^*$.

1980 Mathematics Subject Classification 12J15, 14G30

0271-4132/81/0000-0686/$07.00

Thus we may take as the scenario for the study of finitely generated order-extensions, a pair (T,H), where T is a formally real field and H is a nonempty clopen in $\Omega(T)$. A fundamental question, to which we devote space below, is

$Q_1 = Q_1(\alpha,T,H)$: For α in $\Omega(T)$, does there exist an automorphism σ of T such that $\sigma(\alpha)$ belongs to H ?

1.1. Definition.

The field T is a Q_1-field provided for every non-empty clopen H in $\Omega(T)$ and every order α of T, the answer to Q_1 is "yes"; i.e., for every order α of T there exists an automorphism σ of T such that $\sigma(\alpha) \in H$.

1.2. Examples.

a. Every real closed field, every euclidean field, $\mathbb{Q}(\sqrt{2})$, is a Q_1-field. More generally if for T there exists only one isomorphism class of orders, then T is a Q_1-field.

b. If $T/\mathbb{Q}$ is a simple, real-normal extension - i.e., for every $t \in T$, all the real conjugates of t belong to T (it is assumed that $T \subset \mathbb{R}$), then T is a Q_1-field. The field $\mathbb{Q}[X]/(X^3-3X+1)$ is such a field.

c. If $T/\mathbb{Q}$, for $T \subset \mathbb{R}$, is simple but not real-normal, then T is not a Q_1-field. An example of this type is given by $\mathbb{Q}[X]/(X^3-4X+1)$, which has three isomorphism classes.

d. A field T, whose order space consists of precisely two non-isomorphic orders is surely not a Q_1-field. Such a field can be constructed over the pure transcendental extension $\mathbb{Q}(X)$, in such a way that one of its orders is Archimedian and the other is not - see $[D_2]$.

1.3. Theorem.

Let T be a field such that for every order α of T, (T,α) is dense in its real closure, and assume further that T is a Q_1-field. Then every finitely genrated pure transcendental extension $T(X_1,\dots,X_n) \equiv F$ is also a Q_1-field.

Proof.

We are going to reduce the theorem to the case of a real-closed T. Let Γ be a basic open set in $\Omega(F)$, and let α be an order of F. Take ρ to be the inclusion map $T \to F$. Then $\rho^*(\Gamma)$ is a clopen in $\Omega(T)$. Since T is, by hypothesis, a Q_1-field, there exists an automorphism σ of T such that $\sigma(\rho^*(\alpha)) \in \rho^*(\Gamma)$. Let T_1 and T_2 be the real closures of $(T,\rho^*(\alpha))$ and $(T,\sigma(\rho^*(\alpha)))$, and let σ_{12} be the (unique) order-isomorphism of T_1 onto T_2 which extends σ. Now extend σ_{12} to $\bar{\sigma}_{12}$, an isomorphism from $T_1(X_1,\dots,X_n)$ onto $T_2(X_1,\dots,X_n)$, where the X_i are assumed to be independent variables over T_1 and T_2 as well as over T. By an induction from the case of a single variable, which was proved by Massaza [M], we extend α to an order $\bar{\alpha}$ in $T_1(X_1,\dots,X_n)$, i.e., $\alpha = \zeta_1^*(\bar{\alpha})$, where ζ_1 is the inclusion map of $T(X)$ in $T_1(X)$. Similarly $\Gamma \subset \operatorname{Im} \zeta_2^*$, where ζ_2 is the inclusion map of $T(X)$ in $T_2(X)$. Set $\bar{\Gamma} = \zeta_2^{*-1}(\Gamma)$. Now we have $\sigma\rho^*\alpha \in \rho^*\Gamma$, while $\bar{\sigma}_{12}\bar{\alpha}$ and $\bar{\Gamma}$ are in $\Omega(T_2(X))$.

By the Lemma 1.4 (below), we choose an automorphism $\bar{\tau}$ of $T_2(X)$, whose restriction τ to $T(X)$ is an automorphism of $T(X)$, and which maps $\bar{\sigma}_{12}(\bar{\alpha})$ into $\bar{\Gamma}$. We claim then that $\tau(\sigma(\alpha))$ belongs to Γ. See Figure 1 below.

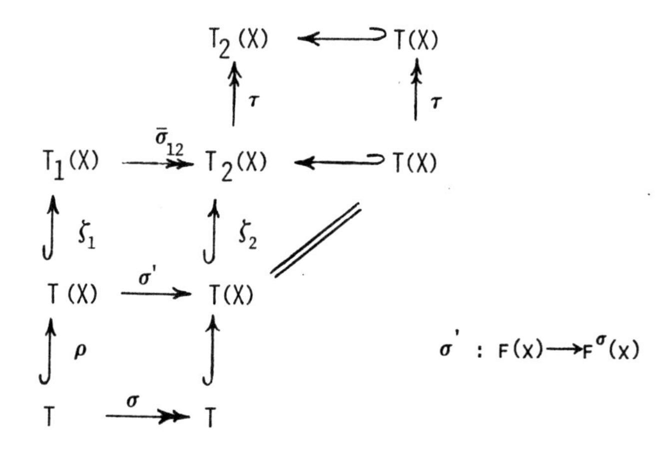

FIGURE 1

The only thing we have left to show is that $\sigma'(\alpha)$ is the image of $\bar{\sigma}_{12}(\bar{\alpha})$ - i.e., $\sigma'(\alpha) = \zeta_2^* \bar{\sigma}_{12}\bar{\alpha}$ - since

$$\Gamma = \bigcup_{i=1}^{m} \Omega_{E_i} \; (T(X)) \implies \bar{\Gamma} = \bigcup_{i=1}^{m} \Omega_{E_i} (T_2(X)) .$$

But the middle square in the diagram is commutative and $\bar{\alpha} = \zeta_1^* \alpha$, so the required equality is valid.

Thus the Theorem will be proved as soon as we prove the following Lemma:

1.4. Lemma.

Let (k,k^+) be an ordered field, let $\bar{k}$ be its real closure and suppose k is dense in $\bar{k}$. Let H be a non-empty clopen in $\Omega(F/k^+)$, where F is the pure transcendental extension $\bar{k}(X_1,\ldots,X_n) = \bar{k}(X)$; and assume that β is any order of F/k . Then there exists an automorphism $\sigma : F \to F$ such that $\sigma|_{k(X)}$ is an automorphism of $k(X)$ and $\sigma(\beta) \in H$.

Before starting the proof of the Lemma we need some geometric constructions and preliminary results.

1.5. More Notations.

Given a clopen H in $\Omega(k(V))$, for a real variety V , we need to distinguish between various strongly open semi-algebraic sets associated with H . Let $S = \{S_1,\ldots,S_m\}$ be a finite set of finite subsets of $k[V]\setminus\{0\}$. Then we define:

$$\Omega_S = \bigcup_{i=1}^{m} \{\alpha ; \alpha \supset S_i\} .$$

$$\hat{\Omega}_S = \bigcup_{i=1}^{m} \{z \in V;\; s_i(z) > 0 \quad \text{for all} \quad s_i \in S_i\} .$$

$$\hat{\Omega}_S^u = \bigcup_{i=1}^{m} \{z \in V;\; s_i(z) \geq 0 \quad \text{for all} \quad s_i \in S_i\} ;$$

the last set is McEnerney's untrimmed closure of $\hat{\Omega}_S$ [McE], which contains the strong closure of $\hat{\Omega}_S$.

1.6. Proposition.

Let $H = \Omega_S$, as in the previous section, let V be non-singular. Then $\hat{\Omega}_S$ is non-empty if and only if Ω_S is non-empty.

Proof.

Assume $H \neq \emptyset$, say α contains S_i. By the rational order theorem [D_4] there exists a rational order β centered in Ω_{S_i}. But then Ω_{S_i}, and hence Ω_S, is not empty. Conversely assume $\hat{\Omega}_S \neq 0$, say $\hat{\Omega}_{S_i}$ is not empty. Since the latter is then a non-empty open subset of V it contains a central point z; we center an order β at z which thus contains S_i, so β belongs to Ω_S.

1.7. Proposition.

Keep all hypotheses except that of non-singularity. Then H is non-empty if and only if $\hat{H} \cap V_c$ is non-empty; and even if k is not dense in $\bar{k}$ then it is still true that if $\hat{H} \neq \emptyset$ then $H \neq \emptyset$. But the converse fails as shown in examples found in, e.g., Dubois [D_1], Schwartz [Sch] and McKenna [McK].

1.8. Proposition.

Assume the hypotheses of 1.6. Then $H = \Omega(F/k^+)$ if and only if $\hat{H}$ is (strongly) dense in $\bar{k}^n$.

Proof.

For each sequence $(e) = (e_1, \ldots, e_m)$ with e_i in S_i, let

$$L(e) = \{z \in V;\ e_i(z) \leq 0,\ 1 \leq i \leq m\},$$

$$L'(e) = \{z \in V;\ e_i(z) < 0,\ 1 \leq i \leq m\}.$$

The union of the (finite) collection of all $L(e)$ is precisely the complement of $\hat{H}$. Suppose $L'(e)$ is not empty. Take an order α centered at $z \in L'(e)$. For all i, then, $-e_i(x) \in \alpha$, so $S_i \not\subset \alpha$, and hence $\alpha \notin H$.

On the other hand, suppose every $L'(e)$ is empty. Then the complement of $\hat{H}$ is contained in the proper subvariety $W = \{z \in V;\ e_i(z) = 0\ ,\ 1 \le i \le m\}$ (recall that the sets S_i always exclude the zero element of $k[x]$). Thus the complement of $\hat{H}$, along with W, is nowhere dense in V, which implies that $\hat{H}$ is dense in V. This argument shows that if $H = \Omega(F)$ then $\hat{H}$ is dense in V.

Conversely, assume H is not all of $\Omega(F)$, say α is an order not in H. It follows that there exists a sequence (e) with $-e_i \in \alpha$ for all i, so that α belongs to $\Delta \equiv \Omega_{\{-e_i\}}$. Thus $\hat{\Delta} = L(e)$, which is plainly open and (by proposition 1.6) also non-empty, is contained in the complement of $\hat{H}$. Hence $\hat{H}$ is not dense, and the proof is complete.

1.9. Proof of Lemma 1.4.

Assume H is a basic non-empty clopen defined by

$$H = \Omega_E(F)\ ,\ E = \{e_1,\dots,e_m\} \subset \dot{F}$$

For any automorphism σ of F, note that

$$\sigma(H) = \{\alpha \in \Omega(F/k\ );\ \sigma E \subset \dot{\alpha}\}$$

Now we are going to find a convenient σ such that, putting $\Delta = \sigma H$, $\hat{H} \cup \hat{\Delta}$ is dense in $\overline{k}^n$. By virtue of Proposition 1.8, this will suffice to prove our Lemma. Applying non-emptiness of H and Proposition 1.6, we select a point t in $\hat{H} \cap k^n$ (using the density of k in $\overline{k}$) and a small ball $\hat{B}$ of radius $r \in \dot{k}^+$, centered at t, such that $\hat{B} \subset \hat{H}$. There is no loss of generality assuming, as we do, that t is the origin:

$$\hat{B} = \{z \in \overline{k}^n\ ;\ r^2 - \sum z_1^2 > 0\} \subset \hat{H}\ .$$

Define the automorphism σ by mapping X_i to $r^2 X_i/\Sigma X_j^2$. Then σ defines an inversion in the spherical boundary of $\hat{B}$. Let $\hat{M}$ be the exterior of the closed sphere. Then $\hat{M}$ is contained in $\hat{\Delta}$, and $\hat{B} \cup \hat{M}$ is dense in $\overline{k}^n$ (only the boundary of $\hat{B}$ is missing). Since $\hat{B}$ is contained in $\hat{H}$ the density of $\hat{B} \cup \hat{\Delta}$ is established, and the proof is complete.

1.10. Remark.

After the above results it is clear that $\mathbb{R}(x)$, $\mathbb{Q}(x)$, $\overline{\mathbb{Q}}(x)$ (bar denoting real closure) are all Q_1-fields. This fact does not add much to our knowledge of $\mathbb{R}(x)$; in fact we know a much stronger result, namely, that all its orders are isomorphic. As to $\mathbb{Q}(x)$ or $\overline{\mathbb{Q}}(x)$ we know that there exists both archimedean and non-archimedean orders; property Q_1 implies then that each of these classes of orders is dense in the Harrison space (see Lam [L] 6.7).

Different types of non-isomorphic orders in $\mathbb{R}(x,y)$ are studied in (Brumfiel [B] 8.12). The Q_1-property for this field implies then the density of each type of order.

2. Real Varieties.

Assume T is a Q_1-field. Let Γ be a non-empty clopen in $\Omega(T)$, let G be a subgroup of Aut T. We say T is a $(\Gamma\text{-}G)$ field provided the family $G^{-1}\Gamma$ of all sets $g^{-1}\Gamma$, as g ranges over G, is a cover of $\Omega(T)$. Thus, a Q_1-field is, for every non-empty clopen Γ, a $(\Gamma\text{-AutT})$ field. If T is a $(\Gamma\text{-}G)$ field for every non-empty clopen Γ, we say it is a Q_1-field. For a (Γ,G) field T we let $D(T,\Gamma,G)$ stand for the minimum cardinality of all subcovers of $G^{-1}\Gamma$. This is finite, since $\Omega(T)$ is compact. For real closed k and a pure transcendental extension $T(X)$ in n variables, the arguments in Section 1 show that for $G = \text{Aut } T$, $D(T,\Gamma,G) \leq 2$, while $D(T,\Gamma,G) = 1$ if and only if $\Gamma = \Omega(T)$, therefore, if and only if $\hat{\Gamma}$ is dense in k^n. In case T is not a (Γ,G) field we define $D(T,\Gamma,G)$ to be infinity.

2.1. Assume $T = k(x_1,\ldots,x_n)$, for algebraically independent x_i. An affine automorphism of T/k is one which induces an automorphism of $k[x]$ (characterization of these is difficult even for $n = 2$; see, for example Nagata [N_1], [N_2] and Hartshorne [H], problem 3.19). A projective automorphism of T/k is one which is induced by an isomorphism of the projective space $\mathbb{P}^n(k)$ of which T is the rational function field; it is therefore an element of PGL(n,k) (see Hartshorne [H]). In the one variable case PGL(1,k) = Aut(T/k).

The notations $G(A)$ and $G(P)$ are used to denote the groups of all affine and all projective automorphisms.

Example.

A Q_1-field need not be a $Q_1^{G(A)}$-field. For example, let $\Gamma = \Omega_{\{1-x^2\}}$ in $\Omega(T)$, $T = \mathbb{R}(X)$, one variable case. Now every order in Γ has a finite center, and affine automorphisms preserve finiteness of center. The group $G(P)$ is better behaved.

2.2. Proposition.

The pure transcendental extension $T = k(X) = k(X_1,\ldots,X_n)$ over a real closed k is a $Q_1^{G(P)}$ field: for every non-empty clopen $\Gamma \subset \Omega(T)$ and every order $\alpha \in \Omega$, there exists $\sigma \in G(P)$ such that $\sigma(\alpha) \in \Gamma$.

Proof.

Let Γ be a non-empty clopen. There exists a positive $r \in k$ and a point t in k^n such that for $f(x) = r^2 - \Sigma(X_i-t_i)^2$, the sphere of radius r and center t is contained in $\hat{\Gamma}$. As in Section 1 we conclude that $\Omega_{\{f\}} \subset \Gamma$. This reduces the proposition to the case of a distinguished clopen $\Gamma = \Omega_{\{f\}}$. By means of translations and dilatations, we may assume $f(X) = 1 - \Sigma X_i^2$. For a member $c \in k$, to be determined later, we define projective automorphisms σ_j , for $j = 1,\ldots,n$, as follows:

$$\sigma_j : \lambda_i \to \frac{\lambda_i}{c\lambda_j} \quad \text{if} \quad i \neq j \ , \ \lambda_j \to \frac{1}{c\lambda_j} \ .$$

Let $\hat{S}_0 = \hat{\Omega}_{\{f\}}$ be the set of all z in k^n such that $f(z) > 0$, let $\hat{S}_i = \hat{\Omega}_{\{\sigma_i f\}}$, $i = 1,\ldots,n$. Then the union of the sets $\hat{S}_i$ is dense in k^n , as we now show. Assume z is outside the union. Then

$$1 - \sum z_i^2 \leq 0$$

$$c^2z_1^2 - z_2^2 - \ldots - z_n^2 - 1 \leq 0$$

$$c^2z_2^2 - z_1^2 - \ldots - z_n^2 - 1 \leq 0$$

$$\ldots$$

$$c^2z_n^2 - z_1^2 - \ldots - z_n^2 - 1 \leq 0 \ .$$

Adding yields

$$1 - \sum z_i^2 + c^2 \sum z_i^2 - (n-1)\sum z_i^2 - n + 1 \le 0 .$$

Now let $c^2 > n$, $c^2 - n > n - 1$ hold (for large c); thus we get:

$$\sum z_i^2 \le \frac{n-1}{c^2 - n} \le 1$$

Thus $1 - \sum z_i^2 > 0$, contrary to choice of z . We conclude that $\hat{S} \cup \sigma_1 \hat{S} \cup \ldots \cup \sigma_n \hat{S} = k^n$. By 1.8 of Section 1, the open set $\bigcup \Omega_{\{\sigma_i f\}}$, where σ_0 = identity, covers $\Omega(T)$.

Corollary.

For the T of the propositon, and for all Γ ,

$$D(T,\Gamma,G(P)) \le n + 1 .$$

2.3. Examples.

a. Case $n = 2$, $k = \mathbb{R}$, $\Gamma = \{\alpha; 1-x_1^2-x_2^2 \in \alpha\}$.

Then for any projective automorphism σ , there exist first degree polynomials L , M , N such that the matrix of coefficients is invertible and such that

$$\sigma(\Gamma) = \{\alpha; L^2 - M^2 - N^2 \in \alpha\}$$

It follows that $L^2 - M^2 - N^2$ is not positive semi-definite on $\mathbb{R}^2$, which implies that $L^2 - M^2 - N^2$ is the ideal of a real irreducible conic in $\mathbb{R}^2$. If in the definition of Γ the circle is replaced by an arbitrary irreducible real conic the argument still applies, which shows that we need more than one automorphism, and hence more than two. Thus $D(T,\Gamma,G(P)) \ge 3$, and, in view of the corollary

$$D(T,\Gamma,G(P)) = 3 .$$

b. For arbitrary n now, with $T = \mathbb{R}(X_1,\ldots,X_n)$, let G be any subgroup of $\mathrm{Aut}\, T$.

Then for all Δ , $D(T,\Delta,G) = 1$ if and only if $\hat{\Delta}$ is dense in $\mathbb{R}^n$, and $D(T,\Gamma,G(A))$ is finite if and only if $\hat{\Gamma}$ is unbounded.

c. Let F be the function field of a real variety V of genus g and dimension d over $\mathbb{R}$, let $T = \mathbb{R}(X_1,\ldots,X_d)$ be a pure transcendental extension of $\mathbb{R}$ of degree d .

For each of the many embeddings $\rho : T \to F$ the set $\mathrm{Im}\rho^*$ is a clopen in $\Omega(T)$, and so $D(T,\mathrm{Im}\rho^*,G)$ may be defined. Now we define m_F^G and M_F^G as follows:

$$m_F^G = \min_\rho D(T,\mathrm{Im}\rho^*,G)$$

$$M_F^G = \sup_\rho D(T,\mathrm{Im}\rho^*,G) \ .$$

Case $d = 1$, $g = 0$.

Then

$m_F^{G(P)} = 1$,	$M_F^{G(P)} = 2$
$m_F^{G(A)} = 1$	$M_F^{G(A)} = \infty$

Case $d = 1$, $g = 1$.

Consider the real curve $y^2 - x^3 + x = 0$, over $\mathbb{R}$, and let F be its function field. The genus is one. By considering the projection into the y-axis we see for all G that $m_F^G = 1$. The projection into the x-axis, which corresponds to the embedding $\mathbb{R}(X) \to \mathbb{R}(X)(\bar{y})$, with $\bar{y}^2 = x^3 - x$, shows that $M_F^{G(P)} = 2$. Also, by the bounded model theorem [D_3], we see that $M_F^{G(A)} = \infty$.

Case arbitrary V , arbitrary genus.

By the bounded model theorem [D_3], we see that $M_F^{G(A)} = \infty$. This bounded model is real-complete in the sense of Coste-Coste-Roy [C-C].

Now consider a real d-dimensional affine variety V with coordinate ring $\mathbb{R}[V]$. We apply Noether's normalization Lemma to choose independent variables $x_1,\dots,x_d$ in $\mathbb{R}[V]$ so that $\rho : \mathbb{R}[x_1,\dots,x_d] \to \mathbb{R}[V]$ is an integral extension. As before we take $T = \mathbb{R}(x_1,\dots,x_d)$, $F = \mathbb{R}(V)$, and extend the embedding to $\rho : T \to F$. As before $\mathrm{Im}\rho^*$ is a clopen in $\Omega(T)$. Now we define $D^{\mathcal{G}}(V/\rho)$ to be $D^{\mathcal{G}}(T/\mathrm{Im}\rho^*)$, and take $m_V^{\mathcal{G}}$, $M_V^{\mathcal{G}}$ as the infimum (supremum) of all $D^{\mathcal{G}}(V/\rho)$ as ρ varies over all integral embeddings. Let V_c be the set of all central points on V. We wish to relate V_c with $H \equiv H_\rho = \mathrm{Im}\rho^*$.

2.4. Lemma.

Let V be a variety with $V = V_c$. Let $S = \{S_1,\dots,S_m\}$, $U = \{U_1,\dots,U_m\}$ be finite sets of finite subsets of $\mathbb{R}[V]\setminus\{0\}$. Then

$$\overline{\hat{\Omega}_S} = \overline{\hat{\Omega}_U} \iff \Omega_S = \Omega_U ,$$

where the bar denotes, as usual, strong closure.

Proof.

Assume $\Omega_S = \Omega_U$. We show that $\hat{\Omega}_S$ is dense in $\hat{\Omega}_U$ which, by symmetry, will prove half of the assertion. Let z be a point in $\hat{\Omega}_U\setminus\hat{\Omega}_S$. Then there exists a sequence $s_1,\dots,s_m$ with $s_i \in S_i$ such that $s_j(z) \le 0$ for all j. If every $s_i(z)$ were actually less than zero then we could center an order at z which would belong thus to $\Omega_U\setminus\Omega_S$ contrary to hypothesis. Hence we have $s_j(z) = 0$ for some j. This shows that $\hat{\Omega}_U\setminus\hat{\Omega}_S$ is contained in a Zariski closed set and is therefore nowhere dense in $\hat{\Omega}_U$.

Conversely, assume $\Omega_U \neq \Omega_S$, say $\alpha \in \Omega_U\setminus\Omega_S$. Then every S_j contains an element $s_j(x)$ such that $-s_j(x) \in \alpha$, which implies that $\hat{\Omega}_U \cap \{z; s_j(z) < 0 , \text{ for all } j\}$ is not empty. This last shows that $\hat{\Omega}_S$ is not dense in $\hat{\Omega}_U$, whence their closures are unequal. The lemma is all proved.

Note. The proof shows that in general, $\Omega_S = \Omega_U$ if and only if

$$\overline{\hat{\Omega}_S \cap V_c} = \overline{\hat{\Omega}_U \cap V_c} .$$

2.6. Remark.

Let $\rho : \mathbb{R}[x_1,\ldots,x\ ] \to \mathbb{R}[V]$ be an arbitrary extension and let $\bar{\rho} : V \to \mathbb{R}^d$ be the corresponding morphism. Now $\bar{\rho}(V_c)$ is semialgebraic in $\mathbb{R}^d$ (see Recio [R] and Brumfiel [B]). We notice that any polynomial which vanishes all over $\bar{\rho}(V_c)$ must be zero, since ρ is injective and V_c is surely Zariski dense. The same applies if we consider any small neighborhood of a point in $\bar{\rho}(V_c)$, i.e., $\bar{\rho}(V_c)$ is trim in the sense of McEnerney. A more precise description follows.

2.7. Theorem.

Let ρ be an integral extension $\mathbb{R}[x_1,\ldots,x_d] \to \mathbb{R}[V]$, for a d-dimensional real affine variety V, and let $\operatorname{Im}\rho^* = H_\rho = \Omega_S \subset \Omega(\mathbb{R}(x))$, where $S = \{S_1,\ldots,S_m\}$ is a finite set of finite subsets of $\mathbb{R}[X_1,\ldots,X_d]$ let $\bar{\rho}$ be the corresponding finite morphism. We have then

$$\bar{\rho}(V_c) = \bar{\hat{\Omega}}_S .$$

Proof.

From Lemma 2.4, we see that $\bar{\hat{\Omega}}_s$ dependends on $H = \Omega_s$. As in Brumfiel ([B], 8.9.4) (see also [R-D]) for this type of modified argument) we see that

$$\rho(V_c) = \{z \in \mathbb{R}^d ; s(z) \geq 0 \ \text{ for all } \ s \in \textstyle\sum\} ,$$

where

$$\textstyle\sum = \{s(x) \in \mathbb{R}[x_1,\ldots,x_d] ; \rho(s) \ \text{ is a sum of squares}\} .$$

Now the polynomials in Γ may be adjoined to S since every order which extends to $\mathbb{R}(V)$ must contain $\sum$. Therefore, the previous lemma shows that $\bar{\hat{\Omega}}_s \subseteq \bar{\rho}(V_c)$. Conversely let $z = \bar{\rho}(y)$, $y \in V_c$, be a point in $\bar{\rho}(V_c)$. There is an order α centered at z which extends to $\mathbb{R}(V)$. In fact consider the multiplicative set M of all polynomials which are strictly positive at z. We know $\rho(M)$ is contained in an order of $\mathbb{R}(V)$ provided

$\sum m_i t_i^2 = 0$, for $m_i \in M$, $t_i \in \mathbb{R}[V]$, implies all t_i are zero; but $\sum m_i t_i^2 = 0$ implies every t_i vanishes over some neighborhood of the central point y , which by Dubois-Efroymson [D-E], implies the t_i are zero. Thus there is an order α which contains $\rho(M)$. If we suppose z to be outside $\overline{\hat{\Omega}}_S$ then we will have, because of the existence of extensible orders centered in any point of $\overline{\rho}(V_c)$, as in Lemma 2.4, that z belongs to a proper Zariski-closed subset of $\mathbb{R}^d$, and therefore, (see 2.6) $z \in \overline{\hat{\Omega}}_S$.

<u>2.8. Corollary</u>. (Adkins [A], Recio-Dubois [R-D].

Assume further that the ρ of the theorem induces a birational morphism. Then $\overline{\rho}(V_c) = \mathbb{R}^d$.

<u>2.9. Remark</u>.

The proof of Theorem 2.7, and hence of the Corollary, applies equally well to the general case of an integral embedding $\rho : \mathbb{R}[W] \to \mathbb{R}[V]$ - the algebraic independence of the x_i was not used - and we find then that $\overline{\hat{\Omega}_S \cap W_c} = \overline{\rho}(V_c)$. In particular, $\overline{\rho}(V_c) = W_c$ in the birational case. For a general embedding we still have $\overline{\rho}(V_c) \subset \overline{\hat{\Omega}_S \cap W_c}$, but the reverse inclusion fails. As an example consider the projection of the hyperbola $xy = 1$ in the x-axis. Here we have $\hat{\Omega}_S = \mathbb{R}$ but $\overline{\rho}(V_c) = \mathbb{R}\setminus\{0\}$. But if V_c is bounded we still have $\overline{\hat{\Omega}_S \cap W_c} = \overline{\rho}(V_c)$ - boundedness implies Brumfiel's [B] semi-integral property. Consider, for example, $\mathbb{R}[X,Y] \to \mathbb{R}[X,Y,Z]/p$, where p is generated by $(1-Z^2)X^2 - X^4 - Y^4$.

In general it is true that if $H_\rho = \Omega_S$, then, no matter what S may be used, $\hat{\Omega}_S \cap W_c \subset \overline{\rho V_c}$. For suppose z belongs to $\hat{\Omega}_S \cap W_c$ and that D is a neighborhood of z which is disjoint from $\overline{\rho}(V_c)$ - say D is the set of all points at which $f(x)$ is strictly positive, $f(x) \in \mathbb{R}[W]$. Let α be an order centered at z . Clearly α extends to $\mathbb{R}(V)$ and it contains $f(x)$, whence $\rho(f)$ belongs to the extension of α . But, since $-\rho(f)$ is positive all over V_c , $-\rho(f)$ is a sum of squares in $\mathbb{R}(V)$ (see Brumfiel [B], [R-D], Schwartz [Sch_2]) and hence $-\rho(f)$ belongs to the extended order, in contradiction to membership of $\rho(f)$ in that order. This proves:

2.10. Theorem.

Let $\rho : \mathbb{R}[W] \to \mathbb{R}$ be an arbitrary extension of real coordinate rings.
Then $\overline{\hat{H}_\rho} \cap W_c = \overline{\bar{\rho}(V_c)}$.

2.11. Remark.

From the above results it is clear that the numbers $m_V^{\mathbf{G}}$ and $M_V^{\mathbf{G}}$ furnish measures of the extent, i.e., the quality of being spread out, of the real variety V . It should be recalled that in the complex case the situation is much simpler: if $\bar{\rho} : V \to W$ is a polynomial map of varieties inducing an embedding $\rho : \mathbb{C}[W] \to \mathbb{C}[V]$, then $\bar{\rho}(V)$ is constructible in W by Chevalley's Theorem, it is Zariski-dense by hypotheses, and so it is strongly dense (cf. Serre's GAGA [S]). This is certainly not the case for real varieties as can be shown by simple examples. Of course, the real counterpart of Chevalley's Theorem is the Tarski-Seidenberg Theorem, but that merely tells us that $\bar{\rho}(V_c)$ is semi-algebraic. In this sense our results may be understood as a concrete implementation of Tarki-Seidenberg's in the given situation, since it is easy, theoretically, to describe the set H_ρ corresponding to an integral extension $\rho : \mathbb{R}[x_1,\ldots,x_d] \to \mathbb{R}[V]$. In fact, ρ may be extended to an algebraic extension $\rho : \mathbb{R}(x) \to \mathbb{R}(V)$, and we choose a primitive integral generator θ for $\mathbb{R}(V)/\mathbb{R}(x)$ with minimal monic polynomial $F(t)$. Then, as in Elman, Lam and Wadsworth's Lemma 4.2 [E-L-W], H_ρ is defined by inequalities which appear in either Sturm's or Sylvester's theorem which guarantees that $F(t)$ has a real root. We give an example in which we may avoid most of the calculations required for either the Sturm or the Sylvester algorithm.

Consider the crimped blimp B of Coste-Coste-Roy [C-C], defined in $\mathbb{R}^3$ by the equation $(1-z^2)x^2 = x^4 + y^4$. It is easy to see that B is an irreducible real surface. The inclusion $\rho : \mathbb{R}[X,Z] \to \mathbb{R}[B]$ is clearly integral, and for the minimal polynomial $F(Y)$ we take $Y^4 + X^4 + X^2(Z^2-1)$. Then the projection $\pi_2(B_c)$ in the XZ-plane of the set B_c of all central points of B is just the strong closure of the set in the XZ-plane defined by the inequality $x^4 + x^2(z^2-1) < 0$, or, equivalently, by the system

$\{x^2 + z^2 < 1 , x \neq 0\}$. In other words, $\pi_2(B_c)$ is the closed unit disk in the XZ-plane.

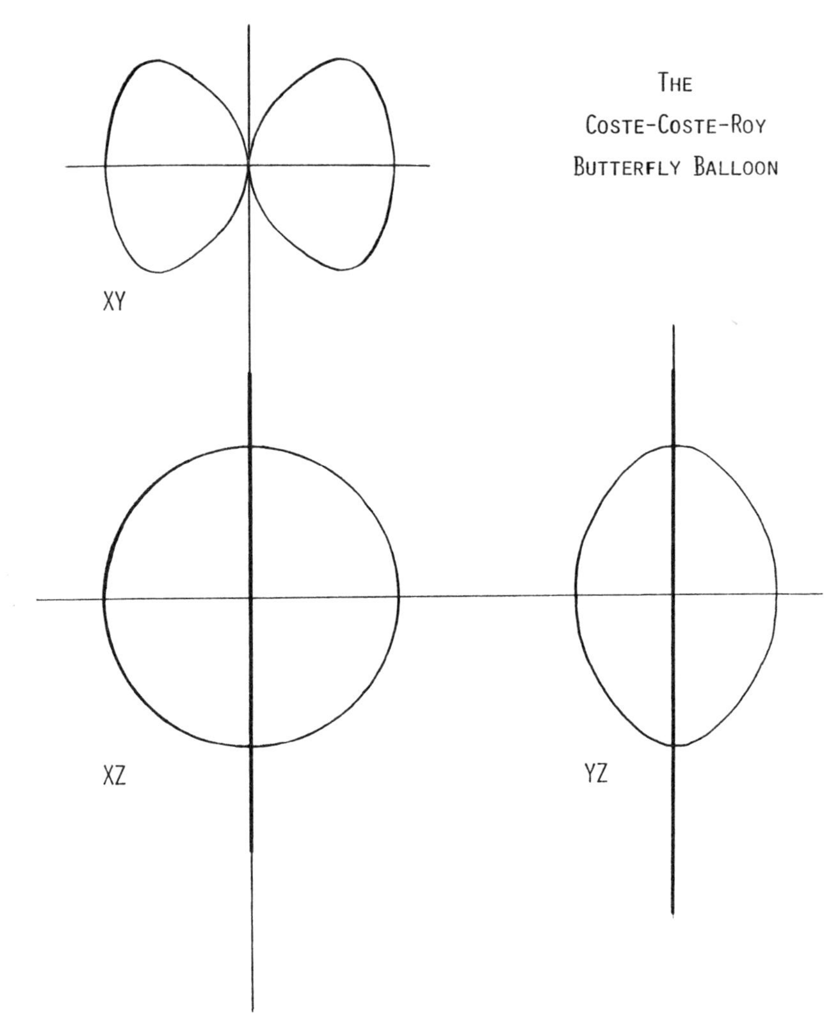

FIGURE 2

NOTE: $\pi_2(B_c)$ is <u>not</u> $\{(x,z) ; x^4 + x^2(z^2-1) \leq 0\}$.

The projection $\pi_3(B_c)$ on the XY-plane satisfies the condition that its closure is equal to the closure of the solution set in the XY-plane of the inequality $x^2 - x^4 - y^4 > 0$; this implies that $\overline{\pi_3(B_c)}$ is equal to the solution set of $x^2 - x^4 - y^4 \geq 0$. Since B_c is bounded and, as always, closed, it is compact and so we even know then that $\pi_3(B_c)$ is itself the solution set of $x^2 - x^4 - y^4 \geq 0$. As for the YZ-projection it is the closure of the solution set of the system:

$$(z^2-1)^2 - 4y^4 > 0 \text{ , and}$$
$$(1-z^2) > 0 \text{ or } -4y^4 > 0 \text{ ,}$$

that is, $\pi_1(B_c) = (z,y);\ z^2 + 2y^2 \leq 1$. See Figure 2.

The essential point of our procedure is that we have changed the "specter" of Tarki's principle from the projection to the closure of an open semi-algebraic set. But even if we don't find the closure we have an approximation to the projection. In any case we have reduced the problem to a lower dimension. Much remains to be done in this case.

3. Hyperplane Sections of Real Varieties.

The techniques of earlier sections are now applied to the subject heading of this section. Let V be a real variety in $\mathbb{R}^n$ of dimension $d > 0$. Many hyperplane sections may well be empty, and the "size" of the set of all such hyperplanes reflects the extent of V . To study this relation we proceed as follows. Let p be the (prime) ideal of V in $\mathbb{R}[X_1,\ldots,X_n]$. Let $\Lambda_0,\ldots,\Lambda_n$ be independent variables and work now in $\mathbb{R}[X,\Lambda] = \mathbb{R}[X_1,\ldots,X_n,\Lambda_0,\ldots,\Lambda_n]$. Let $g = p + (\Lambda_0+\Lambda_1X_1 + \ldots + \Lambda_nX_n)\mathbb{R}[X,\Lambda]$. Let G be the zero set of g in $\mathbb{R}^{2n+1}$, and let $\pi(G)$ be the projection of G on the space $\mathbb{R}^{n+1}$ of the last $n+1$ coordinates. We then have

$$\pi(G) = \{a \in \mathbb{R}^{n+1} ;\ \text{for some } z \text{ on } V ,\ a_0 + \sum_{i=1}^{n} a_i z_i = 0\} ,$$

i.e., the set of all a such that the hyperplane $a_0 + \sum_{i=1}^{n} a_i X_i = 0$ meets V . This includes the <u>odd</u> hyperplane $0 = 0$, and it is the set we wish to study. It is easy to see that $\pi(G)$ is a cone (if $a \in G$ then for all c in $\mathbb{R}$, $(ca_0,\ldots,ca_n) \in G)$ and a semi-algebraic set in $\mathbb{R}^n$ (Tarski-Seidenberg). Some more information is given by the following:

3.1. Proposition (cf. Espino-Recio [E-R]).

With notations given above, $\dim(\mathbb{R}[X,\Lambda]/g) = n+d$, and g is a real prime which contains no polynomials of $\mathbb{R}[\Lambda]$.

Proof (Sketch).

Let m,n be members of $\mathbb{R}[X,\Lambda]$ whose product belongs to g. Divide each of m and n by $\Lambda_0 + \Lambda_1X_1 + \ldots + \Lambda_nX_n$ as polynomials in Λ_0, to obtain

$$m = m' + [\Lambda_0 + \sum \Lambda_iX_i]m''$$

$$n = n' + [\Lambda_0 + \sum \Lambda_iX_i]n'' .$$

Then $m'n'$ belongs to g. Now write m' and n' as polynomials in Λ with coefficients which are functions of X. then for every z on V,

$$m'(\Lambda,z)n'(\Lambda,z) \in (\Lambda_0 + \sum \Lambda_iz_i) ,$$

and hence either $m'(\Lambda,z)$ or $n'(\Lambda,z)$ is zero.
By primality of p, for at least one of m' and n', all coefficients belong to p. Hence m', say, belongs to the extension of p to $\mathbb{R}[X,\Lambda]$, and so to g. The dimension of g is now easily computed by Krull's Hauptidealsatz, while its reality is established by the same techniques as above. Now suppose that m belongs to both g and $\mathbb{R}[\Lambda]$. As above, take any point $z \in V$. Then $m(z)$ is simply m itself and so belongs to $(\Lambda_0 + \sum \Lambda_iz_i)$. As z varies over the infinite set V we obtain infinitely many irreducible factors of m. Thus $m = 0$, and the Proposition is proved.

From the Proposition it follows that there exists an injection $\mathbb{R}[\Lambda] \to \mathbb{R}[X,\Lambda]/g$, so that $\pi(G)$ is Zariski-dense in $\mathbb{R}^{n+1}$. From Recio [R] it is then clear that $\pi(G)$, which contains a non-empty open set in $\mathbb{R}^{n+1}$, has topological dimension $n+1$. Now in the complex case we know that even if a hyperplane misses V a small change in the coefficients will produce a hyperplane which intersects V, hence $\pi(G)$ is strongly dense. But such is not the case for real V as shown in the following examples.

3.2. Examples.

a. Let p be the ideal $(X^2 + Y^2 - 1)$ of the real unit circle in $\mathbb{R}^2$. Then $\pi(G) = \{a \in \mathbb{R}^3 ; a_1^2 + a_2^2 - a_0^2 \geq 0\}$, which is surely not dense in $\mathbb{R}^3$.

b. Let p be the ideal $(Y - X^2)$ in $\mathbb{R}(X,Y)$. Then $\pi(G)$ is the solution set of:

$$a_1^2 - 4a_0a_2 \geq 0 \text{ , and } a_1^2 + a_2^2 > 0$$

or

$$a_1 = a_2 = a_0 = 0 \text{ .}$$

Returning now to the hyperplane situation with which this section begins, recall that $\rho : \mathbb{R}(\Lambda) \to \mathbb{R}(G)$, gives rise, as before, to a clopen set H in $\Omega(\mathbb{R}^{n+1}) = \Omega(\mathbb{R}(\Lambda))$, $H = H_S = \Omega_S = H_\rho = \operatorname{Im}\rho^*$ and thus to the semialgebraic set $\hat{H}_S$ in $\mathbb{R}^{n+1}$. Now if ρ has dimension 1 then G has dimension $n+1$, the extension $\mathbb{R}(\Lambda) \to \mathbb{R}(G)$ is therefore algebraic, and we have via Sturm's (or Sylvester's) theorem a way to "construct" the set S. Our purpose is to study the general relations between the geometric object $\pi(G)$ and the algebraically defined objects H, $\hat{H}_S$, etc. In the preceding section we showed that $\overline{\hat{H}} = \overline{\pi(G_c)}$, where G_c is the set of all central points on G. In this situation this set admits an interesting geometric interpretation. Let $p = (f_1,\ldots,f_s)$, let J_p be its Jacobian matrix $\|\partial f_i/\partial X_j\|$. The Jacobian of g is the $s+1$ by $2n+1$ matrix defined as follows:

$$J_g = \begin{array}{c} s \\ \\ 1 \end{array}\left[\begin{array}{c|c} J_p & 0 \\ \hline \Lambda_1\ldots\Lambda_n & 1 \quad X_1\ldots X_n \end{array}\right]$$
$$\qquad\qquad n \qquad\qquad\quad n+1$$

Let $(p,q) = (p_1,\ldots,p_n,q_0,\ldots,q_n)$ be a point on G; it is regular if $J_g(p,q)$ has rank $n - d + 1$, in which case $J_p(p)$ has rank $n - d$ $(d = \dim V(p))$, so that $p = (p_1,\ldots,p_n)$ is regular on $V(p)$, and the hyperplane, whose equation is $q_0 + \sum q_i x_i = 0$, passes through p and is transversal to the tangent space of $V(p)$ at p. The converse is also true. Thus if we project the set of all regular points of G on $\mathbb{R}^{n+1}$ as above we see that the set of all $a \in \mathbb{R}^{n+1}$ such that the hyperplane whose equation is $a_0 + \sum_{i=1}^{n} a_i X_i = 0$ intersects $V(p)$ transversally at a regular point, is an open semialgebraic set of dimension $n+1$. Now it is easy to see that the set G_c is precisely the set of all points (p,q) in G such that $p \in V_c$, and therefore that $\pi(G_c)$ is the set of all a in $\mathbb{R}^{n+1}$ such that the hyperplane $a_0 + \sum_{i=1}^{n} a_i X_i = 0$ intersects $V(p)$ in a central point. Since, by our earlier results, $\overline{\hat{H}} = \overline{\pi(G_c)} \subseteq \overline{\pi G}$, we naturally inquire whether equality holds: i.e., whether every hyperplane which nearly touches V also nearly touches a central point of V. For the surface in Example 2.10 the answer is clearly "NO"; we take any hyperplane parallel to the XY-plane which cuts the Z-axis at a point far from the origin. But we may still relate $\hat{H}$ with $\pi(G)$ in some way in the special case of hyperplane sections of curves.

3.4. Proposition.

To the above assumptions we add that V is a curve. Then for a proper choice of $S = \{S_1,\ldots,S_m\}$ we have $H = \Omega_S$ and

$$\overline{\pi(G)} \subseteq \hat{H}_S^u \equiv \hat{\Omega}_S^u = \bigcup_{i=1}^{m} \{z; s \in S_i \Rightarrow s(z) \geq 0\} .$$

Proof.

Since $\hat{H}_S^u$ is closed in $\mathbb{R}^{n+1}$ we have only to show that $\pi(G)$ is contained in $\hat{H}_S^u$ for suitable S. It is easy to see that $\pi(G)\backslash\overline{\pi(G_c)}$ is contained in a proper Zariski-closed subset of $\mathbb{R}^{n+1}$, for $\pi(G)\backslash\overline{\pi(G_c)} \subseteq \pi(G\backslash\overline{G_c})$ and $G\backslash\overline{G_c}$ is a semialgebraic set of dimension smaller than $n+1$ (recall that $\dim G = n + d$, which is $n+1$ in this case). Het $h^2 = 0$ be the equation of the Zariski closed set. Multiply every element of S by h^2 to get a new set Q of finite sets. Then $H_Q = H_S$, while the new $\overline{\hat{H}_Q^u}$ clearly contains $\pi(G)$.

3.5. Remark.

This is the best general result we may obtain for curves. In fact, returning to our surface $(1-z^2)x^2 - y^4 - x^4 = 0$ and intersecting it with a well-situated hyperplane we obtain a plane quartic curve which is bounded and has an isolated point outside the curve's convex hull. There exists a line through this point such that nearby lines don't meet any other points of the curve. Therefore $\overline{\pi(G)} \neq \overline{\hat{H}}_S$.

In addition to the number of relations shown above between the clopen H corresponding to the inclusion $\mathbb{R}(\Lambda) \subset \mathbb{R}(G)$ and the geometry of hyperplane sections of V , we offer now another, more quantitative kind of information provided by H .

3.6. Definition.

By $D^G(H/V)$, called the hyperplane number of V (corresponding to a group G of automorphisms), is meant the minimum power of all subcovers of $G^{-1}(H)$.

Note that $\mathbb{R}(\Lambda)$ is a Q_1-field, whence $G^{-1}(H)$ is a cover, and $D^G(H/V)$ is finite. Because of our previous result and the fact that $\hat{H}$ is an unbounded open set ($\overline{\hat{H}}$ is the closure of the cone $\pi(G_c)$) we know that in general $D^G(H/V) \leq 2$, if G is equal to the group of all automorphisms; also $D^{G(P)}(H/V) \leq n + 2$, and $D^{G(A)}(H/V) < \infty$, where $n+1$ is the number of Λ's .

3.7. Proposition.

A necessary and sufficient condition for $D(H/V)$ to be equal to 1 is that every hyperplane should "nearly" intersect V_c ; i.e., for arbitrary $a \in \mathbb{R}^{n+1}$ there exists b arbitrarity close to a such that the hyperplane $b_0 + \sum b_i X_i = 0$ intersects V_c ; or equivalently, the set of all $b \in \mathbb{R}^{n+1}$ such that $b_0 + \sum b_i X_i = 0$ intersects V_c is dense in $\mathbb{R}^{n+1}$.

Proof.

It was shown earlier, at the beginning of Section 2, that $D(H/V) = 1$ if and only if $\hat{H}$ is dense in $\mathbb{R}^{n+1}$. From the above remarks it is clear that $\bar{\hat{H}} = \mathbb{R}^{n+1}$ is equivalent to $\overline{\pi(G_c)} = \mathbb{R}^n$.

For curves the propositon can be made more precise.

3.8. Proposition.

Assume now that $\dim V = 1$. Then $D(H/V) = 1$ if and only if every hyperplane nearly intersects V.

Proof.

One implication is now obvious. So suppose every hyperplane nearly intersects V, i.e., $\overline{\pi(G)} = \mathbb{R}^{n+1}$; thus $\hat{H}^u = \mathbb{R}^{n+1}$. But the difference $\hat{H}^u \backslash \hat{H}$ is contained in a Zariski-closed set in $\mathbb{R}^{n+1}$. Therefore, $\hat{H}^u = \mathbb{R}^{n+1}$ which implies $\bar{\hat{H}} = \mathbb{R}^{n+1}$.

3.9. Remark.

The number $D(H/V)$ is not determined by the function field - it is not a birational invariant - as shown by the examples of the circle and rational cubic, whose fields are the same but whose values for $D(H/V)$ are 2 and 1, respectively. Moreover, the hyperbola has $D(H/V) = 2$. Nor does the number involve parity of the degree, for the curve $V : X^2 + Y^2 - X^2Y^2 = 0$ has even degree but $D(H/V) = 1$.

3.10. More Examples.

a. The Coste-Coste-Roy butterfly balloon, as usual, illustrates an interesting point. Recall the equation: $(1-z^2)x^2 = x^4 + Y^4$. Let $a_0 + a_1X + a_2Y + a_3Z = 0$ be a hyperplane. It misses the Z-axis if and only if $a_0 \neq 0$ and $a_3 = 0$. Therefore every plane nearly cuts the balloon, but many planes don't even come near to striking a simple point. Therefore $\overline{\pi(G)} = \mathbb{R}^4$, while $\bar{\hat{H}}$ is strictly included in $\mathbb{R}^4$, i.e., $D(H/V) > 1$. Thus $\overline{\pi(G)} \neq \hat{H}^u$.

b. By enclosing any bounded variety V in a big sphere we see that $D(H/V) > 1$.

3.11. Proposition.

For all $V \subseteq \mathbb{R}^m$, $D^{G(A)}(H/V) \leq n+1$, where $G(A)$ is the group of all affine automorphisms.

Proof.

Since $\hat{H} = \overline{\pi(G_c)}$ is a cone we may choose a small ball in $\hat{H}$ and let $\hat{C}$ be the cone through the ball with vertex at the origin of the Λ-coordinates. We now show that for some $n+1$ affine automorphisms of $\mathbb{R}[\Lambda]$, say $\sigma_1,\ldots,\sigma_{n+1}$, we have $\bigcup \sigma_j(\bar{\hat{H}}) = \mathbb{R}^{n+1}$, which implies that $\sigma_j(\hat{H})$ is dense in $\mathbb{R}^{n+1}$ (for the σ_i are homeomorphisms of $\mathbb{R}^{n+1}$ on itself) whence $\sigma_j^{-1}H$ is the entire Harrison order space, which suffices to prove the theorem. We suppose for simplicity that the cone $\hat{C}$ is described by inequalities $a_0^2 - \sum_{i=1}^{n} a_i^2 \geq 0$. Consider the automorphisms (b will be determined later in $\mathbb{R}^+$)

$$\sigma_i : \begin{cases} a_j \to a_{[j+i-1]} , & \text{if } j \neq 0 \\ a_0 \to b a_{i-1} & \end{cases}$$

where $[m]$ represents the least residue of m modulo $n+1$. Now if $a = (a_0,\ldots,a_n)$ is outside the union of the $\sigma_i(\hat{C})$, then we have

$$\sum_{i\neq 0} a_i^2 - b^2 a_0^2 > 0$$

$$\cdots$$

$$\sum_{i\neq n} a_i^2 - b^2 a_n^2 > 0 .$$

Adding the inequalities we get

$$n \sum_{i=0}^{n} a_i^2 - b^2 \sum_{i=0}^{n} a_i^2 > 0 .$$

We merely take b large enough and obtain a contradiction.

Final Remark.

Let V be a variety in $\mathbb{R}^n$, let $a_0 + \sum a_i X_i = 0$ be a hyperplane. Using the σ_j constructed in the Proposition we may assert that for some j, the hyperplane $\sigma_j a_0 + \sum_i (\sigma_j a_i) X_i$ nearly intersects V. The number $n+1$ seems not to be the best bound for in the plane circle case two automorphisms is all we need. Sharper estimates and a more thorough study of the geometry involved are clearly desirable.

References

[A] Adkins, W. Communication presented to the special session on Ordered Fields and Real Algebraic Geometry, A.M.S., January 1981.

[B] Brumfiel, G. Partially Ordered Rings and Semi-Algebraic Geometry. Lecture Note Series, London Math Soc. (1979). Cambridge University Press.

[C-C] Coste, M. and Roy, M.-F. La Topologie du spectre reel. Preprint 1980.

[D_1] Dubois, D.W. Note on Artin's solution of Hilbert's 17th problem. Bull. Amer. Math. Soc. 73 (1976), 540-541.

[D_2] Dubois, D.W. Infinite primes and ordered fields, Dissertationes Math. (Warszawa) 69 (1970), 1-40.

[D_3] Dubois, D.W. Real Algebraic curves, University of New Mexico Technical Report No. 227, October 1971.

[D_4] Dubois, D.W. Second note on Artin's solution of Hilbert's 17th Problem. Order Spaces. To appear in Pac. J. Math.

[D-E] Dubois, D.W. and Efroymson, G. Algebraic theory of real varieties I. Studies and Essays Presented to Y.W. Chen on his 60th Birthday (October 1970). Taipeh, Taiwan.

[E-L-W] Elman, R. Lam,T.Y. and Wadsworth, A.R. Orderings under field extensions, J. Reine Angew. Math. 306 (1979), 6-27.

[E-R] Espino, V. and Recio, T. Sobre la seccion hiperplana generica de una variedad algebraica real, Rev. Mat. Hispano Americana, Madrid, 1980.

[H] Hartshorne R. Algebraic Geometry. Springer-Verlag, Berlin, 1977.

[L] Lam, T.Y. The Theory of Ordered Fields, Proceedings of Algebra and Ring Theory Conference. Ed. B. MacDonald. University of Oklahoma 1979.

[M] Massaza, C. Sugli ordinamento di un campo estensione puramente trascendente di un campo ordinato, Rend. Mat. 6, 1968.

[McE] McEnerney, J. Trim stratification of semianalytic sets, Man. Math. 25, 1978.

[McK] McKenna, Kenneth. New Facts about Hilbert's 17th problem. Meodel Theory and Algebra (Mem. Tribute to A. Robinson). Lecture notes in Math 498, Springer-Verlag, 1975.

[N_1] Nagata, M. On automorphism group of K[X,Y]. Lect. in Math. Dept. of Math., Kyoto Univ. Tokyo, 1972.

[N_2] Nagata, M. Polynomial rings and affine spaces. Conference Board of Math. Sci. Regional Conf. Series in Math. No. 37. A.M.S. 1978.

[R] Recio, T. Una descomposicion de un conjunto semialgebraico. Actas de la V reunion de Mathematicos de Expresion Latina, Mallorca, 1977.

[R-D] Recio, T. and Dubois, D.W. A note on Robinson's nonnegativity criterion. Preprint 1980.

[Sch_1] Schwartz, N. Hilbert's 17th problem for parially ordered sets. Preprint.

[Sch_2] Schwartz, N. Communication presented to the special session on Ordered Fields and Real Algebraic Geometry. A.M.S., January 1981.

[S] Serre, J.P. Geometrie analytique et geometrie algebrique, Ann. Inst. Fourier 6, 1956.

[T] Taussky, O. The discriminant matrices of an algebraic number field. J. Lond. Math. Soc. 43, 1968.

DEPARTMENT OF MATHEMATICS
UNIVERSITY OF NEW MEXICO
ALBUQUERQUE, NEW MEXICO

DEPARTAMENTO DE ALGEBRA
UNIVERSIDAD DE MÁLAGA
MÁLAGA

Contemporary Mathematics
Volume 8, 1982

REAL POINTS AND REAL PLACES

Heinz-Werner Schülting[1]

The "real holomorphy ring" of a formally real field K is the intersection of all real valuation rings of K. Various descriptions of this ring can be found in the literature (1), (14). In this paper we give a report on some abstract properties of real holomorphy rings and applications to real algebraic geometry. Real holomorphy rings also appear in connection with axiomatic geometry (1), (18), and they lead to interesting examples in commutative algebra (19). Moreover, there are important applications to the study of sums of n-th powers (2).

1. BASIC PROPERTIES. Let K be a formally real field. We denote the set of $\mathbb{R}$-valued places of K by M_K and the set of real valuation rings of K by S_K. Let X_K be the topological space of orderings of K and Q_K the sums of squares of K. We write M, S, X, Q when no confusion can occur.

PROPOSITION 1 (14), (1), (2o). The following rings $H_1,\dots,H_4$ coincide: $H_1 = \bigcap_{B\in S} B$, $H_2 = \{x\in K \mid \text{ex. } n\in\mathbb{N},\ n\pm x\in Q\}$,

$$H_3 = \mathbb{Q}[\{\tfrac{1}{1+q} \mid q\in Q\}],\quad H_4 = \{\frac{n\cdot x_1}{1+x_1^2+\dots+x_r^2} \mid n,r\in\mathbb{N},\ x_i\in K\}.$$

The ring $H_K=H_1=H_2=H_3=H_4$ is called the <u>real holomorphy ring</u> of K. Overrings of H_K contained in K are called <u>relative real holomorphy rings</u>.

The ring generated by $\{\frac{1}{1+x^2} \mid x\in K\}$ is a Prüfer domain (13). Therefore H_K is a Prüfer domain, too. A consequence of this fact is the following theorem:

THEOREM 2. Let D be a relative real holomorphy ring and E a domain with quotient field $\bar{E}$. Let $h\colon D\to E$ be a ring epimorphism. Then i) $\bar{E}$ is formally real. ii) There is a unique place $\phi\colon K\to\bar{E}$ which extends h.

Proof: (2o), prop. 1.3. The second part is proved in (14), in

---

1980 Mathematics Subject Classification. 12D15, 14G3o.

[1]Supported by Deutsche Forschungsgemeinschaft DFG

the case $D=H_K$.

According to the theorem M is isomorphic to $\mathrm{Hom}(H_K,\mathbb{R})$, the set of ring homomorphisms from H_K to $\mathbb{R}$. We embed this set into the product $\times[-n_x,n_x]$, where x runs through H_K and n_x is chosen such that $n_x \pm x \in Q_K$. We observe that $\mathrm{Hom}(H_K,\mathbb{R})$ is mapped to a closed subset of the above product. Furthermore, points of $\mathrm{Hom}(H_K,\mathbb{R})$ are separated by elements of H_K, hence in this way M becomes a compact Hausdorff space in the induced topology.

Each element f of H_K induces a continuous function $\tilde{f}: M \to \mathbb{R}$, $\phi \mapsto \phi(f)$. Thus, we get a map from H_K to $C(M,\mathbb{R})$, the set of continuous functions from M to $\mathbb{R}$, and according to the Stone-Weierstrass approximation theorem the image of H_K is a dense subset of $C(M,\mathbb{R})$. It is easy to see that the above topology coincides with that which Dubois defines in (14).

2. CONNECTED COMPONENTS OF M. If ϕ is a place in M, the restriction of ϕ to H_K induces a homomorphism $\tilde{\phi}: W(H_K) \to W(\mathbb{R}) \tilde{\to} \mathbb{Z}$. We recall that homomorphisms $W(A) \to \mathbb{Z}$ from the Witt ring of the ring A onto $\mathbb{Z}$ are called signatures of A (16).

THEOREM 3. $\tilde{\phi}$ depends only on the connected component of ϕ. The map:

$$\begin{array}{ccc} \{\text{connected components of } M\} & \to & \{\text{signatures of } H_K\} \\ C & \mapsto & \tilde{\phi} \quad ; \ \phi \in C \end{array}$$

is bijective.

Proof: (20)

The fact that different components can be separated by signatures is not very profound. It follows directly from Stone-Weierstrass: Any open and closed set C can be separated from its complement by a unit of H_K by approximating the characteristic function of C.

Let $H^{\bullet}_{+}$ be the set of totally positive units in H_K and Z_K the factor group $H^{\bullet}/H^{\bullet}_{+}$. $H^{\bullet}_{+}$ consists of the elements $f = (1+g)/n$, $g \in H_K \cap Q$, $n \in \mathbb{N}$ ((20), 2.7). The results of (20), Ch. IV, show that the reduced Wittring $W_{red}(H_K)$ and the set of signatures of H_K have in some sense optimal structures, summarized in the following

THEOREM 4. i) $W_{red}(H_K)$ is generated by one-dimensional forms. ii) The signatures form a space of orderings with stability index ≤ 1 (in the sense of Marshall (17)) with respect to the group Z_K. iii) Let L be the rational function field in n variables over K. Then $W_{red}(H_K)$ is isomorphic to $W_{red}(H_L)$.

As a consequence of part iii) M_K is connected iff M_L is connected (L as above). This is not true for arbitrary algebraic function fields L/K.

3. REAL VARIETIES. If V is a complete non-singular $\mathbb{R}$- variety with a formally real function field F, we have a continuous map $M_F \to V_r$ with V_r denoting the set of rational points of V with the strong topology. L. Bröcker studied the fibers of this map and proved that they are connected (7). This implies that this map induces a bijection between the connected components of V_r and those of M_F, the latter corresponding to the signatures of H_F. Thus we obtain a good explanation of the fact that the number of components of V_r is a birational invariant of non-singular complete $\mathbb{R}$-varieties (10).

Knebusch (16) is dealing with the related problem of associating the signatures of the Witt ring of V to the components of V_r. This problem however is much more difficult.

We now shall consider varieties over an arbitrary real closed field R. In this situation one has to study the set of R-valued places of F/R. But this set hasn't a good tpology. It is the recent definition of the real spectrum of a ring by Coste and Coste-Roy which makes it possible to generalize the above result on varieties over $R = \mathbb{R}$ (11),(12). We use a definition which differs slightly from the original one (see (6)).

DEFINITION. Let B be a commutative ring. A subset α of B is called a prime ordering of B if it satisfies the following conditions: $\alpha+\alpha \subset \alpha$, $\alpha\cdot\alpha \subset \alpha$, $\alpha \cup -\alpha = B$, $-1 \notin \alpha$, $\alpha \cap -\alpha$ is a prime ideal. The sets $D(a_1,\dots,a_n) := \{\alpha \mid \alpha$ is a prime ordering of B, $-a_i \notin \alpha\}$ $a_1,\dots,a_n \in B$, form a basis of open sets of a topology. The corresponding topological space is called the <u>real spectrum of B</u>, denoted by $\mathrm{spec}_r(B)$.

The concept of real spectra includes the space M_K, since the map $M_K \to \mathrm{spec}_r(H_K)$, $\phi \mapsto \phi^{-1}(\mathbb{R}^2)$, defines a homeomorphism from M_K onto the set of closed points of $\mathrm{spec}_r(H_K)$. The real spectrum of a ring is (quasi-) compact (12) and the set of closed points is compact and Hausdorff (8).

Let R be a real closed field and $\widetilde{V}$ a projective non-singular variety over R with a formally real function field F. We can find an open affine subvariety V of $\widetilde{V}$, such that $V_r = \widetilde{V}_r$ (10). Then V is real complete in the sense of (12). In this case, Coste and Coste-Roy give a birational description of the partition into semi-algebraic components of V_r. (A semi-algebraic component of V_r is an open and closed semi-algebraic subset of V_r which is not the disjoint union of two closed semi-algebraic sets. See (9), prop. 8.13.14.) Let $R[V]$ be the coordinate ring of V. Every point $p \in V_r$ induces a prime ordering $\alpha_p := \{f \in R[V] \mid f(p) \geq 0\} \in \mathrm{spec}_r(R[V])$. In this way V_r is identified with a subset of the real spectrum and under this identification the restrictions of the components of $\mathrm{spec}_r(R[V])$ are exactly the semi-algebraic components of V_r (see (12), Th. 5.1).

We consider the ring $H(F/R) := H_F \cdot R$. As an overring of H_K the ring $H(F/R)$ is again a Prüfer domain. Hence $H(F/R)$ is an intersection of valuation rings and all valuation rings containing $H(F/R)$ are real (see Th. 2). It follows that $H(F/R) = \bigcap_{B \supset R \text{ real}} B = \{f \in F \mid \text{ex. } r \in R,\ r \pm f \in Q_F\}$. The relative real holomorphy ring $H(F/R)$ contains $R[V]$, so we obtain the continuous restriction maps

$$X_F = \mathrm{spec}_r(F) \xrightarrow{h_1} \mathrm{spec}_r(H(F/R)) \xrightarrow{h_2} \mathrm{spec}_r(R[V]).$$

Let C be a connected component of $\mathrm{spec}_r(R[V])$ and $\widetilde{C} = h_2^{-1}(C)$. $\widetilde{C}$ is not emty since V is non-singular. Assume $\widetilde{C}$ is not connected, i.e. $\widetilde{C} = C_1 \cup C_2$, $C_1 \cap C_2 = \emptyset$, and C_1, C_2 are closed non-empty sets. Let α be any element of C_i, $i \in \{1, 2\}$, $k(\alpha)$ the quotient field of $H(F/R)/_{\alpha \cap -\alpha}$ and π: $H(F/R) \to k(\alpha)$ the canonical homomorphism. According to theorem 2, π extends to a place $\mu: F \to k(\alpha) \cup \{\infty\}$. Let P be an ordering of F compatible with μ and $\bar{\alpha}$, where $\bar{\alpha}$ denotes the ordering of $k(\alpha)$ induced by $\pi(\alpha)$. Since we have $P \cap H(F/R) \subset \alpha$, α is contained in the closure of $\{P \cap H(F/R)\}$. Hence $P \in h_1^{-1}(C_1)$ and $(h_2 \circ h_1)^{-1}(C)$ is the disjoint union of the closed sets $h_1^{-1}(C_1)$ and $h_1^{-1}(C_2)$. But Coste and Coste-Roy show that this is impossible. We obtain:

THEOREM 5 (Coste, Coste-Roy). The restriction map: $\mathrm{spec}_r(H(F/R)) \to \mathrm{spec}_r(R[V])$ induces a bijection between the connected components of $\mathrm{spec}_r(H(F/R))$ and those of $\mathrm{spec}_r(R[V])$, the latter correspond to the semi-algebraic components of V_r.

In the case of the real numbers $R = \mathbb{R}$, open and closed subsets of M_F can be separated by a unit of $H_F = H(F/R)$, as mentioned above. It follows easily that open and closed subsets of $\mathrm{spec}_r(H(F/R))$ are also separated by $H(F/R)^{\bullet}$. We give an example which shows that this is not true, when R is a non-archimedean real closed field.

EXAMPLE. We construct a function field F over a real closed field R which satisfies the following property: The real spectrum of the relative holomorphy ring $H(F/R)$ possèsses two open and closed sets C_1, C_2, such that there is no f in F with $C_1 \subset D(f)$,

$C_2 \subset D(-f)$. We assume that R^2 contains an infinitely small element a. Let $\lambda : R \to \mathbb{R} \cup \{\infty\}$ be the unique $\mathbb{R}$-valued place of R. There is a canonical extension of λ to a place $\tilde{\lambda} : R(X,Y) \to \mathbb{R}(X,X) \cup \{\infty\}$. Let ϕ_Y be the place of $\mathbb{R}(X,Y)$ which is trivial on $\mathbb{R}(X)$ and which satisfies $\phi_Y(Y) = 0$, and let ϕ_X be the unique real place of $\mathbb{R}(X)$ with $\phi_X(X) = 0$. Consider the real place $\phi := \phi_X \circ \phi_Y \circ \tilde{\lambda}$. The value group of ϕ is $C \times \mathbb{Z} \times \mathbb{Z}$ with a divisible group C, hence there are four orderings of R(X,Y) compatible with ϕ. They correspond to the signs of X and Y. The intersection of these orderings form a fan T ((3), §4). Define $F := R(X,Y)(\sqrt{X^2-a}, \sqrt{Y^2-a}, \sqrt{1-X^2}, \sqrt{1-Y^2})$. First we show that the preordering T' generated by T and $\{\sqrt{X^2-a}, \sqrt{Y^2-a}, \sqrt{1-X^2}, \sqrt{1-Y^2}\}$ is a fan. We have $\phi(1-X^2) = \phi(1-Y^2) = \phi(\frac{X^2-a}{X^2}) = \phi(\frac{Y^2-a}{Y^2}) = 1$. Hence X^2-a, Y^2-a, $1-X^2$ and $1-Y^2$ have even values in the value group of ϕ and they are positive with respect to each ordering compatible with ϕ. Therefore, the extensions of an ordering $P \supset T$ to F correspond to the allotment of signs to $\sqrt{\frac{X^2-a}{X^2}}, \sqrt{\frac{Y^2-a}{Y^2}}$, $\sqrt{1-X^2}$, $\sqrt{1-Y^2}$, as well as the extensions of ϕ. If P is an ordering over T we choose an extension $\tilde{P}$ of P with $\{\sqrt{\frac{X^2-a}{X^2}}, \sqrt{\frac{Y^2-a}{Y^2}}$, $\sqrt{1-X^2}$, $\sqrt{1-Y^2}\} \subset \tilde{P}$. Then the places induced by these orderings coincide. Hence T' is a fan.
It suffices to prove that the real spectrum of H(F/R) is the disjoint union of four open sets each of them containing an ordering over T'. Then one of these open and closed sets cannot be separated from its complement by any function in F.
The elements X and Y are units in H(F/R) since we have $a < |X|, |Y| < 1$ with respect to each ordering of F. Suppose, there is a $\alpha \in \operatorname{spec}_r(H(F/R))$ and a unit f of H(F/R) with $f \in \alpha \cap -\alpha$. The homomorphism $H(F/R) \to H(F/R)/_{\alpha \cap -\alpha}$ is the restriction of some real place ψ with $\psi(f) = 0$. But this is not possible since f is a unit. Therefore the sets $D(\pm X, \pm Y)$ are open and each of these sets contains an ordering over T'.

A last point should be mentioned, which explains why real holomorphy rings contain a lot of informations about birational properties of real varieties.

In the above situation let O_V be the ring of those functions which are regular on V_r and write $S = O_V^{\bullet} \cap R[V]$. Using the compactness of $\operatorname{spec}_r(R[V])$ one can show that O_V is the localization of R[V] at S (see (5)). Then it is easy to conclude that the map

$$\operatorname{spec}_r(R[V]) \to \operatorname{spec}_r(O_V)\ ,\ \alpha \mapsto \alpha_S := \{\frac{f}{g} \mid f, g \in \alpha,\ g \in S\}$$

is well defined, continuous, and it is the inverse of the restriction map $\mathrm{spec}_r(R[V]) \to \mathrm{spec}_r(O_V)$. Hence the real spectra are homeomorphic.

Now we follow an idea of L. Bröcker: Let V be a complete non-singular R-variety. Again we assume that V has real points. Let $B(V)$ be the set of pairs (W,ϕ) with W being a complete non-singular R-variety and $\phi: W \to V$ a birational R-morphism. This set is ordered in the following way: $(W,\phi) > (\tilde{W},\tilde{\phi})$ iff there is an R-morphism from W onto $\tilde{W}$ representing $\tilde{\phi}^{-1} \circ \phi$. In this manner $B(V)$ becomes a directed set (see(21), p. 45) and the rings O_W, $(W,\phi) \in B(W)$, form a direct system in an obvious way.

THEOREM 6. i) $H(F/R) = \varinjlim O_W$

ii) $\mathrm{spec}_r(H(F/R)) = \varprojlim \mathrm{spec}_r(O_W)$

Proof: ii) follows directly from i) and (12), prop. 2.4.

i): A function $f \in H(F/R)$ can be considered as a rational map $\hat{f}: V \to \mathbb{A}^1_R \to \mathbb{P}^1_R$. According to Hironaka ((15), main theorems I, II) we find a complete non-singular R-variety W_o, a birational R-morphism $\phi: W_o \to V$, and an R-morphism f_o, such that $f_o = \hat{f} \circ \phi$. Since f_o and f considered as rational functions coincide we may assume that $\hat{f}: V \to \mathbb{P}^1_R$ is a morphism. If B is a valuation ring of F centered in a point p of V we have: $f \in B$ iff $\hat{f}(p) \in \mathbb{A}^1_R$ iff f is regular in p. But f is contained in every real valuation ring over R. Thus f is regular in every real point of V.

BIBLIOGRAPHY

1. F. Bachmann, Aufbau der Geometrie aus dem Spiegelungsbegriff, Berlin-Göttingen-Heidelberg, 1959.

2. E. Becker, "Summen n-ter Potenzen in Körpern und der reelle Holomorphiering", to appear.

3. E. Becker, E. Köpping, "Reduzierte quadratische Formen und Semiordnungen reeller Körper", Abh. Math. Sem. Univ. Hamburg, 46 (1977), 143-177.

4. L. Bröcker, "Zur Theorie der quadratischen Formen über formal reellen Körpern", Math. Ann.,21o(1974), 233-256.

5. L. Bröcker, "Reelle Divisoren", to appear.

6. L. Bröcker, "Positivbereiche in kommutativen Ringen", to appear.

7. L. Bröcker, "Zusammenhangskomponenten im Funktionenkörper und im Modell", unpublished.

8. L. Bröcker, Manuscript on real spectra and distributions of signs.

9. G. W. Brumfiel, Partially ordered rings and semi-algebraic geometry, London Math. Soc. Lecture Note Series 37, Cambridge U.P., 1979.

10. J. L. Colliot-Thélène, "Formes quadratique multiplicatives et variétés algébriques", Bull. Soc. Math. France, 106 (1978), 113-151.

11. M. Coste, M.-F. Coste-Roy, "Topologies for real algebraic geometry", in A. Kock ed.: Topos theoretic methods in geometry, Various publication series 30, Aarhus Universitet, 1979.

12. M. Coste, M.-F. Coste-Roy, "La topologie du spectre reel", manuscript.

13. A. Dress, "Lotschnittebenen mit halbierbarem rechten Winkel", Arch. Math. (Basel), 16 (1965), 388-392.

14. D. W. Dubois, "Infinite primes and ordered fields", Dissertationes Math. (Rozprawy Mat.), 69 (1970).

15. H. Hironaka, "Resolution of singularities of an algebraic variety over a field of characteristic zero", Ann. of Math., 79 (1964), 109-326.

16. M. Knebusch, "Signaturen, reelle Stellen und reduzierte quadratische Formen", Jber. Deutsch. Math.-Verein., 82 (1980), 109-127.

17. M. Marshall, "The Witt ring of a space of orderings", Trans. Amer. Math. Soc., 258 (1980), 505-521.

18. W. Pejas, "Die Modelle des Hilbertschen Axiomensystems der absoluten Geometrie", Math. Ann., 143 (1961), 212-235.

19. H. W. Schülting, "Über die Erzeugendenanzahl invertierbarer Ideale in Prüferringen", Comm. Algebra, 7 (1979), 1331-1349.

20. H. W. Schülting, "On real places of a field and their holomorphy ring", to appear.

21. I. R. Shafarevich, Lectures on minimal models and birational transformations of two dimensional schemes, Tata Institute of Fundamental Research, Bombay, 1966.

ABTEILUNG MATHEMATIK
UNIVERSITÄT DORTMUND
D-4600 DORTMUND
FED. REP. GERMANY

Contemporary Mathematics
Volume 8, 1982

THE STRONG TOPOLOGY ON REAL ALGEBRAIC VARIETIES

Niels Schwartz

ABSTRACT. For a totally ordered field (k, T) with real closure R and a real algebraic k-variety $V \subset R^n$ it is investigated whether the sets $N(f) = \{x \in V_c; f(x) < 0\}$ with $f \in k[X_1, \ldots, X_n]$ form a basis of the strong topology of the set V_c of central points of V. These sets always form a subbasis of the strong topology, and, if $\dim V > 0$, every strongly open subset of V_c contains some set of the form $N(f)$. But the sets $N(f)$ do not always form a basis of the strong topology: For a variety V of dimension > 0, the sets $N(f)$ form a basis of the strong topology of V_c if and only if for every algebraic extension $k \subset K$ the set of extensions of T to total orders of K has the strong approximation property. These results on the strong topology of V_c lead to the following solution of a generalization of Hilbert's 17th problem: For a variety V of positive dimension and a subset $M \subset V$, the functions $f \in k[V]$ that are positive definite on M are of the form $f = p_1f_1^2 + \ldots + p_rf_r^2$ with $0 < p_i \in k$ and $f_i \in k(V)$ if and only if $M \cap V_c$ is dense in V_c. Finally, a few results on Hilbert's 17th problem for partially ordered fields are proved.

Let k be a field with a total order T, R its real closure. For a real prime ideal $P \subset k[X_1, \ldots, X_n]$ (for real ideals see [6]) the set of zeros in R^n is called a <u>variety</u>. As an ordered field R is a topological field with respect to its interval topology. The product topology on R^n is called the <u>strong topology</u> (in contrast to the Zariski topology). The strong topology on a variety V is the topology induced

---

1980 Mathematics Subject Classification. 13J25, 14G30.

by the strong topology of R^n. The main subject of this paper is to investigate how the strong topology on a variety V can be described by polynomials in $k[X_1, \ldots, X_n]$: For the polynomials $f_1, \ldots, f_r \in k[X_1, \ldots, X_n]$ call $N(f_1, \ldots, f_r) = \{x \in V; \; f_i(x) < 0 \text{ for all } i\}$ the <u>set of negativity</u>. A first (and rather easy) result is that the nonempty sets $N(f_1, \ldots, f_r)$ form a basis of the strong topology of V. The main results deal with the question when the nonempty sets of negativity $N(f)$ of a single polynomial form a basis of the strong topology of V.

Section 1 deals with the variety R^n. For a point $a \in R^n$ let $P(a)$ be the set of points in R^n k-isomorphic to a. It is shown that every point in $P(a)$ has a neighborhood basis of sets of the form $N(f)$ if and only if the set $T_{k(a)}$ of orderings of $k(a)$ extending the ordering T of k has the strong approximation property (SAP) (see [2], [17]). From this it follows that the sets $N(f)$ form a basis of the strong topology of R^n if and only if, for every algebraic extension $k \subset K$, the set T_K of extensions of T to K has the SAP. This, of course, shows that the sets $N(f)$ do not always form a basis of the strong topology of R^n. But the following weaker result does always hold: Given any strong open set $U \subset R^n$, there is some $f \in k[X_1, \ldots, X_n]$ such that $\emptyset \neq N(f) \subset U$.

A field K with a set $\underline{T}$ of total orders is called a <u>hereditary SAP field</u> if for any algebraic extension $K \subset L$ the set $\underline{T}_L$ of orderings of L extending some ordering in $\underline{T}$ has the SAP. A point $a \in R^n$ is called an <u>SAP point</u> or <u>non-SAP point</u> according as $T_{k(a)}$ does or does not have the SAP. The results of Section 1 clearly call for a closer examination of hereditary SAP fields and SAP points. Therefore, Section 2 is devoted to such an examination. The main tool in this section is a valuation theoretic criterion for a field to be a hereditary SAP field. This is a variation of a similar criterion for SAP fields by Prestel (see [18, Satz 2.2] in connection with [8, §3]) and Becker and Köpping ([2, Korollar zu Satz 23]). The main results deal with the behaviour of the hereditary SAP under field extensions. Using this one obtains the following picture of the distribution of SAP points and non-SAP points in R^n: If the ordered field (k, T) is a hereditary SAP field then every point in R^n is an SAP point. If (k, T) is not a hereditary SAP field then, for all $i = 1, \ldots, n$ and all $a_i < b_i$ in R, there is some $c_i \in R$, $a_i < c_i < b_i$ such that the hyperplane $X_i = c_i$ in R^n consists entirely of non-SAP points.

The discussion of arbitrary real algebraic varieties is taken up in Section 3. The results can be roughly summarized by stating that the results of Section 1 carry over to arbitrary varieties. In many places this works only under the following two restrictions: The variety must have positive dimension. And, secondly, one has to work with the set V_c of central points (see [4]) of the variety V instead of with the whole variety.

The investigations of the first three sections were motivated by the following generalization of Hilbert's 17th problem:

> Suppose that $M \subset R^n$ is a subset and $f \in k[X_1, \ldots, X_n]$ is a polynomial which is positive definite on M. For which M does this imply that f is a positive square combination in $k(X_1, \ldots, X_n)$, i.e., $f = p_1f_1^2 + \ldots + p_rf_r^2$ with $0 < p_i \in k$ and $f_i \in k(X_1, \ldots, X_n)$?

It will be shown in Section 4 that the solution of this problem is that M must be dense in R^n. In [15, Theorem 9] and [23] Hilbert's 17th problem was generalized to varieties. The same will be done here with the above version of the problem. Just in the same way as the results of Section 1 carry over to varieties (Section 3) the solution of the Hilbert problem carries over. Finally, there is one more direction in which Hilbert's 17th problem can be generalized, namely, the question can be posed for partially ordered fields ([19], [24]). The paper concludes with a few results in this direction.

1. THE STRONG TOPOLOGY OF R^n.

Let k be a formally real field with a fixed total order T and let R be the real closure of (k, T). The unique total order of R gives an interval topology such that R with this topology is a topological field. The <u>strong topology</u> of R^n is the product of the interval topologies on n copies of R. For $a, b \in R^n$ write $a < b$ if $a_i < b_i$ for all i. Then the open intervals $]a, b[= \{x \in R^n; \ a < x < b\}$ form a basis of the strong topology of R^n. In the special case where $k = \mathbb{Q}$ and R is the field of real algebraic numbers a basis of the strong topology can be given without using irrational numbers: Just take the intervals $]a, b[$ with $a, b \in \mathbb{Q}^n$. This, of course, depends on the fact that $\mathbb{Q}$ is dense in the field of real algebraic numbers. For an arbitrary ordered field k, it is also possible to describe the strong topology of R^n by using only elements from k if intervals are replaced by sets of negativity of polynomials:

DEFINITION. For $f_1, \ldots, f_r \in k[X_1, \ldots, X_n]$, call $N(f_1, \ldots, f_r) = \{x \in R^n;\ f_i(x) < 0\}$ for all i the <u>set of negativity of</u> $f_1, \ldots, f_r$.

THEOREM 1. For an ordered field (k, T) with real closure R the sets of negativity $N(f_1, \ldots, f_r)$ of polynomials in $k[X_1, \ldots, X_n]$ form a basis of the strong topology of R^n.

PROOF. Clearly, it suffices to prove this for $n = 1$. So, let $U \subset R$ be strongly open and pick $a \in U$. Let f be the minimal polynomial of a over k, and choose $0 < \varepsilon \in k$ such that a is the only root of f in $]a - 2\varepsilon, a + 2\varepsilon[$ and $]a - \varepsilon, a + \varepsilon[\subset U$. Let $a = a_0, a_1, \ldots, a_h$ be all the real roots of f. Now choose $0 < \alpha \in k$ small enough that

$$N(f_0) \subset \bigcup_{i=0}^{1}]a_i - \varepsilon, a_i + \varepsilon[$$

where $f_0 = f^2 - \alpha$. For each $i = 1, \ldots, h$, let $n_i \in \mathbb{N}$ be minimal with the property that there exists a polynomial g of degree n_i with a root between a_0 and a_i. Now, for $i = 1, \ldots, h$, let f_i be a polynomial with the following properties:

f_i is of degree n_i.

f_i has no roots in $]a_0 - \varepsilon, a_0 + \varepsilon[\cup]a_i - \varepsilon, ai + \varepsilon[$.

f_i has a root between a_0 and a_i.

$f_i(a_0) < 0$.

From this it follows that $f_i(x) < 0$ for $x \in]a_0 - \varepsilon, a_0 + \varepsilon[$ and $f_i(x) > 0$ for $x \in]a_i - \varepsilon, a_i + \varepsilon[$. Thus, $a \in N(f_0, \ldots, f_1) \subset U$. **

Again considering the special case $k = \mathbb{Q}$, R the field of real algebraic numbers one sees that the sets of negativity $N(f)$ of a single polynomial $f \in \mathbb{Q}[X]$ form a basis of the strong topology of R (let f run through the set of polynomials of the form $(X - a)(X - b)$ with $a, b \in \mathbb{Q}$). This raises the following question: For arbitrary ordered k with real closure R, when do the sets $N(f)$, $f \in k[X_1, \ldots, X_n]$, form a basis of the strong topology of R^n?

Of course, this question is also inspired by the strong approximation property (SAP) for spaces of orderings (see [2], [7], [8], [13], [14], [17]). The connections of spaces of orderings with the strong topology are discussed by Dubois in [4]. There, spaces of orderings and varieties with the strong topology are linked through the centering relation which, in fact, is a mapping in the present situation.

For any formally real extension field $K \supset k$, let T_K be the set of orderings extending the given total order of k. Also, for $a \in R^n$, let $P(a)$ be the set of points in R^n k-isomorphic to a.

THEOREM 2. For an ordered field (k, T) with real closure R and a point $a \in R^n$, the following are equivalent:

(i) The sets $P(a) \cap N(f)$, $f \in k[X_1, \ldots, X_n]$ form a basis of the strong (= discrete) topology of $P(a)$.

(ii) The sets $P(a) \cap N(f)$, $f \in k[X_1, \ldots, X_n]$ form the strong topology of $P(a)$.

(iii) In R^n every $b \in P(a)$ has a neighborhood basis of sets of the form $N(f)$, $f \in k[X_1, \ldots, X_n]$.

(iv) The field $k(a)$ with the set $T_{k(a)}$ of orderings has the SAP.

PROOF. Since $P(a)$ is finite, the equivalence of (i) and (ii) is obvious. Also, (iii) $\longrightarrow$ (i) is clear.

(iii) $\longrightarrow$ (iv): There is a natural bijection between $P(a)$ and $T_{k(a)}$. For $b \in P(a)$, let $T(b)$ be the total order of $k(a)$ corresponding to b. By (iii), for any $b \in P(a)$, there is some $f \in k[X_1, \ldots, X_n]$ such that $\{b\} = P(a) \cap N(f)$. Thus, $f(a) < 0$ $(T(b))$ and $f(a) > 0$ $(T(c))$ for $c \in P(a)$, $c \neq b$. This shows that $T_{k(a)}$ has the SAP (see [2, Section 5], [7, Section 3]).

(iv) $\longrightarrow$ (i): Using the natural bijection of $P(a)$ and $T_{k(a)}$ as in (iii) $\longrightarrow$ iv)), this is seen immediately.

(i) $\longrightarrow$ (iii): Again, let $b \in P(a)$ and choose a neighborhood U of b in R^n. By (i), there is some $f \in k[X_1, \ldots, X_n]$ with $f(b) < 0$ and $f(c) > 0$ for $c \in P(a)$, $c \neq b$. Pick $0 < \alpha \in k$ and choose neighborhoods U_c for $c \in P(a)$ subject to the following conditions:

$$U_b \subset U, \quad \text{and} \quad U_b \cap U_c = \emptyset \quad \text{for} \quad c \neq b.$$

$$\text{For} \quad c \neq b \quad \text{and} \quad x \in U_c, \quad 0 < f(x) < \alpha.$$

Let $U' = \{U_c;\ c \in P(a)\}$.

$P(a) \subset R^n$ is a variety of dimension 0. Let $M \subset k[X_1, \ldots, X_n]$ be the corresponding maximal real ideal ([6]). As a real ideal, M is generated by a single polynomial h_1. For all $i = 1, \ldots, n$, let g_i be the minimal polynomial of b_i in $k[X_i]$. Set $g = g_1^2 + \ldots + g_n^2$ and $h_2 = h_1 + g$. Then $P(a)$ is the set of zeros of h_2 in R^n and there is some $0 < \varepsilon \in k$ such that $h_2(x) > \varepsilon$ for $x \notin U'$. Define $h_3 = h_2 - \varepsilon$. Then $N(h_3) \subset U'$. The absolute minimum of h_3 is $-\varepsilon$, and h_3 takes this value precisely in the points of $P(a)$. Finally, define $h_4 = (f - \alpha)^2 h_3 + \varepsilon\alpha^2$. An easy computation shows that $b \in N(h_4) \subset U_b \subset U$. **

The following corollary is one of the main results of this paper. It is an immediate consequence of Theorem 2.

COROLLARY 3. For an ordered field (k, T) with real closure R the following are equivalent:

(i) The sets $N(f)$, $f \in k[X_1, \ldots, X_n]$ form a basis of the strong topology of R^n.

(ii) For every $a \in R^n$, the field $k(a)$ with the set $T_{k(a)}$ of orderings has the SAP.

(iii) For every formally real algebraic extension $k \subset K$, T_K has the SAP.

DEFINITION. Let K be a formally real field with a set $\underline{T}$ of orderings. For a field extension $K \subset L$, $\underline{T}_L$ denotes the set of orderings of L extending an ordering in $\underline{T}$. K is called a <u>hereditary SAP field</u> if $\underline{T}_L$ has the SAP for every algebraic field extension $K \subset L$. K is called a <u>hereditary non-SAP field</u> if $\underline{T}_L$ does not have the SAP for every finite algebraic extension $K \subset L$.

DEFINITION. Let (k, T) be a totally ordered field with real closure R. A point $a \in R^n$ is called an <u>SAP point</u> or a <u>non-SAP point</u> according as $T_{k(a)}$ does or does not have the SAP.

The notions "hereditary SAP field" and "hereditary non-SAP field" are extremely important in determining the distribution of SAP points and non-SAP points in R^n. This will be taken up in some detail in Section 2.

Clearly, there exist totally ordered fields that are not hereditary SAP fields. The field $\mathbb{Q}(X, Y)$ $(X, Y$ indeterminates$)$ with the total order such that $1 \ll X \ll Y$ ("$\ll$" means infinitely less than) is a case in point. Using the criterion of Becker and Köpping ([2, Korollar zu Satz 23]), it is not hard to see that the set of extensions of the above total order of $\mathbb{Q}(X, Y)$ to total orders of $\mathbb{Q}(\sqrt{X}, \sqrt{Y})$ does not have the SAP. Thus, the equivalent conditions of Corollary 3 do not always hold. But a weaker statement does always hold:

THEOREM 4. Let (k, T) be an ordered field with real closure R. Then, for any nonempty strongly open $U \subset R^n$, there exists some $f \in k[X_1, \ldots, X_n]$ such that $\emptyset \neq N(f) \subset U$.

PROOF. First, suppose the theorem is true for $n = 1$. It will be shown that this implies the claim for arbitrary n: Let $a = (a_1, \ldots, a_n) \in U$ be any point and choose $0 < \Delta = (\delta, \ldots, \delta) \in k^n$ such that the interval $]a - \Delta, a + \Delta[$ is contained in U. By hypothesis, for each $i = 1, \ldots, n$, choose a polynomial $f_i \in k[X_i]$ with $\emptyset \neq N(f_i) \subset]a_i - \delta/2, a_i + \delta/2[$. Then there is some $0 < \varepsilon \in k$ such that $\varepsilon < f_i(x_i)$ for all i and all $x_i \notin]a_i - \delta, a_i + \delta[$. If $f = f_1^2 + \ldots + f_n^2 \in k[X_1, \ldots, X_n]$, then $f(x) > \varepsilon$ for all $x \notin]a - \Delta, a + \Delta[$ and $f(x) = 0$ for some $x \in]a - \Delta, a + \Delta[$. Now $\emptyset \neq N(f - \varepsilon) \subset U$.

Now the theorem will be proved for the case $n = 1$: Pick any $a \in U$ and some $0 < \varepsilon_0 \in k$ such that $]a - \varepsilon_0, a + \varepsilon_0[\subset U$. If $]a - \varepsilon_0, a + \varepsilon_0[\cap k \neq \emptyset$ then there are $\alpha, \beta \in k, \alpha \neq \beta$ in this intersection. In this case, $(X - \alpha)(X - \beta)$ is the desired polynomial. So, suppose that $]a - \varepsilon_0, a + \varepsilon_0[\cap k = \emptyset$. Let f_0 be the minimal polynomial of a over k and let $a_{-r} < \ldots < a_{-1} < a = a_0 < a_1 < \ldots < a_s$ be all the roots of f_0 in R. Choose some $\varepsilon_1 \in k$, $0 < \varepsilon_1 < \varepsilon_0$ such that $a_i + 2\varepsilon_1 < a_{i+1} - 2\varepsilon_1$ for all $i = -r, \ldots, s - 1$. Set $I_i =]a_i - \varepsilon_1, a_i + \varepsilon_1[$ for all $i = -r, \ldots, s$. Now let $0 < \delta_0 \in k$ be small enough that $\emptyset \neq N(f_0^2 - \delta_0) \subset \bigcup_{i=-r}^{s} I_i$. Define n_{-r} as the smallest natural number such that there is a polynomial of degree n_{-r} in $k[X]$ with a root in the interval $]a_{-r} + \varepsilon_1, a_0 - \varepsilon_1[$. Pick $g_1 \in k[X]$ subject to the following conditions:

g_1 is of degree n_{-r}.

g_1 has a root in $]a_{-r} + \varepsilon_1, a_0 - \varepsilon_1[$.

$g_1(a_0) > 0$.

Then $g_1(x) < 0$ for all $x \in I_{-r}$. Pick $0 < \mu_1 \in k$ such that $-\mu_1 < g_1(x)$ for all $x \in I_{-r}$. Define $f_1 = (g_1 + \mu_1)^2(f_0^2 - \delta_0) + \delta_0\mu_1^2$. This polynomial has real roots (since $f_1(a_0) < 0$ and $f_1(a_{-r}) > 0$) all of which are in $\bigcup_{i=-r+1}^{s} I_i$ (since $f_1(x) > \delta_0\mu_1^2 > 0$ for $x \notin \bigcup_{i=-r}^{s} I_i$ and $f_1(x) > 0$ for $x \in I_{-r}$). Next, choose $0 < \delta_1 \in k$ small enough that $\emptyset \neq N(f_1^2 - \delta_1) \subset \bigcup_{i=-r+1}^{s} I_i$. An inductive application of the above argument finishes the proof.

2. HEREDITARY SAP FIELDS

Corollary 3 of Section 1 clearly indicates that it is of some interest to take a closer look at hereditary SAP fields. As it will turn out in this section, this will also provide some insight into the distribution of SAP points and non-SAP points in R^n. The first result is a very useful criterion for a field to have the hereditary SAP. This criterion is inspired (and makes use of) the valuation theoretic characterizations of SAP fields by Becker and Köpping [2, Korollar zu Satz 23] and Prestel [18, Satz (2.2)].

A set $\underline{T}$ of total orders on the formally real field k is called <u>saturated</u> if $T \in \underline{T}$ for any total order T of k with $T \supset \cap \underline{T}$. For a field extension $k \subset K$, $\underline{T}_K$ is the set of total orders of K extending one of the orderings in $\underline{T}$.

THEOREM 5. For the formally real field k, partially ordered by $T = \cap \underline{T}$ with $\underline{T}$ a saturated set of total orders, the following are equivalent:

(i) k with $\underline{T}$ is a hereditary SAP field.

(ii) For every real place $\lambda : k \longrightarrow \overline{k} \cup \{\infty\}$ with valuation v, value group G and $\overline{T} \neq \overline{k}$ ($\overline{T}$ is the image of T under λ in $\overline{k}$) either

(a) $[G : 2G] = 1$ or

(b) $[G : 2G] = 2$ and $\overline{k}$ is hereditarily euclidean.

PROOF. <u>(ii) $\longrightarrow$ (i)</u>: Let $K = k(\alpha) \supset k$ be a finite extension and $T_K = \cap \underline{T}_K$. The criterion of Becker and Köpping ([2, Korollar zu Satz 23]) will be used to prove that $\underline{T}_K$ has the SAP.

Suppose, $w : K^* \longrightarrow H$ is a real valuation with corresponding real place $\mu : K \longrightarrow \overline{K} \cup \{\infty\}$ such that $\overline{T_K} \neq \overline{K}$. Let $v : k^* \longrightarrow G$ be the restriction of w. Then also $\overline{T} \neq \overline{k}$. Since H is torsion free, the mapping $H \longrightarrow 2H : h \longmapsto 2h$ is an isomorphism and, therefore,

$[H:G] = [2H:2G]$. Moreover, $[H:G][G:2G] = [H:2H][2H:2G]$ so that $[H:2H] = [G:2G] \leqslant 2$ (by (ii)). Since $2H \subset w(T_K^*)$ it follows that $[H:w(T_K^*)] \leqslant 2$. In case $[H:w(T_K^*)] = 1$ all is done. So, suppose $[H:w(T_K^*)] = 2$. Then also $[G:2G] = 2$ and (by (b)) $\overline{k}$ is hereditarily eculidean. But then $\overline{K}$ must be euclidean. Hence $2 \leqslant [\overline{K}^* : \overline{T_K}^*] \leqslant [\overline{K}^* : \overline{K}^{*2}] = 2$.

(i) $\longrightarrow$ (ii): Suppose, $v : k^* \longrightarrow G$ is a real valuation with $\overline{T} \neq \overline{k}$ and $[G:2G] \geqslant 4$. Pick $g_1, g_2 \in G \backslash 2G$ such that $g_1 + g_2 \notin 2G$. Let $\overline{T_1}$ be any total order of $\overline{k}$ extending the partial order $\overline{T}$. By the construction of [17, Theorem (7.9)] there is a total order T_1 of k making the place belonging to v order compatible with respect to T_1 on k and $\overline{T_1}$ on $\overline{k}$ and such that $T_1 \in \underline{T}$. Since $v(T_1^*) = G$, it is possible to pick $x_1, x_2 \in T_1$ with $v(x_i) = g_i$. Now let $K = k(\sqrt{x_1}, \sqrt{x_2})$. By the choice of the g_i and x_i, $[K:k] = 4$ and T_1 has four extensions S_1, S_2, S_3, S_4 to total orders of K. Let $S = S_1 \cap S_2 \cap S_3 \cap S_4$. Let $w : K^* \longrightarrow H$ be a real extension of v which is order compatible with S_1 Then $\overline{S} \neq \overline{K}$ with respect to the real place belonging to w. Clearly, $[H:G] = 4$. Now the fundamental inequality for extensions of valuations ([10, 17.5]) shows that w is in fact the only extension of v to K (and hence is order compatible with all S_i, $i = 1, \ldots, 4$) and that $\overline{K} = \overline{k}$ and $\overline{S} = \overline{T_1}$. Eventually, it will turn out that K with the total orders $S_1, \ldots, S_4$ does not have the SAP. But then K with $\underline{T}_K$ cannot have the SAP either, and this part of the proof is complete. To see this, it will be shown that $w(S^*) = G$. By [2, Korollar zu Satz 23], the claim then follows from $[H:G] = 4$. So, assume (by way of contradiction) that $w(S^*) \supsetneqq G$. Then pick $y \in S^*$ such that $w(y) = g \notin G$. Since S is the intersection of all the total orders of K extending T_1, it follows from [1, Satz 2] that $y = t_1 y_1^2 + \ldots + t_r y_r^2$ with $t_i \in T_1$ and $y_i \in K$. Since w is compatible with all the S_i, this shows that $w(y) \geqslant w(t_i y_i^2) = w(t_i) + 2w(y_i)$ for all $i = 1, \ldots, r$. On the other hand, $w(y) \leqslant \max\{w(t_i y_i^2);\ i = 1, \ldots, r\}$. Therefore, $w(y) = w(t_i) + 2w(y_i)$ for some i. The exponent of H/G is 2. Thus, $2w(y_i) \in G$. Also, by the choice of t_i, $w(t_i) \in G$. Now this gives the desired contradiction, since y was chosen so that $w(y) \notin G$. Altogether, this shows that $[G:2G] \leqslant 2$.

Now suppose that $v : k^* \longrightarrow G$ is a real valuation with $\overline{T} \neq \overline{k}$ and $[G:2G] = 2$ and such that $\overline{k}$ is not hereditarily euclidean.

First assume that $\overline{T}$ is not a total order of $\overline{k}$. Since k with $\underline{T}$ is a hereditary SAP field, $\underline{T}$ in particular has the SAP. Then by [2, Korollar zu Satz 23], $[G : v(T^*)] = 1$. Pick $g \in G \backslash 2G$ and $t \in T^*$ with $v(t) = g$. v can be extended uniquely to a valuation $w : k(\sqrt{t})^* \longrightarrow H$ ([10, 17.5]). Then $[H : G] = 2$ and the residue field of w is $\overline{k}$. Since $\overline{T}$ is not a total order of $\overline{k}$ there are total orders $\overline{S}_1$, $\overline{S}_2$ of $\overline{k}$, $\overline{S_1} \neq \overline{S_2}$ containing $\overline{T}$. Again by Prestel's construction ([17, Theorem (7.9)]) there are total orders S_1, $S_2 \in \underline{T}$ of k such that the place belonging to v is order compatible with respect to the orderings S_i of k and $\overline{S_i}$ of $\overline{k}$ $(i = 1, 2)$. Let S_{11}, S_{12} and S_{21}, S_{22} be the extensions of S_1 resp. S_2 to total orders of $k(\sqrt{t})$, and let S be the intersection of all of these. Clearly, w is compatible with every S_{ij} $(i, j = 1, 2)$. Assume (by way of contradiction) that $w(S^*) = H$. Then there is some $s \in S$ with $w(s) = w(\sqrt{t}) \in H \backslash G$. Now $s = t_1x_1^2 + \ldots + t_rx_r^2$ with $t_i \in S_1 \cap S_2$, $x_i \in k(\sqrt{t})$ ([1, Satz 2]). Since w is order compatible this implies $w(s) \geqslant w(t_ix_i^2)$ for all $i = 1, \ldots, r$. On the other hand, $w(s) \leqslant \max\{w(t_ix_i^2);\ i = 1, \ldots, r\}$ so that $w(s) = w(t_ix_i^2)$. But $w(t_ix_i^2) \in G$ (since $[H : G] = 2$) and this contradicts the choice of s. Thus, $w(S) = G$ has been proved. Moreover, the partial order $\overline{S}$ of $\overline{k}$ is not a total order (since $\overline{S} \subset \overline{S_1} \cap \overline{S_2}$). Now, by [2, Korollar zu Satz 23], $k(\sqrt{t})$ with the total orders S_{ij}, $i, j = 1, 2$ does not have the SAP. Then $k(\sqrt{t})$ with $\underline{T}_{k(\sqrt{t})}$ cannot have the SAP either. This contradicts the hypothesis that k with $\underline{T}$ be a hereditary SAP field, and $\overline{T}$ must be a total order of $\overline{k}$.

Since $\overline{k}$ is not hereditarily euclidean by assumption, there exists an extension $\overline{k} \subset \overline{k}(a)$ such that the total order $\overline{T}$ of $\overline{k}$ has two different extensions $\overline{S_1}$, $\overline{S_2}$ to $\overline{k}(a)$. Let $k \subset k(b)$ be an extension with the following properties:

$$[k(b) : k] = [\overline{k}(a) : \overline{k}],$$

v can be extended to a valuation $w : k(b)^* \longrightarrow H$ such that $\lambda_w(b) = a$, where λ_w is the place belonging to w.

Then the value groups G and H are equal ([10, 17.5]). As before, by Prestel's construction ([17, Theorem (7.9)]), there are total orders S_1, $S_2 \in \underline{T}_{k(b)}$ such that w is order compatible with respect to S_i on $k(b)$ and $\overline{S_i}$ on $\overline{k}(a)$. Let $S = \cap \underline{T}_{k(b)}$. Then $\overline{S} \subset \overline{S_1} \cap \overline{S_2}$ so that $\overline{S}$ is not a total order for $\overline{k}(a) = \overline{k(b)}$. But $k(b)$ with $\underline{T}_{k(b)}$ is a hereditary SAP field, and, therefore, this gives the same contradiction as the assumption that $\overline{T}$ is not a total order on $\overline{k}$.

**

For the special case of a field $k = k_1(x, y)$ with x transcendental over k_1, y algebraic over $k_1(x)$, the criterion of Theorem 5 has the following simpler form ([18, Satz (3.6)]). The following statements (i) and (ii) are equivalent:

(i) k is a hereditary SAP field with respect to the set of all total orders.

(ii) k_1 is hereditarily euclidean.

Next, the behaviour of the hereditary SAP under algebraic field extensions will be examined. For this, the following lemma turns out to be quite useful:

LEMMA 6. Let k be a totally ordered field with a saturated set $\underline{T}$ of total orders. Then the following are equivalent:

(i) $\underline{T}$ has the hereditary SAP.

(ii) For every $T \in \underline{T}$, k with the single total order T is a hereditary SAP field.

PROOF. <u>(i) ⟶ (ii)</u>: Suppose that, for some $T \in \underline{T}$, (k, T) is not a hereditary SAP field. Then there is a finite extension $k \subset K$ such that the set T_K of total orders of K does not have the SAP. Then $\underline{T}_K$ cannot have the SAP either. Contradiction.

<u>(ii) ⟶ (i)</u>: Assume (by way of contradiction) that $\underline{T}$ does not have the hereditary SAP. Then there is some real valuation $v : k^* \longrightarrow G$ with $\overline{T} \neq \overline{k}$ (where $T = \cap \underline{T}$) such that $[G : 2G] \geqslant 4$ or $[G : 2G] = 2$ and $\overline{k}$ is not hereditarily euclidean. In case $[G : 2G] \geqslant 4$ or $[G : 2G] = 2$ and $\overline{T}$ is a total order of $\overline{k}$ and $\overline{k}$ is not hereditarily euclidean, it has been shown in the proof of Theorem 5 that there is a total order S of k and a finite extension $k \subset K$ such that the set S_K of total orders of K does not have the SAP. But then (k, S) cannot be a hereditary SAP field. Contradiction.

Now it only remains to consider the case that $[G : 2G] = 2$ and $\overline{T}$ is not a total order of $\overline{k}$. Choose any total order $S \in \underline{T}$ compatible with v. Then there is some $x \in S$ with $\overline{x} = \lambda_v(x) \in \overline{k}\backslash\overline{k}^2$ (where λ_v is the place belonging to v). Let $S_1 \neq S_2$ be the extensions of S to total orders of $k(\sqrt{x})$. By [10, 17.5], v can be extended uniquely to a valuation w of $k(\sqrt{x})$. The value group of w is G, the residue field is $\overline{k}(\sqrt{\overline{x}})$ and w is compatible with S_1 and S_2. By hypothesis, (k, S) is a hereditary SAP field. But then, so is $k(\sqrt{x})$ with the total orders S_1, S_2. Now the proof of Theorem 5 shows that $\overline{S_1 \cap S_2}$ must be a total order on $\overline{k(\sqrt{x})} = \overline{k}(\sqrt{\overline{x}})$. This is impossible since $\sqrt{\overline{x}} > 0$ $(\overline{S_1})$ and $\sqrt{\overline{x}} < 0$ $(\overline{S_2})$. **

THEOREM 7. Suppose, k is a formally real field with a saturated set $\underline{T}$ of total orders, and $k \subset K$ is a finite field extension. Let $\rho : \underline{T}_K \longrightarrow \underline{T}$ be the restriction map. Then the following are equivalent:

(i) $\underline{T}_K$ has the hereditary SAP.

(ii) For all $T \in \rho(\underline{T}_K)$, (k, T) is a hereditary SAP field.

PROOF. (ii) $\longrightarrow$ (i): Obvious, by Lemma 6. (i) $\longrightarrow$ (ii): Assume that there is some $T \in \rho(\underline{T}_K)$ such that (k, T) does not have the hereditary SAP. By Theorem 5, there is some real valuation $v : k^* \longrightarrow G$ with $\overline{k} \neq \overline{T}$ and either

(a) $[G : 2G] \geqslant 4$ or

(b) $[G : 2G] = 2$ and $\overline{k}$ is not hereditarily euclidean.

CASE (a). Let w be an extension of v to a real valuation of K which is order compatible with some $S \in \rho^{-1}(T)$. Let H be the value group of w. Then $[H : 2H] = [G : 2G] \geqslant 4$, so that $\underline{T}_K$ does not have the hereditary SAP by Theorem 5.

CASE (b). As in Case (a), extend v to a valuation w of K with value group H and residue field $\overline{K}$, order compatible with $S \in T_K = \rho^{-1}(T)$. Let $R = \cap T_K$. Then, since T_K has the hereditary SAP (by Lemma 6), $\overline{K}$ is hereditarily euclidean and $\overline{R}$ is the unique total order of $\overline{K}$ (Theorem 5). Since $\overline{K}$ is a finite extension of $\overline{k}$, [14, Proposition 10.4 (d)] shows that $\overline{k}$ must also be hereditarily euclidean. Contradiction. **

COROLLARY 8. Let $\underline{T}$ be a saturated set of total orders on the formally real field k. As a subset of the set of all total orders X_k of k, $\underline{T}$ carries the topology induced by the Harrison topology of X_k ([17, §6]). The following are equivalent:

(i) There exists a finite extension $k \subset K$ such that $\underline{T}_K$ has the hereditary SAP.

(ii) The set $\underline{T}_0 = \{T \in \underline{T};\ (k, T)$ is a hereditary SAP field$\}$ has a nonempty interior in $\underline{T}$.

PROOF. (ii) $\longrightarrow$ (i): There are $a_1, \ldots, a_n \in k$ such that the basic open set $H(a_1, \ldots, a_n) \cap \underline{T} = \{T \in \underline{T};\ a_1, \ldots, a_n < 0\ (T)\}$ of the Harrison topology is contained in $\underline{T}_0$. Let $K = k(\sqrt{-a_1}, \ldots, \sqrt{-a_n})$. Then (with ρ as in Theorem 7) $\rho(\underline{T}_K) \subset \underline{T}_0$ and, by Theorem 7, $\underline{T}_K$ has the hereditary SAP.

<u>(i) $\longrightarrow$ (ii)</u>: By Theorem 7, $\rho(\underline{T}_K) \subset \underline{T}_0$. But $\rho(\underline{T}_K)$ is an open subset of $\underline{T}$ ([9, Theorem 4.4]).

For the case of a finite set $\underline{T}$, statement (ii) of Corollary 8 has a somewhat simpler form: $\underline{T}_0 \neq \emptyset$. The next corollary is a restatement of Corollary 8 for the special case that $\underline{T}$ consists of a single total order. This is closely related to [14, Proposition 10.4].

COROLLARY 9. For the totally ordered field (k, T) the following are equivalent:

(i) (k, T) is a hereditary SAP field.

(ii) There exists a finite extension $k \subset K$ such that K with T_K has the hereditary SAP.

The above results on hereditary SAP fields are complemented by the following theorem which deals with hereditary non-SAP fields. This is a key result for the investigation of the distribution of SAP points and non-SAP points in R^n.

THEOREM 10. Let (k, T) be an ordered field which is not a hereditary SAP field. Then there is a finite extension $k \subset K$ which is a hereditary non-SAP field with respect to T_K.

PROOF. Since (k, T) is not a hereditary SAP field there exists a real valuation $v : k^* \longrightarrow G$ which is order compatible with T (i.e., $\overline{T}$ is a total order on the residue field $\overline{k}$) such that

(a) $[G : 2G] \geqslant 4$ or

(b) $[G : 2G] = 2$ and $\overline{k}$ is not hereditarily euclidean

(Theorem 5).

CASE (a). As in the proof of Theorem 5, pick $g_1, g_2 \in G\backslash 2G$ linearly independent mod $2G$ and $x_1, x_2 \in T$ with $v(x_i) = g_i$ $(i = 1, 2)$. Let $K = k(\sqrt{x_1}, \sqrt{x_2})$, and let T_1, T_2, T_3, T_4 be the different extensions of T to total orders of K. Let $w : K^* \longrightarrow H$ be the unique extension ([10, 17.5]) of v to a valuation of K. Then w is compatible with each T_i. There are $h_1, h_2 \in H$ such that $2h_i = g_i$. In particular, $[H : G] = 4$. The residue fields of v and w are the same.

Now let $K \subset L$ be any finite extension such that the set $(T_K)_L = T_L$ of total orders of L is nonempty. Let $S = \cap\, T_L$. Let $u : L^* \longrightarrow F$ be a real valuation order compatible with some $R \in T_L$, extending v and w. Then L with T_L does not have the SAP if $[F : u(S^*)] \geqslant 4$ ([2, Korollar zu Satz 23]). To show this, define an ascending sequence of subgroups of F by

$$F_t = \{\alpha \in F;\ 2^t\alpha \in H\} \quad (t \in \mathbb{N}_0).$$

Clearly, $F_0 = H$, $F_t \subset F_{t+1}$ for all t, and there is some minimal $m \in \mathbb{N}_0$ with $F_m = F_{1+m}$ (since $[F : H] < \infty$). Then $F_1 = F_{1+s}$ for all $s \in \mathbb{N}$. Let H_t be the image of

$$\phi_t : F_t \longrightarrow H : \alpha \longrightarrow 2^t\alpha \ (t \in \mathbb{N}_0).$$

Then $H = H_0 \supset H_1 \supset \dots$ is a descending sequence, and $H_{1+m} \subset 2H \subset G = w(T_0^*)$. ($T_0 = T_1 \cap T_2 \cap T_3 \cap T_4$. $G = w(T_0^*)$ is seen just as in the proof of Theorem 5.) For $h_1, h_2, h_3, = h_1 + h_2$, let $n_i \in \mathbb{N}_0$ be maximal with the property that $(h_i + G) \cap H_{n_i} \neq \emptyset$. Clearly, $n_i \leqslant 1$ for all i. For $i = 1, 2, 3$, pick $f_i \in F_{n_i}$ such that $2^{n_i}f_i = h_i + g_i'$ for some $g_i' \in G$. It will be shown that two elements of the set $\{f_1, f_2, f_3\}$ are linearly independent in $F \bmod u(S^*)$, which then concludes the proof of $[F : u(S^*)] \geqslant 4$.

In a first step it is shown that $f_1, f_2, f_3 \notin u(S^*)$. Assume (by way of contradiction) that $f_i \in u(S^*)$, say $f_i = u(s)$. By [1, Satz 2], $s = p_1s_1^2 + \dots + p_rs_r^2$ with $0 < p_i \in k$ and $s_i \in L$. By compatibility of u with $R \in T_L$, $u(s) = \max\{u(p_js_j^2);\ j = 1, \dots, r\}$. Assume without loss of generality that $u(s) = u(p_1) + 2u(s_1)$. Since $f_i \in F_{n_i} \setminus F_{n_i-1}$ and $u(p_1) \in G$, $f_i - u(p_1) = 2u(s_1) \in F_{n_i} \setminus F_{n_i-1}$ so that $u(s_1) \in F_{n_i+1} \setminus F_{n_i}$. But then

$$2^{n_i+1}u(s_1) = 2^{n_i}(f_i - u(p_1)) = h_i + g_i' - 2^{n_i}u(p_1) \in (h_i + G) \cap H_{n_i+1},$$

contradicting the choice of n_i.

Now suppose that $n_1 < n_2$ and assume that $f_1 + f_2 \in u(S^*)$. With the same argument as above, $f_1 + f_2 = u(p) + 2u(s)$ for suitable $0 < p \in k$ and $s \in L$. Then

$$2^{n_2}(f_1 + f_2) = 2^{n_2-n_1}(h_1 + g_1') + h_2 + g_2' = 2^{n_2}u(p) + 2^{n_2+1}u(s),$$

where

$$2^{n_2-n_1}(h_1+g_1') \in 2H \subset G, \quad g_2' \in G, \quad 2^{n_2}u(p) \in G.$$

Thus, $h_2 + g = 2^{n_2+1}u(s)$ for some $g \in G$, i.e., $(h_2+G) \cap H_{n_2+1} \neq \emptyset$. This contradicts the choice of n_2.

Finally, suppose that $n_1 = n_2 = n$ and assume that $f_1 + f_2 \in u(S^*)$. As before, it follows that $f_1 + f_2 = u(p) + 2u(s)$ for some $0 < p \in k$ and $s \in L$. This implies

$$2^n(f_1+f_2) = h_1 + h_2 + g_1' + g_2' = 2^n u(p) + 2^{n+1}u(s)$$

with g_1', g_2', $2^n u(p) \in G$. Therefore, $(h_3+G) \cap H_{n+1} \neq \emptyset$ and $n_3 > n$. Now the same argument as in the case "$n_1 < n_2$" shows that f_1 and f_3 are linearly independent mod $u(S^*)$.

CASE (b). Let $\overline{K}$ be a finite extension of $\overline{k}$ admitting two different extensions of $\overline{T}$ to total orders of $\overline{K}$. Choose a field extension k $k \subset K'$ with $[K' : k] = [\overline{K} : \overline{k}]$ and such that v can be extended uniquely to a real valuation $w' : K'^* \longrightarrow G$ with residue field $\overline{K}$. Let $T' = \cap T_{K'}$. Then $G = w'(T'^*) = v(T^*)$. Pick $g \in G \setminus 2G$ and $x \in T$ with $v(x) = g$, and set $K = K'(\sqrt{x})$. Every element of $T_{K'}$ can be extended to a total order of K. w' has a unique extension $w : K^* \longrightarrow H$ which is order compatible with every total in T_K. Clearly, $[H : G] = 2$. As in the proof of Theorem 5, it is easy to see that $w(S^*) = G = 2H$ where $S = \cap T_K)$. The residue field of w is $\overline{K}$, and, by the construction of K and T_K, $[\overline{K}^* : \overline{S}^*] \geqslant 4$.

Now let $K \subset L$ be any finite extension with $T_L = \{R_1, \ldots, R_m\} \neq \emptyset$ and set $R_0 = R_1 \cap \ldots \cap R_m$. Extend w to a real valuation $u : L^* \longrightarrow F$ compatible with some R_i. Again, as in the proof of Theorem 5, $[F : 2F] = 2$ and $F \supsetneqq u(R_0^*) = 2F$. Let $\overline{T_1}$ be the total order induced on $\overline{K}$ by the ordering $\overline{R_i}$ of $\overline{L}$. Pick a total order $\overline{T_2} \neq \overline{T_1}$ of $\overline{K}$ which extends the ordering $\overline{T}$ of $\overline{k}$. Then $\overline{S} \subset \overline{T_1} \cap \overline{T_2}$. Let $\overline{f}$ be the minimal polynomial of some primitive element of $\overline{L}$ over $\overline{K}$. If the degree of $\overline{f}$ is odd, both $\overline{T_1}$ and $\overline{T_2}$ can be extended to $\overline{L}$. If the degree of $\overline{f}$ is even, $\overline{f}$ has at least two different roots in the real closure of $(\overline{K}, \overline{T_1})$. Thus, $\overline{T_1}$ can be extended to $\overline{L}$ in at least two different ways. Thereby it has been shown that, in any case, $\overline{L}$ has two different total orders $\overline{S_1}$, $\overline{S_2}$ extending the ordering $\overline{T}$

of $\bar{k}$. By Prestel's construction ([17, Theorem (7.9)]), there are total orders $S_1, S_2 \in T_L$ on L which make the place belonging to u order compatible with respect to S_i on L and $\overline{S_i}$ on $\bar{L}$. From $S_1 \cap S_2 \supset R_0$ it follows that

$$[\bar{L}^* : \overline{R_0}^*] \geqslant [\bar{L}^* : \overline{S_1 \cap S_2}^*] \geqslant [\bar{L}^* : \overline{S_1}^* \cap \overline{S_2}^*] = 4.$$

By [2, Korollar zu Satz 23], this shows that L with T_L cannot have the SAP. **

Finally in this section, the above general results on hereditary SAP fields and hereditary non-SAP fields can be applied very easily to give an impression of the distribution of SAP points and non-SAP points in R^n, where R is the real closure of the ordered field (k, T). First note that always there are some SAP points, namely at least the elements of k^n.

THEOREM 11. Let (k, T) be an ordered field with real closure R. If k is not a hereditary SAP field, then for all $i = 1, \ldots, n$ and all $a_i, b_i \in R$, $a_i < b_i$ there is some $c_i \in R$, $a_i < c_i < b_i$ such that every point on the hyperplane $\{x \in R^n;\ x_i = c_i\}$ is a non-SAP point.

PROOF. By Theorem 10, there is a finite extension $k \subset K$ such that K with the set T_K of total orders is a hereditary non-SAP field. Then $K' = K(a_i, b_i)$ with $T_{K'}$ is a hereditary non-SAP field as well. Since K' is finite over k, the set of primitive elements of K' over k is dense in K'. Pick any primitive element c_i of K' over k with $a_i < c_i < b_i$. Then, for any $x \in R^n$ with $x_i = c_i$, $k(x_1, \ldots, x_n) \supset K'$ is a finite extension, and, by the definition of hereditary non-SAP fields, $k(x_1, \ldots, x_n)$ does not have the SAP. Thus, x is a non-SAP point. **

3. THE STRONG TOPOLOGY ON REAL ALGEBRAIC VARIETIES

The object of this section is to show that the results obtained so far for the special variety R^n can be extended almost completely to arbitrary real algebraic varieties. In many places, the following restrictions are required: The varieties considered must have positive dimension. And, attention must be restricted from the whole variety to the set of central points ([4]). Because of this second restriction it is very useful to have the following alternative description of the central points at hand ([4, Theorem 6]).

LEMMA 12. Let (k, T) be an ordered field with real closure R. Let $V \subset R^n$ be the set of zeros of the real prime ideal $P \subset k[X_1, \ldots, X_n]$.

Let V_c be the set of central points of V, V_s the set of all points every strong neighborhood of which is Zariski-dense in V, V_1 the strong closure of the set of simple points of V. Then $V_c = V_s = V_1$.

PROOF. $V_1 \subset V_c$: [5, Theorem 3]. $V_1 = V_s$: [5, Remark at the end]. $V_c \subset V_s$: The argument of [4, Theorem 6] works here, since it depends solely on the fact that the sets of negativity $N(f_1, \ldots, f_r)$ form a basis of the strong topology of V, which, of course, carries over from R^n (Theorem 1). **

In Section 1 it was shown that the sets of negativity $N(f)$ of single polynomials form a basis of the strong topology of R^n if and only if (k, T) is a hereditary SAP field. Clearly, if (k, T) is a hereditary SAP field, then the sets $N(f)$ form a basis of the strong topology on any variety. Now the converse of this statement will be examined. This will be done in three steps each of which is of some interest in its own right. The first step is an implicit function theorem:

THEOREM 13. Suppose that (k, T) is an ordered field with real closure R, and that $V \subset R^n$ is the variety belonging to the real prime ideal $P \subset k[X_1, \ldots X_n]$. Suppose $\dim V = r > 0$. Let $a \in V$ be a central point. Then (after renumbering the coordinates if necessary) in every neighborhood U of a there is some $b \in V$ and there are a neighborhood U^r of $(b_1, \ldots, b_r)$ in R^r and neighborhoods U^i of b_i in R $(i = r+1, \ldots, n)$ and continuous functions $\phi_i : U^r \longrightarrow U^i$ $(i = r+1, \ldots, n)$ such that

$$V \cap (U^r \times U^{r+1} \times \ldots \times U^n) = \{(x_1, \ldots, x_r, \phi_{r+1}(x_1, \ldots, x_r), \ldots, \phi_n(x_1, \ldots, x_r); (x_1, \ldots, x_r) \in U^r\}.$$

PROOF. Let $k[x_1, \ldots, x_n]$ be the coordinate ring of V and suppose that $x_1, \ldots, x_r$ are algebraically independent over k. Pick any $y \in k(x_1, \ldots, x_n)$ such that $k[x_1, \ldots, x_n] \subset k[x_1, \ldots, x_r, y]$. Then, for $i = r+1, \ldots, n$, there are polynomials $p_i \in k[X_1, \ldots, X_r, Y]$ such that $x_i = p_i(x_1, \ldots, x_r, y)$. And

$$y = \Sigma \frac{q_{i(r+1) \ldots i(n)}(x_1, \ldots, x_r)}{r_{i(r+1) \ldots i(n)}(x_1, \ldots, x_r)} x_{r+1}{}^{i(r+1)} \ldots x_n{}^{i(n)}$$

with $q_{i(r+1) \ldots i(n)}, r_{i(r+1) \ldots i(n)} \in k[X_1, \ldots, X_r]$. Let F be the irreducible polynomial of y over $k[x_1, \ldots, x_r]$. Now let $U \subset R^n$ be a neighborhood of a which will be fixed for the rest of this proof. Since none of the $r_{i(r+1)\ldots i(n)}$ is in P and by $V_c = V_s = V_1$ (Lemma 12), there is some $c \in V_c$ and a neighborhood U'^r of $c' = (c_1, \ldots, c_r)$ such that $r_{i(r+1)\ldots i(n)}(z') \neq 0$ for all $z' \in U'^r$ and all $(i(r+1), \ldots, i(n))$. Let W be the hypersurface in R^{r+1} determined by F. Then, by the choice of U'^r,

$$\Phi : W_c \cap (U'^r \times R) \longrightarrow V_c \cap (U'^r \times R^{n-r})$$
$$(z', t) \longmapsto (z', p_{r+1}(z', t), \ldots, p_n(z', t))$$

is a bijective mapping. Moreover, Φ is continuous and open. By Lemma 12, there is some open $U' \subset U \cap (U'^r \times R^{n-r})$ such that every point of $V \cap U'$ is simple. Now, by invoking Lemma 12 for W, it is easy to see that there is some open $U'' \subset U'^r \times R$ such that $W \cap U''$ consists entirely of simple points and $\Phi(W_c \cap U'') \subset V_c \cap U'$. Let $\Phi' : W_c \cap U'' \longrightarrow \Phi(W_c \cap U'')$ be the restriction of Φ. Pick $b' \in R^r$ and $\overline{b} \in R$ arbitrarily such that $(b', \overline{b}) \in W_c \cap U''$. Then, by [20, Theorem (7.4)], there is a neighborhood $U^r \subset R^r$ of b' and a neighborhood $\overline{U} \subset R$ of $\overline{b}$ such that $U^r \times \overline{U} \subset U''$ and such that there is a continuous function $\phi : U^r \longrightarrow \overline{U}$ with

$$W \cap (U^r \times \overline{U}) = W_c \cap (U^r \times \overline{U}) = \{(z', \phi(z')); \ z' \in U^r\}.$$

Since Φ' is a homeomorphism one may assume without loss of generality that $\Phi'(W \cap (U^r \times \overline{U})) = V \cap (U^r \times U^{r+1} \times \ldots \times U^n)$ for neighborhoods $U^i \subset R$ of $p_i(b', \overline{b}) = b_i$ $(i = r+1, \ldots, n)$. Now define $\phi_i : U^r \longrightarrow U^i$ by $p_i(z', \phi(z')) = \phi_i(z')$ for $i = r+1, \ldots, n$. This concludes the proof since

$$V \cap (U^r \times U^{r+1} \times \ldots \times U^n) = \Phi'(W \cap (U^r \times \overline{U})) = \{(z', \phi_{r+1}(z'), \ldots, \phi_n(z'); \ z' \in U^r\}.$$

**

This implicit function thecrem in connection with Theorem 11 of Section 2 leads to the following result on the distribution of non-SAP points on varieties:

THEOREM 14. Suppose (k, T) is an ordered field with real closure R and $V \subset R^n$ is a real variety of dimension $r > 0$. If k is not a hereditary SAP field, then the non-SAP points are dense in V_c.

PROOF. Assume that $a \in V_c$ and U is a strong neighborhood of a. By Theorem 13, there is some $b \in V_c \cap U$ and a neighborhood U^r of $(b_1, \ldots, b_r)$ in R^r and a neighborhood $U^{n-r} \subset R^{n-r}$ of $(b_{r+1}, \ldots, b_n)$ such that $U^r \times U^{n-r} \subset U$ and an implicit function $\phi : U^r \longrightarrow U^{n-r}$ such that $V \cap (U^r \times U^{n-r}) = \{(x, \phi(x)); x \in U^r\}$. There are $y, z \in R^r$ with $y < z$ and $]y, z[\subset U^r$. By Theorem 11, there exists some $t_1 \in R$, $y_1 < t_1 < z_1$, such that $k(t_1)$ with the set $T_{k(t_1)}$ of total orders is a hereditary non-SAP field. Choose any $t \in U^r$ with first component t_1. Then $(t, \phi(t)) \in V_c \cap U$ is a non-SAP point.**

Theorem 14 is false for varieties of dimension 0: Let (k, T) be any ordered field which does not have the hereditary SAP. Let $n \in \mathbb{N}$, $n \geq 2$ be minimal with the property that there exists an irreducible polynomial of degree n with a root in the real closure R of k. Let f be such a polynomial of degree n. Then the set T_K of orderings of $K = k[X]/(f)$ will always have the SAP, and the set of zeros of f in R consists of SAP points.

Theorem 14 is also false if the set of central points is replaced by the whole variety: Let k be any ordered field which does not have the hereditary SAP. $X^2 - X^3 + Y^2 \in k[X, Y]$ is an irreducible polynomial which generates a real prime ideal. The curve determined by this polynomial in R^2 has $(0, 0)$ as an isolated point. $(0, 0)$ is, of course, an SAP point. Thus, the statement of Theorem 14 fails at this point.

Now the generalization of Corollary 3 to arbitrary real varieties is almost trivial:

COROLLARY 15. For an ordered field (k, T) with real closure R and a real algebraic variety V of dimension $r > 0$ the following are equivalent:

(i) The sets of negativity $N(f)$ of single polynomials $f \in k[X_1, \ldots, X_n]$ form a basis of the strong topology of V_c.

(ii) (k, T) is a hereditary SAP field.

PROOF. <u>(ii) $\longrightarrow$ (i)</u>: Clear. <u>(i) $\longrightarrow$ (ii)</u>: Suppose, (k, T) is not a a hereditary SAP field. By Theorem 14, there is some $a \in V_c$ which is a non-SAP point. Let $P(a) \subset R^n$ be the set of all points k-isomorphic to a. Clearly, $P(a) \subset V_c$. Now the desired contradiction follows immediately from Theorem 2. **

Finally in this section, it will be shown that the weaker property proved for R^n in Theorem 4 also carries over to arbitrary real varieties. The proof requires the following lemma which shows that the intermediate value theorem of ordinary calculus, which is well known to hold for polynomials over real-closed fields ([12, Lemma, p. 278]), is also true for implicit functions.

LEMMA 16. Let R be a real-closed field, $F \in R[X_1, \ldots, X_n, Y]$ be a polynomial, $(a, b) = (a_1, \ldots, a_n, b) \in V$ (hypersurface determined by F) a point with $F_Y(a, b) \neq 0$. Let $\phi : U^n \longrightarrow U$ be a continuous function from a neighborhood $U^n =]a', a''[$ of a to a neighborhood $U =]b', b''[$ of b such that $V \cap (U^n \times U) = \{(x, \phi(x));\ x \in U^n\}$ and such that $F_Y(x, y) \neq 0$ for $(x, y) \in V \cap (U^n \times U)$ ([20, Theorem (7.4)]). Then $\phi(U^n)$ is an interval or the disjoint union of two intervals.

PROOF. Clearly, it suffices to do the proof for $n = 1$. Assume (by way of contradiction) that there are $y_1 < y_2 < y_3 < y_4 < y_5$ in U such that $y_1, y_3, y_5 \in \phi(U^1)$ while $y_2, y_4 \notin \phi(U^1)$. Pick $x_1, x_3, x_5 \in U^1$ with $\phi(x_i) = y_i$. Without loss of generality, assume that $F(x_1, y_2) > 0$ Since $F(x, y_2) \neq 0$ for all x between x_1, x_3, x_5, it follows that $F(x_1, y_2) > 0$ for all $i = 1, 3, 5$. Since $F_Y(x, y) \neq 0$ for $(x, y) \in V \cap (U^1 \times U)$, y_3 is a single root of $F(x_3, Y)$. Moreover, y_3 is the only root of this polynomial in the interval U. Thus, $F(x_3, y_4) < 0$. And, by the same argument as with y_2, it follows that $F(x_i, y_4) < 0$ for $i = 1, 3, 5$. By hypothesis, y_5 is the only root of $F(x_5, Y)$ in U. But, on the other hand, $F(x_5, y_2) > 0 > F(x_5, y_4)$ so that by the intermediate value theorem for polynomials over real closed fields ([12, Lemma, p. 278]) $F(x_5, y) = 0$ for some $y \in U$ with $y_2 < y < y_4 < y_5$. Contradiction. **

The statement of Lemma 16 is, of course, slightly different from the classical intermediate value theorem in that it allows the image of an interval to be a union of two disjoint intervals. But an immediate consequence of this is that locally the classical theorem does hold:

COROLLARY. In the situation of Lemma 16, for every $c \in U^n$ there is an interval $c \in]c', c''[\subset U^n$ such that $\phi(]c', c''[)$ is an interval.

THEOREM 18. Let (k, T) be an ordered field with real closure R, $V \subset R^n$ a variety of dimension $r > 0$. Then, for every open $U \subset R^n$ with $U \cap V_c \neq \emptyset$, there is some $f \in k[X_1, \ldots, X_n]$ such that $\emptyset \neq N(f) \cap V \subset U \cap V_c$.

PROOF. Without loss of generality, one may assume that the prime ideal P of V does not contain any non-trivial polynomial $p \in k[X_i]$ for any i. By a linear transformation of the variables, P can always be brought into this form.

Let $U \subset R^n$ be open such that $U \cap V_c \neq \emptyset$. Then there is some $a \in U \cap V_c$ and a neighborhood U^r of $(a_1, \ldots, a_r)$ and neighborhoods $U^{r+1}, \ldots, U^n \subset R$ of $a_{r+1}, \ldots, a_n$, respectively, such that $U^r \times U^{r+1} \times \ldots \times U^n \subset U$ and there are continuous functions $\phi_i : U^r \longrightarrow U^i$ $(i = r+1, \ldots, n)$ such that

$$V \cap (U^r \times U^{r+1} \times \ldots \times U^n) = V_c \cap (U^r \times U^{r+1} \times \ldots \times U^n)$$
$$= \{(z, \phi_{r+1}(z), \ldots, \phi_n(z)); \ z \in U^r\},$$

after rearranging the variables if necessary (Theorem 13). By Theorem 4, there is some $f_r \in k[X_1, \ldots, X_r]$ such that $\emptyset \neq N(f_r) \subset U^r$ and $f_r(z) > n\varepsilon$ for $z \notin U^r$ and $f_r(z) > -\varepsilon$ for all $z \in R^r$, where $0 < \varepsilon \in k$ is fixed. Let U_0^r be an interval contained in $N(f_r)$. By Corollary 17, this interval can be chosen so that $\phi_{r+1}(U_0^r) \subset U^{r+1}$ is an interval. By the remark at the beginning of the proof, this interval is not degenerated to a single point. Let $U_0^{r+1} \subset \phi_{r+1}(U_0^r)$ be an open interval. By Theorem 4, choose $f_{r+1} \in k[X_{r+1}]$ such that $\emptyset \neq N(f_{r+1}) \subset U_0^{r+1}$ and $f_{r+1}(z_{r+1}) > n\varepsilon$ for $z_{r+1} \notin U^{r+1}$ and $f_{r+1}(z_{r+1}) > -\varepsilon$ for all $z_{r+1} \in R$. Let $U_1^{r+1} \subset N(f_{r+1})$ be an interval and choose an interval $U_1^r \subset \phi_{r+1}^{-1}(U_1^{r+1})$. Now apply the same argument successively to $r+2, \ldots, n$ to obtain a decreasing sequence $U_0^r \supset U_1^r \supset \ldots \supset U_{n-r}^r$ of intervals and intervals $U_1^{r+1} \subset U^{r+1}, \ldots, U_{n-r}^n \subset U^n$ with $U_{r-1}^r \subset \phi_i^{-1}(U_{i-r}^i)$ and polynomials $f_i \in k[X_i]$ with $f_i(z_i) > n\varepsilon$ for $z_i \notin U^i$ and $f_i(z_i) > -\varepsilon$ for all $z_i \in R$ and $U_{i-r}^i \subset N(f_i)$ for all

$i = r+1, \ldots, n$. Now define $f = f_r + f_{r+1} + \ldots + f_n \in k[X_1, \ldots, X_n]$

Then, for

$$(z, z_{r+1}, \ldots, z_n) \notin U^r \times U^{r+1} \times \ldots \times U^n,\ f(z, z_{r+1}, \ldots, z_n)$$
$$= f_r(z) + f_{r+1}(z_{r+1}) + \ldots + f_n(z_n) > n\varepsilon - (n-r)\varepsilon = r\varepsilon > 0.$$

For

$$(z, z_{r+1}, \ldots, z_n) \in U^r_{n-r} \times U^{r+1}_1 \times \ldots \times U^n_{n-r},$$
$$f_r(z) < 0,\ f_{r+1}(z_{r+1}) < 0,\ \ldots,\ f_n(z_n) < 0.$$

By the construction of the intervals,

$$V \cap (U^r_{n-r} \times U^{r+1}_1 \times \ldots \times U^n_{n-r}) = V_c \cap (U^r_{n-r} \times U^{r+1}_1 \times \ldots \times U^n_{n-r})$$

is nonempty. Altogether, this shows

$$\emptyset \neq N(f) \cap V \subset U \cap V_c,$$

as claimed. **

4. HILBERT'S 17TH PROBLEM

The investigations of the three sections were motivated by a couple of generalizations of Hilbert's 17th problem. Before formulating these, I will fix a few notations for the following discussions: Let (k, T) be an ordered field with real closure R. A polynomial $f \in k[X_1, \ldots, X_n]$ will be called a positive square combination in $k(X_1, \ldots, X_n)$ if $f = p_1 f_1^2 + \ldots + p_r f_r^2$ with $0 < p_i \in k$ and $f_i \in k(X_1, \ldots, X_n)$. The polynomial f is called positive definite on $M \subset R^n$ if $f(x) > 0$ for all $x \in M$.

As a starting point, consider the following version of Hilbert's 17th problem which is, in fact, itself a generalization of the original problem ([21, p. 24]):

(Q1) Let $f \in k[X_1, \ldots, X_n]$ be positive definite on k^n. For which fields k does this imply that f is a positive square combination in $k(X_1, \ldots X_n)$?

For k a subfield of the real numbers, (Q1) was answered in the affirmative by Artin ([1, Satz 6]). Robinson proved that the answer is also "yes" for k any real closed field ([22, Theorem 5.2]). From this it follows by an easy continuity argument that the answer is the same for k any ordered field which is dense in its real closure. On the other hand, there exist fields for which the answer is "no". This was first proved through an example constructed by Dubois ([3]). Finally, the complete answer was provided by McKenna ([16]): The answer is "yes" if and only if k is dense in its real closure.

The problem (Q1) is generalized by the following:

(Q2) Let $f \in k[X_1, \ldots, X_n]$ be positive definite on $M \subset R^n$. For which M does this imply that f is a positive square combination in $k(X_1, \ldots, X_n)$?

Considering McKenna's solution of (Q1), the first guess (which will turn out to be correct) is, of course, that M must be dense in R^n. But before proving this I will first generalize (Q2) to real varieties much in the same way as the results of Section 1 were generalized to varieties in Section 3.

(Q3) Let $V \subset R^n$ be a real variety with coordinate ring $k[X_1, \ldots, x_n]$. Let $f \in k[x_1, \ldots, x_n]$ be positive definite on $M \subset V$. For which M does this imply that f is a positive square combination in $k(x_1, \ldots, x_n)$?

For similar generalizations of Hilbert's 17th problem to varieties, see [15, Theorem 9], [23].

THEOREM 19. For an ordered field (k, T) with real closure R and a real variety $V \subset R^n$ of dimension $r > 0$ and a subset $M \subset V$, the following are equivalent:

(i) If $f \in k[V]$ is positive definite on M then f is a positive square combination in $k(V)$.

(ii) $M \cap V_c$ is dense in V_c.

PROOF. (ii) $\longrightarrow$ (i): Follows from [4, Theorem 2] by continuity.
(i) $\longrightarrow$ (ii): Suppose that $M \cap V_c$ is not dense in V_c. Then there is some open subset $U \subset R^n$ with $U \cap V_c \neq \emptyset$ and $U \cap M = \emptyset$. Then, by Theorem 18, there is some $f \in k[V]$ with $\emptyset \neq N(f) \cap V \subset U \cap V_c$. In particular, $N(f) \cap M = \emptyset$. But then (i) shows that f is a positive

square combination in $k(V)$. Thus, f is positive definite on all of V_c ([4, Theorem 2]), which contradicts $N(f) \cap V_c \neq \emptyset$.

For varieties of dimension 0, the equivalence of Theorem 19 is false in general. If $\dim V = 0$ and $a \in V$, then a is a generic point of V and V consists of those points of R^n which are k-isomorphic to a. Now Theorem 2 shows that the equivalence of Theorem 19 holds precisely when $k(a)$ with the set $T_{k(a)}$ of total orders is an SAP field.

The following corollary of Theorem 19 is a close relative of [4, Theorem 2]:

COROLLARY 20. In the situation of Theorem 19, for $M \subset V$ these are equivalent:

(i) If $f \in k[V]$ is positive definite on M then f is a positive square combination in $k(V)$.

(ii) If $f \in k[V]$ is positive definite on $M \cap V_c$ then f is a positive square combination in $k(V)$.

COROLLARY 21 ([16]). For the ordered field (k, T) these are equivalent:

(i) If $f \in k[X_1, \ldots, X_n]$ is positive definite on k^n then f is a positive square combination in $k(X_1, \ldots, X_n)$.

(ii) k is dense in its real closure.

Starting out from version (Q1) of Hilbert's 17th problem, there is still another direction in which the problem can be generalized. But first a few notations: Let $\underline{T}$ be a saturated set of total orders on the formally real field k. Let $P = \cap \underline{T}$. A polynomial $f \in k[X_1, \ldots, X_n]$ is called <u>positive definite on k^n with respect to P</u> if $f(x) \in P$ for all $x \in k^n$. f is a <u>positive square combination with respect to P</u> in $k(X_1, \ldots, X_n)$ if $f = p_1 f_1^2 + \ldots + p_r f_r^2$ with $p_i \in P$ and $f_i \in k(X_1, \ldots, X_n)$.

(Q4) Let $f \in k[X_1, \ldots, X_n]$ be positive definite on k^n with respect to P. Does this imply that f is a positive square combination in $k(X_1, \ldots, X_n)$ with respect to P?

If k is an algebraic number field and $\underline{T}$ is the set of all total orders of k, then the answer is "yes" ([1, Satz 5]). This is a special case of the following (obvious) fact: If (k, T) is dense in its real closure for all $T \in \underline{T}$, then the answer is "yes". If $\underline{T}$ is finite, the converse of this is also true (see [19] and note that Prestel's argument is applicable in the present situation). Positive definite functions for partially ordered fields are also considered by van den Dries ([24]), although positive definite means something different there.

The next theorem serves to show that Prestel's above-mentioned result is not true for arbitrary saturated sets $\underline{T}$.

THEOREM 22. Let L be a formally real field with a saturated set $\underline{S}$ of total orders. Let $k = L(X)$ and let $\underline{T}$ be the set of extensions of total orders in $\underline{S}$ to total orders of k. Let $P = \cap \underline{T}$. If $f \in k[Y]$ is positive definite on k with respect to P then f is a positive square combination in $k(Y)$ with respect to P.

PROOF. Suppose $f \in k[Y]$ is positive definite with respect to P. One may assume that $f \in L[X, Y]$. Then, for all $g \in L[X]$, $f(X, g(X)) \in P$, i.e., $f(X, g(X))$ is a positive square combination in $L(X)$ with respect to $Q = \cap \underline{S}$. Therefore, $f(a, g(a)) \geq 0$ in R_S for all $a \in R_S$ and all $S \in \underline{S}$ (R_S is the real closure of (L, S)).

Now assume (by way of contradiction) that $f < 0(R')$ for some total order R' of $k(Y)$ with $R' \cap k \in \underline{T}$. Let $S = L \cap R'$. By Dubois' Algebraic Order Theorem ([4]) there exists a total order R on $k(Y)$ such that $S = L \cap R$ and $f < 0(R)$ and the images x, y of X and Y, respectively, under the real place belonging to the natural valuation of $(L(X, Y), R)$ over L are finite and algebraic over L. Thus, $x, y \in R_S$. This shows that there is some $(x, y) \in R_S^2$ with $f(x, y) < 0$ in R_S. By continuity of f, there are $a, b \in R_S^2$ such that $f(c) < 0$ for all $c \in]a, b[\subset R_S^2$. By [11, Lemma 2.2], there exists $c \in]a, b[$ such that $c_2 \in L(c_1)$, i.e., $c_2 = g(c_1)$ for some $g \in L[X]$. But then $f(c_1, g(c_1)) < 0$ in R_S for some $c_1 \in R_S$, which is impossible. **

Now, in Theorem 22, let $L = \mathbb{R}$ (real numbers), let $\underline{S}$ consist of the unique total order of $\mathbb{R}$. Let k and $\underline{T}$ be as in Theorem 22. Then every polynomial $f \in k[Y]$ which is positive definite with respect to P is a positive square combination with respect to P in $k(Y)$. But none of the (k, T), $T \in \underline{T}$ is dense in its real closure. This shows that Prestel's above-mentioned result does not extend to arbitrary saturated sets $\underline{T}$.

Certainly, any result extending Prestel's theorem in some way to infinite $\underline{T}$ will have to involve topological spaces of orderings. This is illustrated by the next theorem. The statement and the proof of the theorem require the following notations: Let $\underline{X}$ be the set of orderings of the field k. $H(a_1, \ldots, a_r) = \{T \in \underline{X};\ a_1, \ldots, a_r < 0(T)\}$ are the basic sets of the Harrison topology of $\underline{X}$. Since all the following discussions will take place only inside a fixed saturated subset $\underline{T} \subset \underline{X}$ the basic sets of the induced topology of $\underline{T}$ will also be denoted $H(a_1, \ldots, a_r)$. Note that saturated sets are always closed and that the sets $H(a_1, \ldots, a_r)$ are always saturated.

THEOREM 23: Let k be a formally real field with a saturated set $\underline{T}$ of total orders. The following are equivalent:

(i) If $f \in k[X]$ is positive definite on k with respect to $\cap \underline{T}$ then f is a positive square combination in $k(X)$ with respect to $\cap \underline{T}$.

(ii) There exists a set S^* of saturated closed and open subsets of $\underline{T}$ such that $\underline{T} = \cup S^*$ and for all $\underline{S} \in S^*$ the following holds: If $f \in k[X]$ is positive definite with respect to $\cap \underline{S}$ then f is a positive square combination in $k(X)$ with respect to $\cap \underline{S}$.

(iii) For all saturated closed and open sets $\underline{S} \subset \underline{T}$ the following holds: If $f \in k[X]$ is positive definite with respect to $\cap \underline{S}$ then f is a positive square combination in $k(X)$ with respect to $\cap \underline{S}$.

(iv) For all $a_1, \ldots, a_r \in k$ the following holds: If $f \in k[X]$ is positive definite with respect to $\cap H(a_1, \ldots, a_r)$ then f is a positive square combination in $k(X)$ with respect to $\cap H(a_1, \ldots, a_r)$.

PROOF. (i), (ii) and (iv) are special cases of (iii). (ii) is a special case of (iv).

<u>(ii) $\longrightarrow$ (i)</u>: Assume (by way of contradiction) that there is some $f \in k[X]$ which is positive definite with respect to $\cap \underline{T}$, but which is not a positive square combination in $k(X)$ with respect to $\cap \underline{T}$. Then there is some $T \in \underline{T}$ such that f is not positive definite on the real closure R_T of (k, T). By (ii), there is some $\underline{S} \in S^*$ such that $T \in \underline{S}$. Then f is positive definite with respect to $\cap \underline{S}$ without being a positive square combination in $k(X)$ with respect to $\cap \underline{S}$. Contradiction.

<u>(iv) $\longrightarrow$ (iii)</u>: This is a special case of the implication (ii) $\longrightarrow$ (i). For, any saturated closed and open $\underline{S} \subset \underline{T}$ is covered by the sets $H(a_1, \ldots, a_r) \subset \underline{S}$.

<u>(i) $\longrightarrow$ (iv)</u>: Assume that there are $a_1, \ldots, a_r \in k$ and that there is a polynomial $f(X)$ which is positive definite with respect to $\cap H(a_1, \ldots, a_r)$ without being a positive square combination in $k(X)$ with respect to $\cap H(a_1, \ldots, a_r)$. Hence, there exists a $T \in H(a_1, \ldots, a_r)$ such that f is not positive definite on the real closure R_T of (k, T).

Let $f = \alpha_d X^d + \ldots + \alpha_0$. For every $S \in H(-a_d)$ there is some $x_S \in k$ such that $y_{S,\varepsilon} = x_S^2 - (\varepsilon_0\alpha_0 + \ldots + \varepsilon_d\alpha_d)/a_1 > 0\ (S)$ for all $\varepsilon \in \{+1, -1\}^{d+1}$. Then

$$\bigcup_{S \in H(-a_1)} H(-y_{S,\varepsilon};\ \varepsilon)$$

is an open cover of $H(-a_1)$. By compactness of $H(-a_1)$, there are $S_1, \ldots, S_m \in H(-a_1)$ such that $H(-y_{S_1,\varepsilon};\ \varepsilon), \ldots, H(-y_{S_m,\varepsilon};\ \varepsilon)$ form a cover of $H(-a_1)$. Finally, choose $x_0 \in k$ so that $-a_1x_0^2 > \varepsilon_0\alpha_0 + \ldots + \varepsilon_d\alpha_d\ (T)$ for all $\varepsilon \in \{+1, -1\}^{d+1}$. Define $y_0 = x_0^2 + x_{S_1}^2 + \ldots + x_{S_m}^2$ and $g_1 = X^2 + a_1y_0$. Then $f_1 = f(g_1(X))$ is positive definite on each real closure R_S of (k, S) $(S \in H(-a_1))$ and on k with respect to $\cap H(a_1, \ldots, a_r)$, but not on R_T.

By iteration of this argument, one obtains a polynomial $f_r \in k[X]$ which is positive definite on R_S for all $S \in H(-a_1) \cup \ldots \cup H(-a_r)$ and on k with respect to $\cap H(a_1, \ldots, a_r)$, but not on R_T. Since $\underline{T} = H(a_1, \ldots, a_r) \cup H(-a_1) \cup \ldots \cup H(-a_r)$ it follows that f_r is positive definite on k with respect to $\cap \underline{T}$. But the fact that f_r is not positive definite on R_T implies that f_r cannot be a positive square combination in $k(X)$ with respect to $\cap \underline{T}$. **

Theorem 23 is stated and proved only for polynomials in one indeterminate. The construction used in the proof is not applicable in the case of several indeterminates. Therefore the next theorem serves to extend Theorem 23 to polynomials in any finite number of indeterminates:

THEOREM 24. Let k be a formally real field with a saturated set $\underline{T}$ of total orders. Then the following are equivalent:

(i) If $f \in k[X]$ is positive definite with respect to $\cap \underline{T}$ then f is a positive square combination in $k(X)$ with respect to $\cap \underline{T}$.

(ii) If $f \in k[X, Y]$ is positive definite with respect to $\cap \underline{T}$ then f is a positive square combination in $k(X, Y)$ with respect to $\cap \underline{T}$.

PROOF. (i) is a special case of (ii). $\underline{(i) \longrightarrow (ii)}$: Suppose, $f(X, Y)$ is positive definite on k with respect to $\cap \underline{T}$. Then, for $g \in k[X]$, $f(X, g(X))$ is positive definite on k with respect to $\cap \underline{T}$. By (i), $f(X, g(X))$ is a positive square combination in $k(X)$ with respect to

$\cap \underline{T}$. This shows that $f(X, Y)$ is positive definite with respect to $\cap \underline{S}$ on $k(X)$, where $\underline{S}$ is the set of extensions of the total orders in $\underline{T}$ to total orders of $k(X)$. By Theorem 22, f is a positive square combination in $k(X, Y)$ with respect to $\cap \underline{S} : f = g_1 f_1^2 + \ldots + g_r f_r^2$ with $g_i \in \cap \underline{S}$ and $f_i \in k(X, Y)$ for all i. But the g_i are positive square combinations with respect to $\cap \underline{T}$ in $k(X)$ ([1, Satz 2]). Now, by replacing the g_i in the above representation of f by positive square combinations with respect to $\cap \underline{T}$, one obtains the desired representation for f. **

By induction, the equivalence of Theorem 24 can be extended to any finite number of indeterminates.

Prestel has informed me that Theorem 24 has been known to him for some time and that he has mentioned this result repeatedly in colloquium talks. But I feel justified to present this result here since Prestel's result is unpublished and since Prestel's proof is different from the one presented here.

Note that Theorem 24 gives yet another proof of Robinson's result that the answer to Hilbert's 17th problem (version (Q1)) is "yes" for real closed fields ([22]): For, if k is real-closed and $\underline{T}$ consists of the unique total order of k and $f \in k[X]$ is positive definite with respect to $\underline{T}$, then it is well known that f is a sum of squares in $k[X]$. Now Robinson's result follows from an inductive application of Theorem 24.

Finally, another corollary of Theorem 23 and Theorem 24 is Prestel's result referred to before Theorem 22 ([19]):

COROLLARY 25. Let k be a formally real field with a finite saturated set $\underline{T}$ of total orders. Then these are equivalent:

(i) If $f(X_1, \ldots, X_n)$ is positive definite on k^n with respect to $\cap \underline{T}$ then f is a positive square combination in $k(X_1, \ldots, X_n)$ with respect to $\cap \underline{T}$.

(ii) For each $T \in \underline{T}$, (k, T) is dense in its real closure.

PROOF. (ii) $\longrightarrow$ (i) is trivial by [1, Satz 2]. (i) $\longrightarrow$ (ii) follows immediately from Theorem 23, Theorem 24 and Corollary 21. **

BIBLIOGRAPHY

1. E. Artin: Über die Zerlegung definiter Funktionen in Quadrate. Abh. Math. Sem. Univ. Hamb. 5, 100-115 (1927).

2. E. Becker, E. Köpping: Reduzierte quadratische Formen und Semiordnungen reeller Körper. Abh. Math. Sem. Univ. Hamb. 46, 143-177 (1977).

3. D. W. Dubois: Note on Artin's Solution of Hilbert's 17th Problem. Bull. AMS 73, 540-541 (1967).

4. D. W. Dubois: Second Note on Artin's Solution of Hilbert's 17th Problem. Order Spaces. Preprint.

5. D. W. Dubois: Real Commutative Algebra I. Places. Rev. Mat. Hisp.-Amer. (4) 39, 57-65 (1979).

6. D. W. Dubois, G. Efroymson: Algebraic Theory of Real Varieties Varieties I. In: Studies and Essays Presented to Yu-Why Chen on his 60th Birthday. Taipei, Taiwan (1970).

7. R. Elman, T. Y. Lam: Quadratic Forms over Formally Real Fields and Pythagorean Fields. Amer. J. of Math. 94, 1155-1194 (1972).

8. R. Elman, T. Y. Lam, A. Prestel: On some Hasse Principles over Formally Real Fields. Math. Z. 134, 291-301 (1973).

9. R. Elman, T. Y. Lam, A. R. Wadsworth: Orderings under Field Extensions. J. reine agnew. Math. 306, 7-27 (1979).

10. O. Endler: Valuation Theory. Universitext. Springer 1972.

11. P. Erdös, L. Gillman, M. Henriksen: An Isomorphism Theorem for Real-Closed Fields. Ann. of Math. 61, 542-554 (1955).

12. N. Jacobson: Lectures in Abstract Algebra. III. Van Nostrand 1964.

13. M. Knebusch, A. Rosenberg, R. Ware: Structure of Witt Rings and Quotients of Abelian Group Rings. Amer. J. of Math. 94, 119-155 (1972).

14. T. Y. Lam: The Theory of Ordered Fields. To appear in: Proceedings of the Algebra and Ring Theory Conference (Ed.: B. McDonald), University of Oklahoma, 1979. P Preprint.

15. S. Lang: The Theory of Real Places. Ann. of Math. 57, 378-391 (1953).

16. K. McKenna: New Facts about Hilbert's 17th Problem. Springer Lecture Notes 498, 220-230 (1975).

17. A. Prestel: Lectures on Formally Real Fields. IMPA, Rio de Janeiro 1975.

18. A. Prestel: Quadratische Semi-Ordnungen und quadratische Formen. Math. Z. 133, 319-342 (1973).

19. A. Prestel: Sums of Squares over Fields. Atas da 5 escola de álgebra. IMPA, Rio de Janeiro 1978.

20. A. Prestel, M. Ziegler: Model Theoretic Methods in the Theory of Topological Fields. J. reine angew. Math. 299/300, 318-341 (1978).

21. Proceedings Symp. Pure Math. 28: Mathematical Developments Arising from Hilbert Problems. AMS 1976

22. A. Robinson: On Ordered Fields and Definite Functions. Math. Ann. 130, 257-271 (1955).

23. A. Robinson: Further Remarks on Ordered Fields and Definite Functions. Math. Ann. 130, 405-409 (1956).

24. L. van den Dries: Artin-Schreier Theory for Commutative Regular Rings. Ann. Math. Logic 12, 113-150 (1977).

MATHEMATISCHES INSTITUT
DER UNIVERSITÄT MÜNCHEN
THERESIENSTRAßE 39
8 MÜNCHEN 2, W. GERMANY

Contemporary Mathematics
Volume 8, 1982

THE SQUARE CLASS INVARIANT FOR PYTHAGOREAN FIELDS

Daniel B. Shapiro[1] and T.-Y. Lam[1]

INTRODUCTION.

Let q be a (nonsingular) quadratic form over a field K, of char. $\neq 2$. For $a \in \dot{K}$ let $m_q(a)$ be the largest integer m such that $m\langle a\rangle < q$. Equivalently, if $q \cong m\langle a\rangle \perp q'$ where q' does not represent a, then $m = m_q(a)$. This map $m_q : \dot{K}/\dot{K}^2 \longrightarrow \mathbb{Z}$ is the square class invariant of q, as defined by A. Solow [S1]. She examined the classification problem: if $m_\phi = m_\psi$ for forms ψ, ϕ over K, then must we have $\phi \cong \psi$? We will consider this problem in the case K is a (real) pythagorean field.

One precursor to this question is a classification result using value sets: if K is a pythagorean field and if ϕ, ψ are both (pure parts of) n-fold Pfister forms, then $D_K(\phi) = D_K(\psi)$ implies $\phi \cong \psi$. This is proved in [EL, p. 1183] and [BK, p. 158]. However, the value set $D(\phi)$ alone, or with the other invariants $\dim \phi$, $\det \phi$, and $c(\phi)$, is never enough to classify all forms over K. For instance, every isotropic form ϕ has $D_K(\phi) = \dot{K}$.

The invariant m_ϕ is a more sensitive measure of represented values, since $D(\phi) = \{a \in \dot{K} \mid m_\phi(a) \geq 1\}$. Solow proved that m does classify forms for certain types of pythagorean fields K:

(1) [S1] If K is SAP and the square class group $\dot{K}/\dot{K}^2$ is finite, then m classifies forms;

(2) [S1] If K is superpythagorean, then m classfies forms;

(3) [S2] If K has a complete discrete valuation with residue class field k, and if m classfies forms over k, then m classifies forms over K.

In this work we show that Solow's results include, up to equivalence, all the pythagorean fields with finite square class group for which m

---

1980 Mathematics Subject Classification. 10C04, 12D15, 12J15, 13K05.

[1]Supported in part by the National Science Foundation.

0271-4132/81/0000-0682/$04.50

classifies forms. Furthermore, everything can be extended to pythagorean fields which have only finitely many real-valued places.

THEOREM. Suppose K is a pythagorean field with only finitely many real-valued places. Then m classfies forms over K if and only if K is equivalent to an iterated power series field over k, where k is some SAP pythagorean field.

Here "equivalence" is in the sense of Harrison and Cordes: two fields are equivalent iff their Witt rings are isomorphic. This theorem is proved in Remark 13. It is deduced from a more general result about abstract spaces of orderings.

Our presentation and proofs rely heavily on Marshall's beautiful theory of abstract spaces of orderings [M1], [M2], [M3], [M4]. Craven's work on reduced Witt rings [C1], [C2] is closely related. We are grateful to A. Wadsworth for providing the proofs for the infinite SAP case.

THE SQUARE CLASS INVARIANT AND ABSTRACT SPACES OF ORDERINGS

We assume familiarity with the notations and results of Marshall's papers [M1], [M2], [M3], [M4]. If (X,G) is a space of orderings we will write $f \cong g$ (over G) for isometric forms, and we use $f \perp g$ for orthogonal sum. Also, $D(f) = D_G(f)$ denotes the set of elements of G represented by the form f (over G).

Let (X,G) be a space of orderings and f a form over G. Define the <u>square class invariant</u> $m_f : G \longrightarrow \mathbb{Z}$ as follows: for $a \in G$ let $m_f(a)$ be the largest integer m such that $f \cong m\langle a\rangle \perp f'$, (over G) for some form f'. (Of course in this generality, G no longer stands for "square classes".)

PROPOSITION 1.

(1) If $f \cong g$, then $m_f = m_g$.

(2) $m_f(a) = m_{af}(1)$.

(3) If $f \cong r\langle a\rangle \perp s\langle -a\rangle \perp f'$, where f' does not represent a or $-a$, then $m_f(a) = r$ and $m_f(-a) = s$.

(4) $m_{r\mathbb{H} \perp f}(a) = r + m_f(a)$. If m_f = constant function, then f is hyperbolic.

(5) $m_{n \times f}(a) = n \cdot m_f(a)$.

PROOF. Straightforward from the definitions. For (3) note that if $r\langle a\rangle < s\langle -a\rangle \perp f'$, then $r\langle a\rangle < f'$. □

REMARK. If m classfies anisotropic forms over (X,G), then m classifies all forms over (X,G). This is easily proved from property (4) above.

Marshall considered two basic ways to build spaces of orderings from smaller ones. First [M2, Def. 3.6] there is the group extension $(X,G) = (X',G') \times H$. Here $G = G' \oplus H$, and $X = X' \times \chi(H)$ is the set of all extensions of X' to G. Second [M2, Def. 2.6] there is the direct sum $(X,G) = (X_1,G_1) \oplus (X_2,G_2)$. This means that $G = G_1 \oplus G_2$ and $X = X_1 \cup X_2$ (disjoint union), where $\sigma \in X_i$ is defined to act trivially on G_j for $i \neq j$.

As proved in [J, Th. 2 and Th. 3], each of these types of decomposition can be concretely realized in the category of pythagorean fields, using 2-henselian valuations and euclidean closures. However, in the present paper we will not use this correspondence.

We will analyze the invariant m for those spaces (X,G) which can be built from SAP spaces using the direct sum and group extension operations a finite number of times. (See Definition 10.) This category S of spaces includes the category C consisting of all spaces of finite chain length [M4, Th. 1.6]. C can also be characterized as the category of all spaces $(X_K, \dot{K}/\dot{K}^2)$, where K is a pythagorean field having only finitely many real-valued places [M4, Th. 1.5], [Cl, Th. 2.1].

We begin by discussing the behavior of m_f for each of these two types of decomposition.

PROPOSITION 2. Suppose $(X,G) \cong (X',G') \times H$, and let f be any form over G. Then $f \cong x_1f_1 \perp \ldots \perp x_sf_s$, where the x_i are distinct elements of H and the forms f_i are defined over G'. If f is isotropic, then for some i, f_i is isotropic.

PROOF. [M1, Lem. 4.9] and [M2, Rem. 3.7]. □

If f is anisotropic, then it follows that these forms f_i are unique up to isomorphism. Consequently, the Witt ring is $W(X,G) \cong W(X',G')[H]$, a group ring.

COROLLARY 3. Let $f = x_1f_1 \perp \ldots \perp x_sf_s$ be an anisotropic form over (X,G), as above. Let $a \in G$. If $a \in D_G(f)$ then $a = x_ia'$, for some i and some $a' \in D_{G'}(f_i)$. In fact, $m_f(a) = m_{f_i}(a')$.

PROOF. $a \in D_G(f)$ iff $f \perp \langle -a \rangle$ is isotropic over G. By Proposition 2 we must have $a = x_ia'$ for some $i = 1,\ldots,s$ and some $a' \in G'$, and $f_i \perp \langle -a' \rangle$ is isotropic over G'. Now suppose $f \cong m\langle a \rangle \perp g$, for some form g over G. By the uniqueness of the decomposition we have $m\langle a' \rangle < f_i$. The converse is clear, so $m_f(a) = m_{f_i}(a')$. □

COROLLARY 4. Suppose (X,G) is a group extension of (X',G'). Then m classifies forms over (X,G) iff m classfies forms over (X',G').

PROOF. "if." Let f, g be anisotropic forms over G with $m_f = m_g$. Since $D(f) = D(g)$ we can express $f = x_1 f_1 \perp \dots \perp x_s f_s$ and $g = x_1 g_1 \perp \dots \perp x_s g_s$, (the same x_i's here). For fixed i and any $a' \in G'$ let $a = x_i a'$. Then

$$m_{f_i}(a') = m_f(a) = m_g(a) = m_{g_i}(a').$$

By hypothesis this implies $f_i \cong g_i$ over G'. This is true for all i, so that $f \cong g$ over G.

"only if." Let f', g' be anisotropic forms over G' with $m_{f'} = m_{g'}$ (over G'). View f', g' as forms over G and apply Corollary 3 to conclude that $m_{f'} = m_{g'}$ over G. By hypothesis, this implies $f' \cong g'$ over G, and hence over G'. □

This corollary corresponds to a generalization of the pythagorean case of [S2, Th. 4]. The general case of this theorem of Solow can be extended to include any 2-henselian valuation.

A space (X,G) is a <u>fan</u> iff (X,G) is a group extension of (X',G') where $|X'| = 1$. Fans are defined in [M3, Def. 2.1], originating from [BK, Satz 20]. If $|X'| = 1$, then m certainly does classify forms. Hence, Corollary 4 implies that m classifies forms over every fan (X,G). This corresponds to [S1, Th. 219] since a pythagorean field K is superpythagorean iff $(X_K, \dot{K}/\dot{K}^2)$ is a fan.

Now suppose that the space (X,G) is a direct sum; say $(X,G) \cong (X_1,G_1) \oplus \dots \oplus (X_r,G_r)$. For instance, these (X_i,G_1) might be the components of (X,G), as in [M1, §3] and [M4, §2]. Then $X = X_1 \cup \dots \cup X_r$, (disjoint union), and $G \cong G_1 \times \dots \times G_r$. If $\mathbf{a} \in G$, we write $a = (a_1, \dots, a_r)$ for the components $a_i \in G_i$. Similarly for a form f over G we write $f = (f_1, \dots, f_r)$, where f_i is a form over G_i (of the same dimension). For Witt rings, there is a canonical ring embedding

$$W(X,G) \rightarrowtail \prod_{i=1}^{r} W(X_i,G_i),$$

whose image consists of those r-tuples $(f_1, \dots, f_r)$ of forms having the same dimension modulo 2 [M4, Rem. 2.9].

If $f = (f_1, \dots, f_r)$ is a form over (X,G), it follows that f represents a over G iff f_i represents a_i over G_i, for all i. We can generalize this as follows:

PROPOSITION 5. Let $(X,G) \cong (X_1,G_1) \oplus \dots \oplus (X_r,G_r)$ as above, and let $f = (f_1, \dots, f_r)$ be a form over G and $a = (a_1, \dots, a_r) \in G$. Then

$$m_f(a) = \min\{m_{f_i}(a_i) \mid 1 \leq i \leq r\}.$$

PROOF. Suppose m is this minimum, and express $f_i \cong m\langle a_i\rangle \perp g_i$, for some form g_i over G_i. Then $f \cong m\langle a\rangle \perp g$, where $g = (g_1, \dots, g_r)$. By minimality, there exists i where $a_i \notin D(g_i)$. Therefore $a \notin D(g)$ and by definition we conclude $m_f(a) = m$. □

COROLLARY 6. If m classifies forms over (X,G) then m classifies forms over (X_i,G_i) for all $i = 1, \dots, r$.

PROOF. An exercise for the reader. □

THE CLASSIFICATION THEOREM

If (X,G) satisfies a certain finiteness condition we will find necessary and sufficient conditions for m to classify forms over (X,G). We begin by examining SAP spaces.

Following Marshall's notation [M1, §2], if $a_1, \dots, a_m \in G$ define the Harrison basic set

$$X(a_1, \dots, a_m) = \{\sigma \in X \mid \sigma(a_i) = 1 \text{ for } i = 1, \dots, m\}.$$

These sets form a basis for the topology of X. Each of these basic sets is clopen (closed and open). The space (X,G) satisfies SAP (Strong Approximation Property) if every clopen subset $Y \subseteq X$ is of the form $Y = X(a)$, for some $a \in G$.

PROPOSITION 7. The following statements about a space (X,G) are equivalent:

(1) (X,G) satisfies SAP.

(1') For every $a,b \in G$, $X(a,b) = X(c)$ for some $c \in G$.

(2) There is a SAP pythagorean field K with $(X,G) \cong (X_K, \dot{K}/\dot{K}^2)$.

(3) Every fan in X is trivial (i.e., contains at most 2 elements).

(4) Every totally indefinite form over X is isotropic.

(4') Every form $\langle 1,a,b,-ab\rangle$ is isotropic.

(4") There is no ternary form f over X with $m_f \leq 1$.

(5) If $Y = X(a)$, then Y is a direct summand of X.

(Many further equivalent conditions for SAP spaces have appeared in the literature.)

PROOF. (1) $\longleftrightarrow$ (1') is easy [P, p. 93]. The equivalences (1) $\longleftrightarrow$ (4) $\longleftrightarrow$ (4') are essentially proved in [RW, Th. 3.1]. For (4') $\longleftrightarrow$ (4") note that if $f = \langle 1,a,b\rangle$ then $m_f(x) > 1$ for some x iff f represents ab. (2) $\longleftrightarrow$ (3) is proved in [M3, §5], and (2) $\longrightarrow$ (1) is clear.

(4') $\longrightarrow$ (3): If there is a nontrivial fan in X, then there exist distinct $\sigma_1,\sigma_2,\sigma_3 \in X$ with $\sigma_4 = \sigma_1\sigma_2\sigma_3 \in X$. Choose $a \in G$ with signs $\sigma_1,\sigma_2,\sigma_3,\sigma_4$ equal to $(+,+,-,-)$ and choose $b \in G$ with signs $(+,-,+,-)$. Such a, b can always be found, as in [P, pp. 145-146].) Then by (4'), $\langle 1,a,b\rangle \cong \langle ab,c,c\rangle$ for some $c \in G$. But then c must have signs $(+,+,+,-)$ contrary to the dependence of the σ_i's.

(1') $\longleftrightarrow$ (5): If $Y = X(a)$, then $(Y,G/\Delta)$ is a subspace of (X,G), where $\Delta = Y^{\perp} = D(\langle 1,a\rangle)$ as in [M1, Lem. 2.1]. Let $Y' = X - Y$ and $\Delta' = Y'^{\perp}$. Then Y is a direct summand of X iff $(X,G) \cong (Y,G/\Delta) \oplus (Y',G/\Delta')$, iff the natural map $G \longrightarrow (G/\Delta) \oplus (G/\Delta')$ is surjective, iff $G = \Delta\Delta'$. This occurs iff for every $b \in G$, there exists $c \in G$ with $\sigma(c) = 1$ for $\sigma \in Y$ and $\sigma(c) = \sigma(b)$ for $\sigma \in Y'$; (for then $b = c(cb) \in \Delta\Delta'$). Equivalently, $Y \cup X(b) = X(c)$ for some $c \in G$. □

There are some further equivalent conditions for SAP involving the invariant m. If $f = \langle a_1, \dots, a_n\rangle$ is a form over X and $\sigma \in X$, define $f_\sigma = \langle \sigma(a_1), \dots, \sigma(a_n)\rangle$ to be the form induced on the singleton subspace $\{\sigma\}$. Then $m_{f_\sigma}(1)$ equals the number of a_i with $\sigma(a_i) = 1$, that is, the number of entries of f which are positive at σ. Similarly, $m_{f_\sigma}(-1)$ is the number of entries of f negative at σ. To simplify notation, define $m_{f,\sigma}(a) = m_{f_\sigma}(\sigma(a))$. It is easy to see that $m_{f,\sigma}(a) = \frac{1}{2}(\dim f + \sigma(af))$, so it is well defined.

PROPOSITION 8. The following statements are equivalent for a space (X,G):

(1) (X,G) satisfies SAP.

(2) For every form f over X and every $a \in G$,

$$m_f(a) = \min\{m_{f,\sigma}(a) \mid \sigma \in X\}.$$

(3) For every form f over X,

$$\dim f = \max\{m_f(a) + m_f(b) \mid a,b \in G \text{ and } a \neq b\}.$$

The finite case of this proposition is a restatement of Proposi-2.10 and Theorem 2.16 of [S1]. The argument for the general case was kindly provided by A. Wadsworth. If X is finite, the equivalence of (1) and (2) is quickly settled by Proposition 5.

PROOF: (1) $\longrightarrow$ (2). Certainly $m_f(a) \leq m_{f,\sigma}(a)$ for every σ. We may scale to assume $a = 1$. Suppose $m = m_f(1)$ and $f \cong m\langle 1\rangle \perp f'$. Then $f' \perp \langle -1\rangle$ is anisotropic, so SAP implies that there exists $\sigma \in X$ where $f' \perp \langle -1\rangle$ is definite. Hence f' is negative definite at σ, so that $m_{f,\sigma}(1) = m$.

(2) $\longrightarrow$ (1). Suppose f is totally indefinite, that is, f_σ is indefinite (isotropic) for every $\sigma \in X$. For any $a \in G$, then $m_{f,\sigma}(a) \geq 1$ so that (2) implies $m_f(a) \geq 1$. In particular, f represents 1 and -1 over G, so f is isotropic.

(3) $\longrightarrow$ (1). If (X,G) does not satisfy SAP, then by Proposition 7 (4"), there is a ternary form f with $m_f(a) \leq 1$ for all $a \in G$. This contradicts (3).

(1) $\longrightarrow$ (3). Let $n = \dim f$. For any distinct $a,b \in G$, we have $ab \neq 1$ so there is some $\sigma \in X$ with $(ab) = -1$. Then

$$m_f(a) + m_f(b) \leq m_{f,\sigma}(a) + m_{f,\sigma}(b) = m_{f,\sigma}(1) + m_{f,\sigma}(-1) = n.$$

On the other hand, to see that n is attained, choose $a \in G$ such that $\sigma(af) \geq 0$ for all $\sigma \in X$. (Such a exists by SAP since $C = \{\sigma \in X \mid \sigma(f) \geq 0\}$ is clopen.) Then for every $\sigma \in X$,

$$m_{f,\sigma}(a) \geq n/2 \geq m_{f,\sigma}(-a).$$

Let $r = m_f(a)$. Then by (2) we have

$$r = \min\{m_{f,\sigma}(a) \mid \sigma \in X\} \geq n/2.$$

Now choose $c \in G$ such that $\sigma(c) = 1$ if $m_{f,\sigma}(a) > r$ and $\sigma(c) = -1$ if $m_{f,\sigma}(a) = r$. (Such c exists by SAP.) For every $\sigma \in X$,

$$\text{if } \sigma(c) = 1, \text{ then } m_{f,\sigma}(ca) = m_{f,\sigma}(a) > r \geq n/2;$$

$$\text{if } \sigma(c) = -1, \text{ then } m_{f,\sigma}(ca) = m_{f,\sigma}(-a) = n - r \leq n/2.$$

Therefore (2) implies that $m_f(ca) = n - r$. Hence, $n = m_f(a) + m_f(b)$, where $b = ca$ and $ab = c \neq 1$. □

THEOREM 9. If (X,G) satisfies SAP, then m classifies forms over (X,G).

The finite case of this theorem is in [S1, Th. 2.17]. The extension to the general case is due to A. Wadsworth.

PROOF. Suppose f, g are forms over X with $m_f = m_g$. By Proposition 8 (3) we know the dimensions are equal. Say $n = \dim f = \dim g$. Replacing f by af and g by ag for suitable $a \in G$ (using SAP), we may assume that $\sigma(f) \geq 0$ for every $\sigma \in X$. Then $m_{f,\sigma}(1) \geq n/2$ for every σ, and Proposition 8 (2) implies that $m_g(1) = m_f(1) \geq n/2$. Then $m_{g,\sigma}(1) \geq n/2$ as well.

Let $F_i = \{\sigma \in X \mid m_{f,\sigma}(1) = i\}$ and $G_i = \{\sigma \in X \mid m_{g,\sigma}(1) = i\}$. Letting $\ell = [(n+1)/2]$, we have

$$X = \bigcup_{i=\ell}^{n} F_i,$$

a disjoint union of clopen subsets, and similarly

$$X = \bigcup_{i=\ell}^{n} G_i.$$

To show $f \cong g$, we must prove $\sigma(f) = \sigma(g)$ for all $\sigma \in X$. That is, we need to prove $F_i = G_i$ for each i with $\ell \leq i \leq n$. Since these are disjoint unions, it suffices to show that $F_i \cap G_j = \emptyset$ whenever $i \neq j$.

Suppose for some i, j we have $C = F_i \cap G_j \neq \emptyset$. Then by SAP there exists $b \in G$ with $C = \{\sigma \in X \mid \sigma(b) = -1\}$. Note that i, j lie between n and $n/2$, so $n - i$ and $n - j$ lie between 0 and $n/2$. Take any $\sigma \in X$.

If $\sigma \notin C$, then $m_{f,\sigma}(b) = m_{f,\sigma}(1) \geq n/2$ and $m_{g,\sigma}(b) = m_{g,\sigma}(1) \geq n/2$.

If $\sigma \in C$, then $m_{f,\sigma}(b) = m_{f,\sigma}(-1) = n - i \leq n/2$ and $m_{g,\sigma}(b) = m_{g,\sigma}(-1) = n - j \leq n/2$.

Then Proposition 8 (2) implies that $m_f(b) = n - i$ and $m_g(b) = n - j$. Since $m_f = m_g$ we conclude that $i = j$. This completes the proof. □

It quickly follows that if (X,G) is a group extension of a SAP space, then m classifies forms over (X,G). We can prove the converse for spaces belonging to the following category S:

DEFINITION 10. In the category of all spaces of orderings, let S be the smallest subcategory such that:

(i) S contains every space satisfying SAP;

(ii) If $X_1, X_2 \in S$ then $X_1 \oplus X_2 \in S$;

(iii) If X is a group extension of $X' \in S$, then $X \in S$.

Note that S contains the category C of all spaces of finite chain length [M4, Th. 1.6]. Also S is contained in the category E defined in [M2, Rem. 5.12].

The spaces in S are exactly those spaces which can be built from SAP spaces by using the direct sum and group extension operations a finite number of times. Using this description the following useful properties of S are easily proved:

(1) If X is a group extension of X', then $X \in S$ iff $X' \in S$.

(2) If $X = X_1 \oplus X_2$, then $X \in S$ iff $X_1, X_2 \in S$.

(3) If $X \in S$ then either $|X| = 1$, or X is a proper group extension, or X is a nontrivial direct sum.

It follows that if $X \in S$ is not decomposable as a direct sum, then X is a group extension of a space $X' \in S$ where either $|X'| = 1$ or X' is a nontrivial direct sum.

CLASSIFICATION THEOREM 11. Let (X,G) be a space in the category S. Then m classifies forms over (X,G) if and only if (X,G) is a group extension $(X',G') \times H$ where (X',G') is a space satisfying SAP.

PROOF. The "if" part follows immediately from Theorem 9 and Corollary 4. Conversely suppose m classfies forms over (X,G). We first consider the case X is decomposable.

CLAIM. If (X,G) is decomposable (nontrivially) as a direct sum, then (X,G) satisfies SAP.

Indeed, suppose that X does not satisfy SAP but is decomposable: $(X,G) \cong (X_1, G_1) \oplus (X_2, G_2)$. Switching summands if necessary, we may assume that (X_1, G_1) does not satisfy SAP. Then by Proposition 7 (4") there is a ternary form f_1 over X_1 with $m_{f_1} \leq 1$. Let $g_1 = f_1$ over X_1 and define forms $f_2 = \langle 1,1,-1\rangle$ and $g_2 = \langle 1,-1,-1\rangle$ over X_2. Then $f = (f_1,f_2)$ and $g = (g_1,g_2)$ are forms over X. Clearly $f \not\cong g$ since $f_2 \not\cong g_2$ over X_2. We will show that $m_f = m_g$, and hence m does not classify forms over X. First note that $D_{G_i}(f_i) = D_{G_i}(g_i)$ for both i, and therefore $D_G(f) = D_G(g)$. But by Proposition 5 we have $m_f \leq 1$ and $m_g \leq 1$, and it follows that $m_f = m_g$. This proves the claim.

Now assume that (X,G) is not decomposable. Since (X,G) is in S, it must be a group extension $(X',G') \times H$ where either $|X'| = 1$ or X' is decomposable. Since m classifies forms over X, Corollary 4 implies that m classifies forms over X'. By the claim above, we conclude that (X',G') satisfies SAP. □

The smallest possible space where m does not classify forms has $|X| = 5$ and $|G| = 16$. It is $X \cong X_1 \oplus X_2$ where X_1 is a 4-element fan and $|X_2| = 1$.

COROLLARY 12. If (X,G) is in S, then m classifies ternary forms over (X,G) iff m classfies all forms over (X,G).

PROOF. The proofs of Theorem 11 and Corollary 4 still remain valid under the weaker ternary hypothesis. □

It is worth noting that over any space (X,G), ternary forms are classified by the two invariants m and det. Indeed, suppose f, g are ternary with $D(f) = D(g)$ and $\det f = \det g$. After scaling we have $\det f = \det g = 1$, so that $\phi = \langle 1 \rangle \perp f$ and $\psi = \langle 1 \rangle \perp g$ are 2-fold Pfister forms. Since $D(f) = D(g)$ it follows that $4 \times \phi \cong \phi \otimes \psi \cong 4 \times \psi$. This implies $\phi \cong \psi$ and hence $f \cong g$. (Compare [BK, p. 158].)

Larger examples of (anisotropic) forms f, g with $m_f = m_g$ but $f \ncong g$ can be easily constructed over some spaces (X,G). Using the idea in the proof of Theorem 11, we need only find an n-dimensional form f_1 over a space (X_1, G_1) with $m_{f_1} \leq 1$. For instance, if (X_1, G_1) is a group extension of (X_1', G_1') by H where $|H| \geq n$, such an f_1 is quickly found.

REMARK 13. Let us now return to the original question about pythagorean fields. Let K be a pythagorean field with only finitely many real-valued places. Then by [Cl, Th. 2.1] and [M4, Th. 1.6], the space $(X,G) = (X_K, \dot{K}/\dot{K}^2)$ has finite chain length, hence belongs to the category $C \subseteq S$. Suppose that (X,G) is a group extension of (X',G') by H. Then (X', G') also has finite chain length and therefore [Cl, Lem. 2.4], there is a pythagorean field k with $(X',G') \cong (X_k, \dot{k}/\dot{k}^2)$. Then the Witt rings are $W(K) \cong W(X,G) \cong W(X',G')[H] \cong W(k)[H]$, a group ring. Let L be the iterated power series field over k, where the number of variables equals the cardinality of an $\mathbb{F}_2$-vector space basis of H. (So if $|H| = 2^r$, use r variables.) Then $W(L) \cong W(k)[H]$, so that

K and L are equivalent. Here we follow Harrison and Cordes ([Co] and [L, p. 294, Ex. 13]), and define two fields to be <u>equivalent</u> if their Witt rings are isomorphic. These remarks and Theorem 11 complete the proof of the Theorem stated in the Introduction.

m, DIMENSION, AND OTHER INVARIANTS

We have found examples of spaces (X,G) with forms f, g having $m_f = m_g$ but $f \not\cong g$. for each of these examples we had $\dim f = \dim g$. In fact, this equality of dimesnions must always occur, at least if X is in S. This is a consequence of the next theorem.

THEOREM 14. Suppose (X,G) is a space in the category S, and f, g are anisotropic forms over (X,G) with $m_f(a) \geq m_g(a)$, for all $a \in G$. Then $\dim f \geq \dim g$.

PROOF. First, suppose (X,G) satisfies SAP. Then by Proposition 8, there exist distinct $a,b \in G$ with $\dim g = m_g(a) + m_g(b)$. By another application of Proposition 8, we have $\dim g \leq m_f(a) + m_f(b) \leq \dim f$.

Next suppose $(X,G) = (X_1,G_1) \oplus (X_2,G_2)$ is a direct sum where the theorem holds over both spaces (X_i,G_i). Let $r = \max\{m_g(a) \mid a \in G$ and choose b with $m_g(b) = r$. Then $m_{g_i}(b_i) \geq r$. Also $m_f(b) \geq r$, so that $m_{f_i}(b_i) \geq r$. Switching indices if necessary, we have $m_{g_1}(a_1) \leq r$ for all $a_1 \in G_1$. For otherwise there exists $a_i \in G_i$ with $m_{g_i}(a_i) \geq r$, for each i, and Proposition 5 implies $m_g(a) > r$, where $a = (a_1,a_2) \in G$.

CLAIM. $m_{f_1}(a_1) \geq m_{g_1}(a_1)$ for all $a_1 \in G_1$. If $m_{f_1}(a_1) \geq r$ the claim is clear. Suppose $m_{f_1}(a_1) < r$ and let $a' = (a_1,b_2) \in G$. Then

$$m_{f_1}(a_1) = \min\{m_{f_1}(a_1), m_{f_2}(b_2)\} = m_f(a') \geq m_g(a')$$

$$= \min\{m_{g_1}(a_1), m_{g_2}(b_2)\} = m_{g_1}(a_1).$$

This proves the claim. Applying the theorem over (X_1,G_1) then yields $\dim f = \dim f_1 \geq \dim g_1 = \dim g$.

Finally, suppose $(X,G) = (X',G') \times H$ is a group extension where the theorem holds over (X',G'). Express $f = x_1f_1 \perp \dots \perp x_sf_s$, where the $x_i \in H$ are distinct and the f_i are forms defined over G'. Since

$D(g) \subseteq D(f)$ we may assume that $g = x_1 g_1 \perp \cdots \perp x_t g_t$, where $t \leq s$. Fix i with $1 \leq i \leq t$. For any $a' \in G'$ let $a = x_i a'$. Since f, g are anisotropic we have from Corollary 3 that $m_{f_i}(a') = m_f(a) \geq m_g(a) = m_{g_i}(a')$. The theorem over (X', G') then implies that $\dim f_i \geq \dim g_i$. Adding, we conclude $\dim f \geq \dim g$. □

The classical invariants of a quadratic form q over a (pythagorean) field K are the dimension $\dim q$, the determinant $\det q$, the Witt invariant [L, p. 120] $c(q)$ and the signatures $\sigma(q)$. Certainly the signatures alone completely determine the form q. This is Pfister's Local-Global Principle [L, p. 240] and is the first motivation for studying spaces of orderings. The other three classical invariants do not in general determine q, even if the square class invariant is used as well:

PROPOSITION 15. Suppose K is a pythagorean field with finite chain length, and m does not classify forms over K. Then the invariants m_q, $\dim q$, $\det q$, and $c(q)$ do not always determine the form q over K.

PROOF. By hypothesis there exist forms ϕ, ψ over K with $m_\phi = m_\psi$ but $\phi \not\cong \psi$. By Theorem 14 we must have $\dim \phi = \dim \psi$. (Alternatively, we can use Corollary 12 to find such forms ϕ, ψ which are both ternary.) Let $\Phi = 2 \times \phi$ and $\Psi = 2 \times \psi$. Then

$$\dim \Phi = \dim \Psi = 2 \dim \phi, \quad \det \Phi = \det \Psi = 1, \quad c(\Phi) = c(\Psi) = [-1,-1]$$

(use formulas in [L, p. 121]), and $m_\Phi = m_\Psi = 2m_\phi$. However $\Phi \not\cong \Psi$, since otherwise we would have $\phi \cong \psi$, since $W(K)$ is torsion free. □

This result answers negatively a question raised by Snapper in [Sn, p. 151].

Returning to the abstract situation, we ask whether $m_f \leq m_g$ always implies $f < g$ (f is a subform of g), over some spaces (X,G). This fails whenever $|X| \neq 1$, for then we can choose $f = \langle 1,a \rangle$ where $a \in G, a \neq \pm 1$ and choose $g = \langle 1,-1,-1 \rangle$. However, what if f and g are required to be anisotropic?

PROPOSITION 16. Suppose (X,G) is in the category S. Then (X,G) satisfies the following property:

(*) whenever f, g are anisotropic forms over (X,G) with $m_f \leq m_g$, then $f \leq g$, if and only if (X,G) is a group extension of (X',G') where $|X'| \leq 2$.

For the proof we employ an easy lemma (whose proof is left to the reader).

LEMMA 17. Suppose (X,G) is a group extension of (X',G'). Then (X,G) satisfies property (*) iff (X',G') satisfies (*). □

PROOF OF PROPOSITION 16. If $|X'| \leq 2$ the property (*) can be checked directly, so the "if" part follows from the lemma. Suppose (X,G) satisfies (*). Then certainly m classfies forms over (X,G), so Theorem 11 implies that (X,G) is a group extension of a SAP space. By the lemma, it suffices to show that a SAP space (X,G) with $|X| \geq 3$ cannot satisfy property (*). Since $|X| \geq 3$, there exist clopen sets U, V in X with $\emptyset \subsetneq U \subsetneq V \subsetneq X$. Since X satisfies SAP, there exist $a,b \in G$ with $U = X(a) = \{\sigma \in X \mid \sigma(a) = 1\}$ and $V = X(b)$. Since $X(a) \subseteq X(b)$ we have $D(\langle 1,b \rangle) \subseteq D(\langle 1,a \rangle)$, by [M1, Lem. 2.1]. Let $f = \langle 1,b \rangle$ and $g = \langle 1,a,ab \rangle$ so that $D(f) \subseteq D(g)$. Since $b \neq 1$ we have $m_f \leq 1$ and therefore $m_f \leq m_g$. By using any $\sigma \in U$, we see that f and g are anisotropic. Finally, computing the forms at any $\sigma \in V - U$ we conclude that $f \not\sim g$. □

BIBLIOGRAPHY

[BK] E. Becker and E. Köpping, Reduzierte quadratische Formen und Semiordnungen reeler Körper, Abh. Math. Sem. Univ. Hamburg 46 (1977) 143-177.

[Co] C. Cordes, The Witt group and the equivalence of fields with respect to quadratic forms, J. Algebra 26 (1973) 400-421.

[C1] T. Craven, Characterizing reduced Witt rings of fields, J. Algebra 53 (1978) 68-77.

[C2] __________, Characterizing reduced Witt rings II, Pac. J. Math. 80 (1979) 341-349.

[EL] R. Elman and T.-Y. Lam, Quadratic forms over formally real fields and Pythagorean fields, Amer. J. Math. 94 (1972) 1155-1194.

[J] B. Jacob, On the structure of Pythagorean fields, J. Algebra (to appear).

[L] T.-Y. Lam, The Algebraic Theory of Quadratic Forms, W. A. Benjamin, Reading, Massachusetts, 1973.

[M1] M. Marshall, Classification of finite spaces of orderings, Can. J. Math. 31 (1979) 320-330.

[M2] __________, Quotients and inverse limits of spaces of orderings, Can. J. Math. 31 (1979) 604-616.

[M3] __________, The Witt ring of a space of orderings, Trans. Amer. Math. Soc. (to appear).

[M4] __________, Spaces of orderings IV, Can. J. Math. (to appear).

[P] A. Prestel, Lectures on Formally Real Fields, Monografias de Math. 22, IMPA, Rio de Janeiro, 1975.

[RW] A. Rosenberg and R. Ware, Equivalent topological properties of the space of signatures of a semilocal ring, Publ. Math. Debrecen 23 (1976) 283-289.

[S1] A. Solow, The square class invariant for quadratic forms and the classification problem, Lin. and Multilin. Algebra (to appear).

[S2] __________, The square class invariant for quadratic forms over local fields, preprint, 1979.

[Sn] E. Snapper, Quadratic spaces over finite fields and codes, C. R. Math. Rep. Acad. Sci. Canada 1 (1979) 149-152.

DEPARTMENT OF MATHEMATICS
OHIO STATE UNIVERSITY
COLUMBUS, OHIO 43210

and

DEPARTMENT OF MATHEMATICS
UNIVERSITY OF CALIFORNIA
BERKELEY, CALIFORNIA 94720

Contemporary Mathematics
Volume 8, 1982

COHERENT ALGEBRAIC SHEAVES IN REAL ALGEBRAIC GEOMETRY

Alberto Tognoli

INTRODUCTION.

Let (X, O_X) be a real affine variety and $F \longrightarrow X$ an algebraic vector bundle. F is called A-<u>coherent</u> if the total space F is an affine variety.

The following result is known (see [1]):

THEOREM. Let X be a C^∞ compact manifold; then X is diffeomorphic to an algebraic variety (X_a, O_X) such that any topological vector bundle $F \longrightarrow X_a$ is equivalent (as vector bundle) to an A-coherent vector bundle $F_a \longrightarrow X_a$.

For any compact affine real variety (Y, O_Y) we have: if two A-coherent vector bundles $F' \longrightarrow Y$, $F'' \longrightarrow Y$ are topologically equivalent, then they are algebraically equivalent.

From the above results one could hope to reduce the algebraic classification of vector bundles to the topological one.

The purpose of this paper is to discuss some counterexamples that prove the bounds of this principle.

In fact we have (see [6]):

1) For any $m > 1$ there exists an algebraic bundle $F \longrightarrow \mathbb{R}^n$ such that F is not A-coherent and it is not algebraically trivial.

2) There exists a regular, compact affine variety X of dimension 2 and a topological line bundle $F \longrightarrow X$ such that F has no algebraic structure.

There exists a regular, compact, connected affine variety Y of dimension 5 and a topological line bundle $F \longrightarrow Y$ that has no algebraic structure.

1980 Mathematics Subject Classification. 12J15, 14G30, 58A07

0271-4132/81/0000-0684/$03.50

§1. DEFINITIONS AND WELLKNOWN FACTS

In the following we shall consider algebraic varieties defined on $\mathbb{R}$ or $\mathbb{C}$ endowed with the Zariski topology.

Let $(X, \mathcal{O}_X)$ be a closed algebraic subvariety of $\mathbb{R}^n$; we shall call complexification of X the smallest closed algebraic subvariety $\tilde{X}$ of $\mathbb{C}^n$ that contains X.

Given a real affine variety $(X, \mathcal{O}_X)$ we shall call complexification of X any complexification associated to an embedding $X \hookrightarrow \mathbb{R}^n$.

It is easy to verify that (see [2]):

i) The complexification $\tilde{X}$ of X is uniquely determined, up to isomorphisms, as a germ near X.

ii) Any regular rational function $f : X \longrightarrow \mathbb{R}$ is the restriction of a regular rational function defined on a neighbourhood of X in $\tilde{X}$.

It is known ([2]) that if K is a non-algebraically closed field, then the projective space is an affine variety.

Let $(X, \mathcal{O}_X)$ be an affine variety defined on the field K, i.e., a closed subvariety of K^n; an algebraic sheaf $\mathcal{F}$ is called A-coherent if there exists an exact sequence:

$$\mathcal{O}_X^m \longrightarrow \mathcal{O}_X^p \longrightarrow \mathcal{F} \longrightarrow 0.$$

An algebraic vector bundle $F \longrightarrow X$ is called A-coherent if the associated sheaf is A-coherent.

We shall use the following results (see [2], [1]):

THEOREM 1. *Let K be any field and $(X, \mathcal{O}_X)$ an affine variety defined on K. The following conditions are equivalent:*

i) $\mathcal{F} \longrightarrow X$ is an A-coherent sheaf.

ii) There exists a coherent sheaf $\hat{\mathcal{F}} \longrightarrow \operatorname{Spec} \Gamma(\mathcal{O}_X)$ such that $\hat{\mathcal{F}}|_X = \mathcal{F}$, where $\Gamma(\mathcal{O}_X)$ = ring of rational regular functions.

THEOREM 2. *Let K be any field and $(X, \mathcal{O}_X)$ an affine variety defined on K.*

The functor $\Gamma : \mathcal{F} \longrightarrow \Gamma(\mathcal{F})$ of the global sections, on X, is exact on the category of A-coherent sheaves.

THEOREM 3. *Let $(X, \mathcal{O}_X)$ be an affine variety defined on $\mathbb{R}$ and $F \longrightarrow X$ an algebraic vector bundle.*

Then the following conditions are equivalent:

i) F is A-coherent.

ii) The total space F is an affine variety.

iii) <u>There exists an algebraic map</u> $\phi : X \longrightarrow G_{n,p}$, $G_{n,p}$ = <u>Grassmanian manifold, such that</u> $F \simeq \phi^*$ (<u>tautological bundle</u>) <u>where</u> $\simeq$ <u>is an algebraic isomorphism</u>.

iv) <u>There exists an algebraic vector bundle</u> $F' \longrightarrow X$ <u>such that</u> $F \oplus F'$ <u>is algebraically isomorphic to the trivial bundle</u>.

Let (X, O_X) be a real affine variety and $F \longrightarrow X$ an A-coherent vector bundle.

Let us suppose $\| \ \|$ to be a continuous norm defined on the fibers of F. For any compact K of X (considered with the Euclidean topology) and any section $\gamma : k \longrightarrow F$ let $\|\gamma\|_k = \sup_{n \in k} \|\gamma(n)\|$. Under these assumptions we have:

COROLLARY 1. For any continuous section $\gamma : X \longrightarrow F$ of the A-coherent vector bundle $F \longrightarrow X$, any compact set $K \subset X$ and $\varepsilon > 0$ there exists an algebraic section $\gamma' : X \longrightarrow F$ such that:

$$\|\gamma - \gamma'\|_K < \varepsilon$$

PROOF. The sheaf $\mathcal{F}$ associated to F has a resolution

$$O_X^m \longrightarrow O_X^p \longrightarrow \mathcal{F} \longrightarrow 0.$$

The functor Γ is exact: hence the continuous section γ can be written

$$\gamma = \sum_{i=1}^{t} \alpha_i \gamma_i$$

where α_i are continuous functions and γ_i algebraic sections.

It is now enough to apply the Weierstrass Approximation Theorem to the α_i to obtain the conclusion of the corollary.

REMARK 1. Let (X, O_X) be an affine real variety and F', F'' two A-coherent vector bundles on X.

From Theorem 3 it is easy to deduce that the vector bundle $\mathrm{Hom}(F', F'')$ is A-coherent.

From the corollary we deduce that, if X is compact, we have:

$$F' \overset{\text{topol.}}{\simeq} F'' \Longrightarrow F' \overset{\text{alg.}}{\simeq} F''.$$

REMARK 2. The result of the corollary may be refined considering the C^r topology on the set of sections or the relative case (see [3]).

REMARK 3. Let (X, O_X) be a real affine variety and $F \longrightarrow X$ a topological vector bundle. Let $\phi : X \longrightarrow G_{n,p}$ be a continuous map such that $F \simeq \phi^*$ (tautological bundle).

From Theorem 3 we see that the following problems are equivalent:

1) To find on F a structure of an A-coherent vector bundle.

2) To find an algebraic map $\phi' : X \longrightarrow G_{n,p}$ such that ϕ' is homotopic to ϕ.

We end this paragraph giving some examples that justify the definition of A-coherent vector bundle.

Let us consider $P \in \mathbb{R}[X_1, \ldots, X_n]$, $n > 1$, such that P is irreducible and $\{P = 0\}$ has two connected components.

EXAMPLE.
$$P = X_1^2(X_1-1)^2 + \sum_{i>1} X_i^2.$$

We have $\{P = 0\} = (0, \ldots, 0) \cup (1, 0, \ldots, 0)$.

Let us now consider on the covering $\mathbb{R}^n - (0, \ldots, 0)$, $\mathbb{R}^n - (1, 0, \ldots, 0)$ the cocycle $g_{12} = 1/p$.

It is possible to verify (see [2]) that:

1) The line bundle $F \longrightarrow \mathbb{R}^n$ defined by g_{12} is algebraic but not A-coherent. Hence F is not affine and F is not algebraically trivial.

2) The cocycle g_{12} is not trivial hence $H^1(\mathbb{R}^n, O_{\mathbb{R}^n}) \neq 0$ if $n > 1$.

3) The algebraic global sections of F do not generate the stalk in any point of $\mathbb{R}^n$.

REMARK 4. It is known ([4]) that if (X, O_X) is an affine real variety and F is an A-coherent sheaf, we have $H^q(X, F) = 0$, if $q > 1$.

§2. THE EXAMPLES

Let $C_{\xi,\eta}$ be the plane real curve given by the equation:

$$X^4 + Y^4 - 2\eta X^2 + \xi = 0,$$

where $\eta^2 > \xi > 0$. It is an easy verification that $C_{\xi,\eta}$ is compact, irreducible, regular and it has two connected components separated by the line $X = 0$. Fix one of these curves, say $C = C_{\bar{\xi},\bar{\eta}}$ and let $C = C_0 \cup C_1$ be the decomposition into connected components.

Let

$$S^1 = \{(z, w) \in \mathbb{R}^2 \mid z^2 + w^2 - 1 = 0\}$$

and

$$W = W_0 \cup W_1 = C \times S^1 \subset \mathbb{R}^2 \times \mathbb{R}^2, \quad W_i = C_i \times S^1, \quad i = 0, 1.$$

Choose $\kappa_0 \in S^1$ and let $D = C_0 \times \{\kappa_0\} \subset W_0$.

We shall denote by $[D] \xrightarrow{\pi} W$ the line bundle induced on W by the smooth divisor D. Clearly $[D]$ has a C^∞ structure. Now we have:

THEOREM I. <u>The line bundle</u> $[D] \xrightarrow{\pi} W$ <u>has no algebraic structure</u>.

The proof of Theorem I consists of several steps. We begin with:

LEMMA 1. The line bundle $[D] \longrightarrow W$ has no structure of an A-coherent bundle.

PROOF. Let us suppose the contrary, then $[D]$ has an A-coherent structure.

From the definition of $[D]$ there exists a C^∞ section $\gamma : W \longrightarrow [D]$ such that $D = \{\kappa \in W \mid \gamma(\kappa) = 0\}$ and γ is transverse to the zero section.

Now we can apply the corollary of §1 and approximate γ by an algebraic section $\gamma' : W \longrightarrow [D]$ such that $D' = \{\gamma' = 0\}$ is isotopic to D and very near to D.

Clearly D' is a regular algebraic curve and, if D' is near enough to D, the map $\gamma : D' \longrightarrow \gamma(D') \subset C$ is an analytic isomorphism, where $p : W \longrightarrow C$ is the natural projection.

In fact in these hypotheses we have $p(D') = C_0$.

The following fact is known (see [5]):

(P) Let X be a regular affine variety and $\phi : X \longrightarrow \mathbb{R}^n$ an algebraic map such that $\phi : X \longrightarrow \phi(X)$ is an analytic isomorphism. Under these hypotheses there exists an algebraic closed subvariety Y of $\mathbb{R}^n$ such that: $Y \supset \phi(X)$, $\dim Y - \phi(X) < \dim X$.

It is now clear that we cannot have $p(D') = C_0$ because the smallest algebraic variety containing C_0 is $C = C_0 \cup C_1$ and $\dim C_0 = \dim C_1 = 1$. So we have proved that $[D]$ has no A-coherent structure.

LEMMA 2. Let (X, O_X) be an affine real regular variety and $F \xrightarrow{\pi} X$ an algebraic line bundle.

There exists an algebraic closed subvariety $S \subset X$ such that:

i) $\pi^{-1}(X-S) \longrightarrow X - S$ is A-coherent.

ii) $\dim S \leq \dim X - 2$.

PROOF. X is regular, so we may prove the lemma for the irreducible components of X. In the following we shall suppose X irreducible.

Let us now fix complexification $(\tilde{X}, O_{\tilde{X}})$ of (X, O_X). We can now find an open covering $\mathfrak{U} = \{U_i = X - T_i\}$ of X and an algebraic cocycle $g_{ik} : U_i \cap U_k \longrightarrow \mathbb{R}^* = \mathbb{R} - \{0\}$ that defines the bundle F.

It is known (see [2]) that we can find open neighbourhoods $\tilde{U}_{ik}$ of $U_{ik} = U_i \cap U_k$ in $\tilde{X}$ and regular rational functions

$$\tilde{g}_{ik} : \tilde{U}_{ik} \longrightarrow \mathbb{C}^n = \mathbb{C} - \{0\} \quad \text{such that} \quad \tilde{g}_{ik}|_{U_{ik}} = g_{ik}.$$

From the fact that X, and hence $\tilde{X}$, is irreducible we deduce that the $\tilde{g}_{ik}$ satisfy the compatibility conditions (where they are defined).

To find a complexification of the cocycle g_{ik} and hence of F it is sufficient to find open neighbourhoods $\tilde{U}_i$ of U_i in $\tilde{X}$ such that $\tilde{U}_i \cap \tilde{U}_k = \tilde{U}_{ik}$. From the example given in Remark 3 of §1, we know that this is, in general, impossible.

Now we wish to find a closed subvariety S of X such that the cocycle

$$g'_{ik} = g_{ik}|_{U_{ik}-S}$$

can be complexified in the sense just explained and $\dim S < \dim X - 1$. We remark that it is enough to prove that, given a pair i, k of indexes there exists a closed subvariety S_{ik} of X and two open sets $\tilde{U}'_i$, $\tilde{U}'_k$ of $\tilde{X}$ such that:

$$\tilde{U}'_i \cap \tilde{U}'_k \supset U_{ik} - S_{ik}, \quad \tilde{U}'_i \cap \tilde{U}'_k \subset \tilde{U}_{ik}, \quad \dim S_{ik} < \dim X - 1.$$

In fact in this case it is enough to consider $\bigcup_{i,k} S_{ik}$.

Let now i, k be fixed, then we may write $g_{ik} = h/f$ where h, f are regular functions on X and $h(\kappa) \neq 0$, $f(\kappa) \neq 0$ if $\kappa \in U_{ik}$. From the fact that X is regular we know that $\Gamma(O_X)$ is a factorial ring.

Let $h = h_1, \ldots, h_j$, $f = f_i, \ldots, f_t$ be a decomposition of h, f into a product of prime factors.

Now we remark the following: let ϕ a prime element in $\Gamma(O_X)$; then if the algebraic variety $X_\phi = \{\kappa \in X \mid \phi(\kappa) = 0\}$ is reducible, we have $\dim X_\phi < \dim X - 1$.

In fact let us suppose:

(1) $\dim X_\phi = \dim X - 1$.

If $\tilde{\phi}$ is an extension of ϕ to $\tilde{X}$ then $\tilde{X}_\phi = \{\kappa \in \tilde{X} \mid \tilde{\phi}(\kappa) = 0\}$ is a complex hypersurface of $\tilde{X}$.

From (1) we deduce that there exists an irreducible component $\tilde{X}'_\phi$ of $\tilde{X}_\phi$ such that $\tilde{X}'_\phi$ is the complexification of $\tilde{X}'_\phi \cap \tilde{X}$. Let ψ be an equation of $\tilde{X}'_\phi$; we may suppose ψ has real coefficients and we know that ψ divides ϕ. From the fact that ϕ is prime it follows that $X_\phi = \tilde{X}'_\phi \cap X$.

$\tilde{X}'_\phi$ is irreducible and it is the complexification of X_ϕ hence X_ϕ is irreducible. The remark is now proved.

From the remark we deduce that there exists a closed subvariety S_{ik} of X such that:

1) The subvarieties $\{h_j = 0\} \cap (X - S_{ik})$, $j = 1, \ldots, s$, $\{f_j = 0\} \cap (X - S_{ik})$, $j = 1, \ldots, t$ are irreducible.

2) $\dim S_{ik} < \dim X - i$.

In fact S_{ik} is the union of the reducible varieties of the form

$$\{h_j = 0\} \cap X \quad \text{or} \quad \{f_j = 0\} \cap X.$$

Let us now consider the cocycle:

$$g_{ik} = \frac{h_1, \ldots, h_s}{f_1, \ldots, f_t} : U_{ik} - S_{ik} \longrightarrow \mathbb{R}^n.$$

All the subvarieties $\{h_j = 0\}$, $\{1_j = 0\}$ of $X - S_{ik}$ are irreducible and hence contained in $T'_k = X - (S_{ik} \cup U_k)$ or in $T'_i = X - (S_{ik} \cup U_i)$.

Let now $\tilde{T}'_i$, $\tilde{T}'_k$ be two complexifications of T'_i, T'_k in $\tilde{X}$ and $U'_i = U_i - S_{ik}$, $U_k = U_k - S_{ik}$.

It is now clear, from the construction, that the cocycle $g_{ik}|_{U'_i \cap U'_k}$ can be complexified to an open covering of the form:

$$\{\tilde{X} - \tilde{T}'_i \cup \tilde{S}_{ik},\ \tilde{X} - \tilde{T}_k \cup \tilde{S}_{ik}\}.$$

The assertion is now proved for i, k and hence, as remarked above, we can complexify the line bundle.

It is known (see [2]) that an algebraic vector bundle can be complexified if and only if it is A-coherent.

The lemma is now proved.

PROOF (of Theorem I). Let us now suppose that the line bundle $[D] \xrightarrow{\pi} W$ is algebraic.

From Lemma 2 there exists a finite set $A = \{\alpha_1, \ldots, \alpha_q\}$ of W such that $\pi^{-1}(W - A) \longrightarrow W - A$ is A-coherent.

The space $W - A$ is compact, in the Zariski topology, hence we find a finite number of algebraic sections $\gamma_1, \ldots, \gamma_t$ of F that generate the stalk of $\pi^{-1}(W - A)$ in any point.

So we deduce, as in Corollary 1 of §1, that any continuous section $\gamma : W - A \longrightarrow \pi^{-1}(W - A)$ can be approximated, on compact sets, by algebraic sections.

From the approximation property just proved, it follows that, given a set of small balls $B_1 \ni \alpha_1, \ldots, B_q \ni \alpha_q$ in W, there exists an algebraic section γ_α of $\pi^{-1}(W - A)$ such that

$$\{g_\alpha = 0\} \cap (W - (\bigcup_i B_i)) = \check{D}$$

is analytic regular curve isotopic to D and very near to D.

Let $\hat{D}$ be the smallest algebraic subvariety of W containing $\check{D}$. By construction $\hat{D}$ is an algebraic irreducible curve composed of the component $\check{D}$ and some other pieces $\hat{D}_i = \hat{D} \cap B_i$, $i = 1, \ldots, q$. Let $p : W = C \times X^1 \longrightarrow C$ be the natural projection and $\tilde{W}$, $\tilde{C}$, $\tilde{\hat{D}}$, $\hat{D} \subset \tilde{\hat{W}}$, some complexifications of W, C, $\hat{D}$.

If $\tilde{p}$ is the complexification of p we have that:

(1) $\tilde{p} : \hat{D} \longrightarrow C$ is ramified covering.

C and $\hat{D}$ are irreducible, hence the degree of the covering is well defined.

From the fact that $p : \check{D} \longrightarrow p(\check{D})$ is an analytic isomorphism (this is true because $\check{D}$ is very near to D) we deduce that the degree of the covering (1) is odd.

The degree of the map $p|\hat{D}_i$ is clearly even, hence the $\hat{D}_i$ are contained in the ramification of (1) or $p(\hat{D}_i) \subset C_0$.

Finally we have deduced that the $p(\hat{D}_i) \cap C$ are finite sets and we can use the argument (P) of the Lemma 1 and derive the desired contradiction.

The theorem is now proved.

REMARK 5. If (X, O_X) is a real, affine, compact curve then any topological line bundle $F \longrightarrow X$ has an A-coherent structure (see [2]) hence the minimal dimension for any counterexample is two.

THEOREM II. *There exists a regular, compact, connected real affine variety* X, *of dimension* 5 *and a line bundle* $F \longrightarrow X$ *such that* F *has no algebraic structure*.

PROOF. By $W = W_0 \cup W_1 = C_0 \times S^1 \cup C_1 \times S^1$, $D = C_0 \times \{x_0\}$, we shall denote the non-connected surface of $\mathbb{R}^n$ defined before Theorem I and the curve D on W_0.

Set $Y = (S^1 \times S^1) \times (S^1 \times S^1) \times S^1 = V \times V \times S^1$.

Let us fix $w_0 \in V$, $s_0 \neq s_1$, $s_i \in S^1$ and denote:

$$V_0 = V \times \{w_0\} \times \{s_0\} \subset Y, \quad V_1 = \{w_0\} \times V \times \{s_1\} \subset Y.$$

In the following we shall identify $\mathbb{R}^4$, $\mathbb{R}^4 \supset W$, and $\mathbb{R}^4 \times \{0\} \subset \mathbb{R}^4 \times \mathbb{R}^N$.

Fix two diffeomorphisms $f_i : V_i \longrightarrow W_i$, $i = 0, 1$; if N is big enough there exists (by Whitney) an embedding $f : Y \longrightarrow \mathbb{R}^4 \times \mathbb{R}^N$ which extends the f_i (that is, $f|_{V_i} = (f_i, 0)$).

Put $Y' = f(Y)$ and assume that $f_0(E) = 1)$ where

$$E = S^1 \times \{x_0\} \sim S^1 \times \{x_0\} \times \{w_0\} \times \{s_0\} \subset Y.$$

The normal bundle of W in Y' is trivial, hence, by Theorem 3.2 of [1], we can approximate Y' by a regular algebraic variety X such that: $X \supset W$, X is very near to Y' and there is a diffeomorphism $q : Y' \longrightarrow X$.

The variety X shall be the base of our line bundle.

Let us now denote by $[E] \longrightarrow V$ the line bundle associated to the divisor E. $[E]$ has a C^∞ structure. We remark that $[E]$ coincides with $f^*([D])$.

Let $p : V \times V \times S^1 \longrightarrow V$ be the projection on the first factor (the first factor contains E).

We shall denote by F' the line bundle $p^*([E]) \longrightarrow Y$. The line bundle $F \xrightarrow{\pi} X$ of the theorem is now:

$$F \overset{\text{def}}{=} (g \cdot f)^*(F').$$

We must verify that $F \xrightarrow{\pi} X$ has no algebraic structure.

Now we remark that X is a variety that contains the surface W and F is a line bundle that extends the bundle $[D]$ of Theorem I. So the existence of an algebraic structure on F should imply that there exists an algebraic structure on $[D] \longrightarrow W$ and this is impossible by Theorem I. The theorem is now proved.

REMARK 6. The relevance of the example of Theorem II is in the fact that the base space X is connected.

We believe that it should be possible to find counterexamples in a lower dimension (for example 3).

BIBLIOGRAPHY

[1] R. Benedetti, A. Tognoli: "On real algebraic vector bundles." Bull. Sc. Math. II serie 104 (1980), pp. 89-112.

[2] A. Tognoli: "Algebraic geometry and Nash functions." Institutiones mathematicae. Vol. III. Acad. Press, London and New York.

[3] R. Benedetti, A. Tognoli: "Approximation theorems in real algebraic geometry." To appear on Supplemento G.N.S.A.G.A.

[4] R. Silhol: "Nullité de certaines groupes de cohomologie des variétés sur un corps non algébriquement clos." Bollettino U.M.I.

[5] A. Tognoli: "Su una congettura di Nash." Ann. Sc. N. S. di Pisa 27 (1973), pp. 167-185.

[6] R. Benedetti, A. Tognoli: "Counterexamples in real algebraic geometry." To appear.

DEPARTMENT DE MATHEMATIQUES
UNIVERSITE DE TOURS

and

INSTITUTO DI MATEMATICA
UNIVERSITA DI FERRARA
FERRARA

Contemporary Mathematics
Volume 8, 1982

DIGGING HOLES IN ALGEBRAIC CLOSURES A LA ARTIN—II[1]

Antonio J. Engler[2] and T. M. Viswanathan[2]

ABSTRACT: The authors study fields which are maximal with respect to the exclusion of three elements in algebraically closed fields and present all ordered fields appearing in this situation. The absolute Galois groups are determined as suitable semidirect products with explicit group action. The fields in question are characterized in simple ways and these in turn show how examples may be constructed at will.

The idea of descending from the top of an algebraically closed field to convenient subfields of Ω is due to Artin (and Schreier). In this way Artin constructed two families of fields k so that every finite extension of k is cyclic. (See Exs. 3 and 4, p. 230 of Lang [6].) This method of digging holes gives interesting examples of (profinite) Galois groups.

Let S be a finite set of elements of Ω and k a subfield Ω maximal with respect to disjunction from S. In [3] we presented results of Artin and Quigley [9] and of McCarthy [7], when S contains one and two elements respectively and studied the situation, when S contains three elements. There we dealt with the situation, when k is a nonreal field. In this work, we complete our study by dealing with the remaining case when k is a real field. Once again, we can describe the absolute Galois group completely. Again we use the u-invariant. Our results give specific examples for the theory in Ware [12] and in Neukirch [8]. We also show that these fields are characterized in simple ways and also by their absolute Galois groups.

1980 Mathematics Subject Classification. 12D15, 12A55, 10C05.

[1] Presented to the Society under the title 'Ordered Fields and Valuation Theory'.

[2] Part of this work was done while the first author was at the Universität Konstanz and the second author at the Universität Regensburg, both under the CNPq-GMD convention.

0271-4132/81/0000-0803/$03.50

The ordered fields studied here again fall into the two classes studied widely in the literature and elucidated in [12] from our point of view. They are either of class C in the sense that their Witt rings are group rings or their Hasse numbers are ≤ 2. Even though our fields are rather near the algebraic closure, they may be surprisingly complicated as shown by Theorem 4. This result is a far-reaching generalization of one due to Ribenboim [10, 11] in the number fields case. We should point out that part of the situation here has already been considered by Bredikhin et al. [2]. Some questions are presented at the end, as they naturally arise in our study.

Recall that k is a 3-*maximal field* if it is maximal with respect to the exclusion of three elements in the algebraic closure Ω of k and k is not a 2-maximal field in the obvious sense. As is standard G_k will denote the absolute Galois group $\mathrm{Gal}(\Omega|k)$; $G_k(\sqrt{-1}$ will be the Galois group $\mathrm{Gal}(\Omega|k(\sqrt{-1})) = G_{k(\sqrt{-1})}$. Recall a main result from [3]:

THEOREM 1. *Let* k *be a* 3-*maximal field. Then we have the following:*

i) *Either every finite extension of* k *has cyclic Galois group;*

ii) *Or* k *is a perfect field,* $\Omega|k$ *is a* 2-*extension in the sense that* Ω *is the quadratic closure of* k *and* $|k^{\cdot}/k^{\cdot 2}| = 4$.

PROPOSITION 1. *Let* k *be a* 3-*maximal real field. Then* $\Omega|k$ *is a* 2-*extension and* k *has either one or two orderings*.

PROOF. Clearly k is not real closed, as it is 3-maximal. Also we are not in situation i) of Theorem 1, as otherwise a real closure R of k would be a normal extension of k and so would have many nontrivial automorphisms. Thus we are in situation ii) and so $|k^{\cdot}/k^{\cdot 2}| = 4$. If $k = \pm k^{\cdot 2} \cup \pm ak^{\cdot 2}$, then $P = k^{\cdot 2} \cup ak^{\cdot 2}$ and $Q = k^{\cdot 2} \cup -ak^{\cdot 2}$ are the only possible orderings of k. This proves the result.

Three maximal real fields with one and two orderings can be easily distinguished. They are more extensively characterized in Theorems 2 and 3.

PROPOSITION 2. *Let* k *be a* 3-*maximal real field. Then the following conditions are equivalent*:

1) k *has two orderings*.

2) k *is pythagorean*.

3) *There exists a quadratic extension of* k *which is* 1-*maximal*.

PROOF. 1) $\Longrightarrow$ 2). Write $k = k^{\cdot 2} \cup \pm ak^{\cdot 2}$ as a disjoint union. If k were not pythagorean, then we can choose a to be a sum of two squares of the form $1+c^2$ with $c \in k$. Then $P = k^{\cdot 2} \cup ak^{\cdot 2}$ will be the unique ordering of k, a contradiction. Thus k is pythagorean.

2) $\Longrightarrow$ 3). Using the exact sequence of Theorem 3.4, p. 202 of [5], we see that if $F = k(\sqrt{-1})$, then $|F^{\cdot}/F^{\cdot 2}| = 2$. Since $\Omega|F$ is a 2-extension, then clearly F is 1-maximal.

3) $\Longrightarrow$ 1). This follows from the next proposition.

PROPOSITION 3. _Let_ k _be a 3-maximal real field_. _Then the following conditions are equivalent_:

1) k _has a unique ordering_.

2) k _is not pythagorean_.

3) _Every quadratic extension of_ k _has four or more square classes_.

PROOF. 1) $\Longrightarrow$ 2). Write as above $k^{\cdot} = \pm k^{\cdot 2} \cup \pm ak^{\cdot 2}$ as a disjoint union. If k were pythagorean, then $P = k^{\cdot 2} \cup ak^{\cdot 2}$ and $Q = k^{\cdot 2} \cup -ak^{\cdot 2}$ will give two orderings of k, a contradiction.

2) $\Longrightarrow$ 1). This follows from Propositions 1 and 2.

1) $\Longrightarrow$ 3). Choose a above as a sum of two squares of the form $a = 1 + c^2$, $c \in k$. Then the three quadratic extensions of k are $F_1 = k(\sqrt{-1})$, $F_2 = k(\sqrt{a})$ and $F_3 = k(\sqrt{-a})$. By p. 202 of [5], $|F_1^{\cdot}/F_1^{\cdot 2}| = 4$. The same holds for F_3. Now the form $X^2 - aY^2$ represents 1, $-a$ and $-c^2$; hence the norm $N_{F_2/k}$ represents all the four square classes of R and so $|F_2^{\cdot}/F_2^{\cdot 2}| = 8$.

3) $\Longrightarrow$ 1). For if not R will have two orderings and by Proposition 2, there will be a quadratic extension of k with only two square classes.

We proceed now to calculate the Galois group in each case.

PROPOSITION 4. _Let_ k _be a 3-maximal field with two orderings_. _Then the Galois group_ G_k _is the dihedral group of type_ (2, 2); _that is_, G_k _is generated by two elements_ τ _and_ σ, _where_ $O(\tau) = 2^{\infty}$, $O(\sigma) = 2$ _and_ $\sigma\tau\sigma = \tau^{-1}$.

Moreover, the maximal abelian extension of k _has the Klein 4-group_ $\mathbb{Z}/2\mathbb{Z} \times \mathbb{Z}/2\mathbb{Z}$ _as its Galois group_.

PROOF. From Proposition 2, we get $G_{k(\sqrt{-1})} \simeq \mathbb{Z}_2$, since $k(\sqrt{-1})$ is 1-maximal. Thus G_k is the semidirect product of $\mathbb{Z}_2$ by a group of order 2. Since $G_{k(\sqrt{-1})}$ is abelian, we get by Becker's Lemma (p. 87

of [1]) that $\sigma\tau\sigma^{-1} = \tau^{-1}$ for every $\tau \in G_k(\sqrt{-1})$. In particular this holds for any generator τ of $G_k(\sqrt{-1})$.

As for the second part, we claim that the maximal abelian extension $k_{ab} = k(\sqrt{-1}, \sqrt{a})$ where a is given by Proposition 2. Clearly $k(\sqrt{-1}, \sqrt{a}) \subseteq k_{ab}$. Now $k(\sqrt{-1})$ contains all the 2^nth roots of unity, since k is pythagorean. It then follows by Theorem 3 of [9] that $\Omega = k(\sqrt{-1}, a^{1/2^\infty})$. If $k(\sqrt{-1}, \sqrt{a}) \neq k_{ab}$, then k_{ab} will contain $a^{1/4}$ (Theorem 2 of [9] or using a fact on cyclic extensions). Thus $k(a^{1/4})$ will be a normal extension of k and so will contain $\sqrt{-1}$. On the other hand, $k(a^{1/4})$ is an ordered field, giving rise to a contradiction. Thus $k_{ab} = k(\sqrt{-1}, \sqrt{a})$.

REMARK 1. In [2], the above fact that $\Omega = k(\sqrt{-1}, a^{1/2^\infty})$ is explicitly shown and the Galois group computed more explicitly.

PROPOSITION 5. *Let k be a 3-maximal field with a unique ordering. Then $G_k(\sqrt{-1})$ is the free pro-2-group on two generators and G is the semidirect product of this group by a group $\{1,\sigma\}$ of order 2. Moreover we can choose generators τ and ϕ for $G_k(\sqrt{-1})$ such that* $\sigma\tau\sigma = \phi$ and $\sigma\phi\sigma = \tau$.

Also the maximal abelian extension of k has Galois group $\mathbb{Z}_2 \times \mathbb{Z}/2\mathbb{Z}$.

PROOF. Write $F = k(\sqrt{-1})$ and let $u(F)$ be the u-invariant of F. We have $|F^\cdot/F^{\cdot 2}| = 4$. By Kneser's Theorem (Theorem 4.4, p. 317 of [5]) $u(F) = 2$ or 4. If $u(F) = 4$, write L to indicate the field obtained from F by adjunction of all the roots of unity. Then by [3], $\mathrm{Gal}(\Omega|L)$ will be abelian. Since L is obtained from k also in the same way, we conclude from Theorem 4.1 of [12] that k is a field of class C and so pythagorean, a contradiction. Thus $u(F) = 2$ and by Theorem 3.1 of [12], $\mathrm{Gal}(\Omega|F)$ is a free pro-2-group. The number of free generators is 2, since $|F^\cdot/F^{\cdot 2}| = 4$.

Clearly G_k is the semidirect product of $G_k(\sqrt{-1})$ by a group $\{1,\sigma\}$ of order 2.

To describe the group action, we write $k^\cdot = \pm k^{\cdot 2} \cup \pm ak^{\cdot 2}$ as a disjoint union with $a = 1 + c^2$, $c \in k$. We indicate by i one root of $X^2 + 1$ in Ω. Now $F(\sqrt{a})$, $F(\sqrt{c+i})$ and $F(\sqrt{c-i})$ are the three minimal extensions of F where we choose the square-root in such a way that $\sigma(\sqrt{c+i}) = \sqrt{c-i}$ and $\sqrt{c-i}\,\sqrt{c+i} = \sqrt{a}$. Choose now two free generators τ and ϕ of $G(\sqrt{-1})$. Since $\text{Fix}\{\tau,\phi\} = k(i)$, none of the elements $\sqrt{a}$, $\sqrt{c+i}$ and $\sqrt{c-i}$ can be fixed simultaneously by τ and ϕ. From this it easily follows that each one of τ,ϕ and $\tau\phi$ fixes exactly one of $\sqrt{a}$, $\sqrt{c+i}$ and $\sqrt{c-i}$ and moves the other two. Replacing τ by ϕ or $\tau\phi$ if necessary, we may assume that τ fixes $\sqrt{c+i}$ and similarly that ϕ fixes $\sqrt{a}$ and $\tau\phi$ fixes $\sqrt{c-i}$.

Consider $\sigma\tau\sigma \in G(\sqrt{-1})$. Since $\sqrt{a} \in \text{Fix}\,\sigma$, we get $\sigma\tau\sigma(\sqrt{a}) = -\sqrt{a}$ and $\sigma\tau\sigma(\sqrt{c+i}) = \sigma\tau(\sqrt{c-i}) = -\sqrt{c+i}$. Now τ moves $\sqrt{c-i}$ so that the two automorphisms τ and $\sigma\tau\sigma$ between them move all the three elements $\sqrt{a}$, $\sqrt{c+i}$ and $\sqrt{c-i}$. Hence $\text{Fix}(\langle\tau,\sigma\tau\sigma\rangle) = k(\sqrt{-1})$, by the 3-maximality of $k(\sqrt{-1})$. Hence $G(\sqrt{-1})$ is generated (topologically) by τ and $\sigma\tau\sigma$. Thus we may choose $\phi = \sigma\tau\sigma$ and so clearly $\sigma\phi\sigma = \tau$.

To prove the second part, we see that every proper ordered extension of k must contain $\sqrt{a}$. Since the pythagorean closure of k is a Galois extension of k, it follows by the Artin-Quigley Theorem [9] that $(\text{Pyth}\ k/k)$ has $\mathbb{Z}_2$ as its Galois group. Now $(\text{Pyth}\ k)(\sqrt{-1}) \subseteq k_{ab}$. On the other hand Pyth k is maximal with respect to the exclusion of $\sqrt{-1}$ within k_{ab}. Hence $k_{ab}|(\text{Pyth}\ k)$ is procyclic. If it were infinite, Pyth k will admit a cyclic extension of degree 4, impossible by a Theorem of Diller and Dress [11]. Hence $k_{ab}|\text{Pyth}\ k$ is of degree 2, and $k_{ab} = (\text{Pyth}\ k)(\sqrt{-1})$. Clearly the corresponding Galois group is $\mathbb{Z}_2 \times \mathbb{Z}/2\mathbb{Z}$.

We can characterize the fields we are studying in a simple manner and also by means of their Galois groups. Condition 4) of the theorems below provide simple ways of constructing examples of such fields.

THEOREM 2. *Let* k *be a real field. Then the following conditions are equivalent:*

1) k *is a 3-maximal field with two orderings.*

2) k *is a 3-maximal field and* k *is pythagorean.*

3) k *is a 3-maximal field, having a quadratic extension which is 1-maximal.*

4) k *has exactly two orderings and* k *is a field maximal with respect to these two orderings.*

5) k *is pythagorean, has exactly two orderings and every polynomial of odd degree over* k *has a root in* k.

6) *The Galois group* $G_k(\sqrt{-1})$ *is a procyclic 2-group of order* 2^∞.

7) G_k *is the (topological) semidirect product of the free pro-2-group on one generator by a group of order* 2.

8) G_k <u>is the infinite dihedral group of type</u> $(2^\infty,2)$ <u>with the direct product topology</u>.

PROOF. The equivalence of 1), 2) and 3) is given by Proposition 2.

2) $\Longrightarrow$ 4). Write $k^{\cdot} = \pm k^{\cdot 2} \cup \pm ak^{\cdot 2}$ as a disjoint union. Then $P = k^{\cdot 2} \cup ak^{\cdot 2}$ and $Q = k^{\cdot 2} \cup -ak^{\cdot 2}$ give the two orderings of k. Now any proper extension $L|k$ must contain one of the fields $k(\sqrt{-1})$, $k(\sqrt{a})$ and $k(\sqrt{-a})$. And obviously at most one of the orderings P and Q of k can be extended to these fields and hence to L. This shows the maximality property in 4).

4) $\Longrightarrow$ 5). This is Theorem 1 of [2].

5) $\Longrightarrow$ 1). This is contained in the proof of the sufficiency part in Theorem 1 of [2].

1) $\Longrightarrow$ 6). This is contained in the proof of Proposition 4.

6) $\Longrightarrow$ 7). This is evident since $k = (k(\sqrt{-1}) \cap (\text{a real closure of } k))$.

7) $\Longrightarrow$ 1). Let H be the free pro-2-group on one generator in question and F its fixed field. Let G_k be the semidirect product of H by $\{1,\sigma\}$. Let $R = \mathrm{Fix}\{\sigma\}$. Then $k = F \cap R$ is an ordered field. Now $F|k$ is a Galois extension and $FR = \Omega$. Hence $[F:k] = 2$. Since $|F^{\cdot}/F^{\cdot 2}| = 2$, it is easily seen that $|k^{\cdot}/k^{\cdot 2}| = 4$. Thus k is a 3-maximal real field. Since F is a 1-maximal quadratic extension of k, we get from Proposition 2 that k has two orderings.

1) $\Longrightarrow$ 8). This is Proposition 4.

8) $\Longrightarrow$ 7). This is evident.

REMARK 2. It is proved in [2] that the field k of Theorem 2 has the following hereditary property: Every finite real extension of k is 3-maximal with two orderings. This is seen from (6) above.

REMARK 3. The algebraic closure of k is easily described by $\Omega = k(a^{1/2^\infty}, \sqrt{-1})$ for a suitable $a \in k$. See [2] or [3]. In fact this description of the algebraic closure together with 3-maximality characterizes these fields. These fields are of class C, as seen by Condition 6) above. (See p. 233 of [12] and Condition 3 of Theorem 4.1 there.) Membership in class C together with 3-maximality characterizes these fields.

REMARK 4. If k is a 3-maximal field with two orderings, then there are two real and one non-real quadratic extensions of k. The non-real quadratic extension has Galois group $\mathbb{Z}_2$; the other two are again 3-maximal fields with two orderings (Remark 2).

THEOREM 3. *The following conditions on a real field* k *are equivalent:*

1) k *is a* 3-*maximal field with a unique ordering.*

2) k *is a* 3-*maximal field and* k *is not pythagorean.*

3) k *is a* 3-*maximal field admitting no quadratic extension with two square classes.*

4) *There exists an element* $a \notin k^{\cdot 2}$ *such that* a *is a sum of squares and* k *is maximal with respect to the exclusion of the positive* $\sqrt{a}$ *in some real closure of* k.

5) k *is not pythagorean and* $G_k(\sqrt{-1})$ *is a free pro-2-group on two (free) generators.*

6) *The Galois group* G_k *is the topological semidirect product of a free pro-2-group* $H = \langle\tau,\phi\rangle$ *by a group* $\langle 1,\sigma\rangle$ *of order* 2 *with group action defined by* $\sigma\tau\sigma = \phi$ *and* $\sigma\phi\sigma = \tau$.

The proof involving Condition 4 will require the following lemma.

LEMMA 1. *Let* R *be a real closed field and* k *a subfield of* R *maximal with respect to the exclusion of* $\sqrt{a}$ *with* $a \in k \cap R_+$. *Then* k *is a* 3-*maximal real field.* k *has a unique ordering if and only if* a *is not a sum of squares in* k.

PROOF. Since every polynomial of odd degree has a root in k, $\Omega|k$ is a 2-extension. Thus to show that k is a 3-maximal field, it is enough to show that $|k^{\cdot}/k^{\cdot 2}| = 4$. Let $b \in k^{\cdot}$, but $b \notin k^{\cdot 2} \cup ak^{\cdot 2}$. Then $R(\sqrt{b}) = R(\sqrt{-1})$ and $-b \in R^{\cdot 2}$ whence $k(\sqrt{-b}) \subseteq R$. Thus $-b \in k^{\cdot 2} \in ak^{\cdot 2}$ so that $k^{\cdot} = \pm k^{\cdot 2} \cup \pm ak^{2}$. Thus k is a 3-maximal field. The final statement follows from Proposition 3.

PROOF OF THEOREM 3. The equivalence of conditions 1), 2) and 3) is given by Proposition 3.

2) $\Longrightarrow$ 4). Since k is not pythagorean, then we can write $k^{\cdot} = \pm k^{\cdot 2} \cup \pm ak^{\cdot 2}$ with a as a sum of squares of the form $a = 1 + c^2$ with $c \in k$. Thus every real closure R of k contains $k(\sqrt{a})$. Moreover the other two quadratic extensions of k are not ordered. Hence k is maximal with respect to the exclusion of $\sqrt{a}$ in every real closure of k.

4) $\Longrightarrow$ 1). This follows from Lemma 1.

1) $\Longrightarrow$ 5). This follows from Proposition 5.

5) $\Longrightarrow$ 2). We only need to show that k is a 3-maximal field. Now, since $G_k(\sqrt{-1})$ is a free pro-2-group on two generators, we see that $k(\sqrt{-1})$ is a 3-maximal field. It follows from p. 202 of [5] that $|k^{\cdot}/k^{\cdot 2}| = 4$ or 8. Moreover $|k^{\cdot}/k^{\cdot 2}| = 8$ if and only if k is

pythagorean, which is not the case here. Thus $|k^{\cdot}/k^{\cdot 2}| = 4$ and k is 3-maximal.

1) $\Longrightarrow$ 6). This is again given by Proposition 5.

6) $\Longrightarrow$ 2). Let F be the fixed field of H. It is easily seen that $[F:k] = 2$, since F is a Galois extension of k. Now F is a 3-maximal field. Now G_k is a pro-2-group generated by τ, ϕ and σ. In fact two generators will do, for example, σ and τ. Hence $|k^{\cdot}/k^{\cdot 2}| = 4$, since the number of generators is given by the dimension of $|k^{\cdot}/k^{\cdot 2}|$ as a $\mathbb{Z}/2\mathbb{Z}$-vector space. Thus k is a 3-maximal field, since $|\frac{k^{\cdot}}{k^{\cdot 2}}| \neq 2$. k can not be pythagorean, since in that case no quadratic extension of k could admit the free pro-2-group on two generators as a Galois group (Remark 4, Condition 6) of Theorem 2 and Theorem 3.1 of [12]).

Ribenboim [10, 11] has studied towers of infinite algebraic number fields with interesting properties. We will now prove a far-reaching generalization of his Example 2, p. 355 [10]. Our result shows that his example is simply due to the fact that the field of rational numbers is a non-pythagorean real field.

THEOREM 4. _Let_ k _be a non-pythagorean ordered field. Then there exists a doubly infinite sequence of ordered fields_

$$k \subseteq k_1 \subset P_1 \subset k_2 \subset P_2 \subset \cdots \subset k_n \subset P_n \subset \cdots \Omega$$

satisfying the following properties:

i) _Each_ k_i _has a unique ordering._

ii) _Each_ P_i k _is pythagorean and has an infinite number of orderings._

iii) _For each_ $i = 1, 2, \ldots, P_i/k_i$ _is an infinite procyclic extension of degree_ 2^{∞}.

iv) Ω/k_1 _is of degree_ 2^{∞}.

v) _For each_ $i = 1, 2, \ldots, \mathrm{Gal}(\Omega/k_i(\sqrt{-1}))$ _is a free pro-2-group on two generators._

vi) _For each_ $i = 1, 2, \ldots, \mathrm{Gal}(\Omega/P_i(\sqrt{-1}))$ is a free _pro-2-group with infinite basis._

PROOF OF THEOREM 4. Since k is not pythagorean, there is a sum of 2 squares, say, $a = b^2 + c^2$ which is not a square in k (a,b,c in k). Fix a real closure R of k in Ω and let k_1 be a subfield of R maximal with respect to the exclusion of $\sqrt{a}$. By Lemma 1, k_1 is a 3-maximal real field and moreover k_1 is not pythagorean, since $a \in k_1^{\cdot 2} + k_1^{\cdot 2}$. Thus by Lemma 1, k_1 has a unique ordering.

Let P_1 be the pythagorean closure of k_1. Then P_1/k_1 is a Galois extension and $k_1(\sqrt{a})/k_1$ is the unique minimal subextension of P_1/k_1. It follows immediately that P_1/k_1 is a procyclic extension of degree 2^∞. It is well-known that P_1/k_1 has an infinite number of orderings, if it is a proper extension. These facts prove ii) and iii) for the fields k_1 and P_1 and iv) follows from Proposition 1. v) follows from Proposition 5 for $i = 1$. vi) is clear, since a subgroup of a free pro-2-group is again a free pro-2-group and the cardinality of a basis is given by the $\mathbb{Z}/2\mathbb{Z}$-dimension of $P_1^\cdot/P_1^{\cdot 2}$ which is clearly infinite.

To complete the proof of the theorem, we need to show that the argument can be repeated. We claim that the field P_1 is not hereditorily pythagorean in the sense of Becker [1]. For if it were, $\mathrm{Gal}(\Omega/P_1(\sqrt{-1}))$ would be abelian by Theorem 1, p. 86 of [1]; also $\mathrm{Gal}(P_1(\sqrt{-1})/k_1(\sqrt{-1})$ is a subgroup of $\mathrm{Gal}(P_1/k_1)$ and so abelian. Hence $\mathrm{Gal}(\Omega/k_1(\sqrt{-1}))$ is metabelian. It would then follow from Theorems 4.5 and 4.2 of [12] that k_1 is a field of class C and so pythagorean, a contradiction. Hence P_1 is not hereditarily pythagorean and there exists an overfield F_2 of P_1 which is ordered and not pythagorean. We can now repeat the argument with F_2 instead of k. This completes the proof of Theorem 4.

We will now indicate some questions that naturally arise in our study.

QUESTION 1. Let F be a field of characteristic 0 containing $\sqrt{-1}$, Ω its algebraic closure and R a real closed subfield of Ω. Let $k = F \cap R$. When can we say that $[F : k] = 2$? The answer should enable us to construct examples of such k.

QUESTION 2. Do there exist real pythagorean fields k such that $\mathrm{Gal}(\Omega/k(\sqrt{-1}))$ is the free pro-2-group on two generators?

QUESTION 3. Ribenboim [10, 11] also constructs an infinite descending tower of algebraic number fields with special properties. Does his result also hold for any non-pythagorean real field? See his Example 2 and Theorem 4 above for comparison.

ACKNOWLEDGEMENTS. The authors wish to thank A. Prestel and P. Ribenboim for several discussions on the problems studied here. Their special thanks are due to O. Endler and J. Neukirch for their encouragement and support and to M. Knebusch for permitting the typing of the manuscript at the Universität Regensburg.

BIBLIOGRAPHY

[1] Becker, E., Hereditarily-Pythagorean Fields and Orderings of Higher Level. Monografias de Matematica No. 29, IMPA, Rio de Janeiro, 1978.

[2] Bredikhin, S. V., Ershov, Yu. L., and Kalnei, V. E., "Fields with two linear orderings", Math. Notes. U.S.S.R., 7-8 (1970), 319-325.

[3] Engler, A. J., and Viswanathan, T. M., "Digging holes in algebraic closures a la Artin—I".

[4] Geyer, W.-D., "Unendliche algebraische Zahlkörper, über denen jede Gleichung auflösbar von beschränkter Stufe ist", J. Number Theory 1 (1969), 346-374.

[5] Lam, T. Y., The Algebraic Theory of Quadratic Forms. Benjamin, New York, 1973.

[6] Lang, S., Algebra. Addison-Wesley, Reading, Massachusetts, 1965.

[7] McCarthy, P. J., "Maximal fields disjoint from certain sets", Proc. Amer. Math. Soc. 18 (1967), 347-351.

[8] Neukirch, J., Über gewisse ausgezeichnete unendliche algebraische Zahlkörper. Bonner Math. Schr. 25, 1965.

[9] Quigley, F., "Maximal subfields of an algebraically closed field not containing a given element", Proc. Amer. Math. Soc. 13 (1962), 562-566.

[10] Ribenboim, P., "On orderable fields", Math. Nachrichten, 40 (1969), 343-355.

[11] Ribenboim, P., "L'Arithmetique des Corps". Hermann, Paris, 1972.

[12] Ware, R., "Quadratic forms and profinite 2-groups", J. Algebra 58 (1979), 227-237.

[13] Ware, R., "When are Witt rings group rings?", Pacific J. Math. 49 (1973), 279-284.

[14] Ware, R., "When are Witt rings group rings? II", Pacific J. Math. 76 (1978), 541-564.

DEPARTAMENTO DE MATEMATICA
UNIVERSIDADE ESTADUAL DE CAMPINAS
CAMPINAS, S. P.
13.100 BRASIL

and

INSTITUTO DE MATEMATICA PURA E APLICIADA
RUA LUIS DE CAMOES - 68
RIO DE JANEIRO
20060 BRASIL

ABCDEFGHIJ–AMS–898765432